FRED HOYLE'S UNIVERSE

FRED HOYLE'S
UNIVERSE

Jane Gregory

OXFORD
UNIVERSITY PRESS

OXFORD

UNIVERSITY PRESS

Great Clarendon Street, Oxford OX2 6DP

Oxford University Press is a department of the University of Oxford.
It furthers the University's objective of excellence in research, scholarship,
and education by publishing worldwide in

Oxford New York

Auckland Cape Town Dar es Salaam Hong Kong Karachi
Kuala Lumpur Madrid Melbourne Mexico City Nairobi
New Delhi Shanghai Taipei Toronto

With offices in

Argentina Austria Brazil Chile Czech Republic France Greece
Guatemala Hungary Italy Japan Poland Portugal Singapore
South Korea Switzerland Thailand Turkey Ukraine Vietnam

Oxford is a registered trade mark of Oxford University Press
in the UK and in certain other countries

Published in the United States
by Oxford University Press Inc., New York

British Library Cataloguing in Publication Data
Data available

Library of Congress Cataloging in Publication Data
Data available

ISBN 0–19–850791–7

1

Typeset by RefineCatch Limited, Bungay, Suffolk
Printed in Great Britain by Clays Ltd, St Ives plc

Contents

Acknowledgements

The research for this book has its roots in my postgraduate studies, and my PhD thesis 'Fred Hoyle and the Popularisation of Cosmology' (University of London, 1998). The people who helped with that deserve thanks here too.

More recently, Hoyle's colleagues have generously been the source of information, entertainment and enlightenment, and meeting them was a privilege. Their names are given alongside their contributions.

Of the many libraries that were useful, three deserve special mention. The Caltech Institute Archives at the California Institute of Technology is a remarkable resource, and proved invaluable. Shelley Erwin, Bonnie Ludt and Jay Labinger enabled my very productive time there. The National Archive at Kew, still known to many as 'the PRO' from its days as the Public Record Office, is a national treasure. The Library of St John's College, Cambridge holds Hoyle's personal papers, and Jonathan Harrison has done sterling work in cataloguing this extensive collection in the short time since Hoyle's death. He was always helpful and hospitable in the face of my frequent requests.

Sincere thanks are due to four people who read draft manuscripts: Jenny Campbell, Alison Goddard, Andrew Gregory (no relation) and Colin Gregory (my father). Each brought a different critical perspective to bear on the book, and helped me to see it with fresh eyes.

Two colleagues generously shared unpublished research with me. Carolyn Little's work on radioastronomy in Australia provided important insights, and Allan Jones' on-going study of the early days of science broadcasting added valuable context.

I am grateful to Cambridge University for allowing me access to administrative records, and to Jacqueline Cox for facilitating this research. Thanks too to Jen Cole and Simon Lock for assistance with Chapter 18, to Andrew Gregory, Ofer Lahav and Steve Miller for sharing their technical expertise, and to Bruce Armbruster and John Faulkner for opening doors in California. Jacqueline Cox, Carol Dyhouse, Shelley Erwin, Jonathan Harrison, Geoffrey Hoyle and Angela Milner responded mercifully to last-minute cries for help. Lawyers Carter-Ruck gave constructive and sensitive advice.

Small but hugely empowering research funds were awarded to this project by Birkbeck College (my former employer) and by the Royal Society.

At Oxford University Press, this book was in the charge of Michael Rodgers until his recent retirement. I am grateful to him for his advice and support before, during and since.

Jane Gregory

1
Coming to light

Schooldays and scientific ambition; undergraduate life at Cambridge; becoming a mathematician; graduate studies; marriage to Barbara

As autumn turned into the bitter, fog-bound winter of 1930, a small volume of thoughts on astronomy was selling fast in the bookshops of Britain. Its author, Sir James Jeans, a distinguished and well-known mathematician, had abandoned Cambridge University for the Surrey countryside to commit himself to two passions: music, and the popularization of science. His little book, *The Mysterious Universe*, was the text of recent lecture, and had been rushed into print to take advantage of the interest the lecture had aroused. Jeans himself was enthralling radio audiences every Tuesday evening on the BBC, and the *Daily Mirror* – a publication stuffed with advertisements for radio sets – marked out his talks as broadcasting highlights. The *Mirror*'s editor noted that 'thousands who are not mathematicians, physicists or chemists do yet endeavour to stare unblinded at the latest shafts of light cast on man's destiny and place in the world by discoveries about atoms, electrons and wavelengths.'[1] His readers wrote in to ask how scientists could say anything at all about electrons when they could not even see them.

Much of the interest in Jeans' book focused on one particularly startling suggestion:

> Fifty years ago, when there was much discussion on the problem of communicating with Mars, it was desired to notify the supposed Martians that thinking beings existed on the planet Earth, ... [It] was proposed to light chains of bonfires in the Sahara, to form a diagram illustrating the famous theorem of Pythagoras. . . . To most of the inhabitants of Mars such signals would convey no meaning, but it was argued that mathematicians on Mars, if such existed, would surely recognise them as the handiwork of mathematicians on earth. . . . So it is with the signals from the outer world of reality. . . . [F]rom the intrinsic

evidence of his creation, the Great Architect of the universe now begins to appear as a pure mathematician.[2]

The talk of Martians was ordinary enough. But God, a mathematician? Churchmen were divided: from pulpits across the country, Jeans was denounced for this blasphemy – unless, that is, he was congratulated for opening a space within science for religious belief: as the *Church Times* noted, 'the most up-to-date cosmology, then, if it does not exactly postulate a creator, at least leaves logical room for such a conception, and that very obviously.'[3] A letter to *The Times* pointed out that others had already suggested God the Poet and God the Lover: were we now, by adding God the Mathematician:

> . . . setting up a fresh pantheon containing a variety of departmental deities . . .? Either we ask who made the mathematician, or we take refuge in the old myth that the earth is supported on a giant who stands upon a tortoise that rests upon an elephant, and the elephant's legs reach all the way down. There seems to be no escape from the conclusion that every avenue of thought leads to a mystery at the end.[4]

The science journal *Nature* was more interested in how Jeans' original lecture, with 'the clarity and suggestiveness to which we have grown accustomed in his welcome utterances' carried his readers from the old mechanical universe into the new world of quantum physics. *Nature* noted that Jeans' views 'would not meet with general acceptance, and indeed it was one of the merits of the lecture that it was provocative of far more thought than it expressed.'[5] When astrophysicist Herbert Dingle of Imperial College London reviewed the book, he chastised Jeans for his submission to the new physics, and concluded with a warning about the limits of scientific enquiry:

> We can follow Sir James Jeans through the maze of present-day physics, and do so gladly, charmed by the facile mastery of his exposition, which is so obviously the child of clarity and depth of thought. But when he attempts to discuss the status of physical conceptions in the world of realities, in an utterance which will be accepted as the authentic voice of science by thousands who seek guidance in matters of philosophy and religion, we feel strongly that he is darkening counsel, not by words without knowledge, but, much more dangerously, by knowledge without equivalent balance of judgement. Physics has much to say for the present time; there is no need for it to speak for other departments of thought and feeling as well.[6]

Fervent correspondences were transacted through the winter, about gods and physics, among scientists and in the press, and the fires were stoked every Tuesday evening by the radio talks from London. And miles away, in a northern public library, a muddy teenage truant, sheltering from the weather and awaiting the next instalment of this saga, added Jeans to his list of heroes.

Fred Hoyle was 15 years old when Sir James Jeans stirred up the establishment with his tall tale of God the Mathematician, and he enjoyed the whole affair thoroughly.[7]

Missing school in order to explore the latest science was typical of Fred Hoyle's childhood, which teetered between the kind of rebellion that could have ruined him for life, and a conscientiousness and application that would bring him many rewards.[8] His parents' lives similarly were full of contradictions: his mother, Mabel Pickard, had left her job in a textile mill in the small town of Bingley, Yorkshire, to study music at the Royal Academy in London, and had worked as a singer and music teacher. She had returned to Bingley in 1911 to marry Ben Hoyle, a working man whose fortunes were to swing them in and out of the middle class. Fred, their first child, was born in June 1915, and Ben, who was well over 30 by then, had just enough time to meet his new son before he was conscripted into the Machine Gun Corps and dispatched to France. Machine gunners were swiftly consumed in the searing gunfights across the trenches, and his family thought they would never see him again – every day, the arrival of the postman was anticipated with dread. But the compact, cheerful, sprightly Gunner Hoyle would fight on and dodge the bullets for long enough to return to his family and resume his trade as a merchant of textiles – he bought the substandard cloth that the mills did not want to sell to their big customers, and sold it on to tailors.

Home, 4 Milnerfield Villas, was a sturdy stone house, rather suburban in style, at the top of a hill in Gilstead, then just outside Bingley but now on its outskirts. The house still stands, though the address has been changed. It was adjacent to Milnerfield Estate, the grand residence of the managing director of one of the textile mills, and had once been home to one of the higher-ranking servants there. It was a rather ambitious investment for the Hoyles, but it gave them splendid views across the Milnerfield Estate and up to the Yorkshire moors. Hoyle later reckoned that he could have walked to Edinburgh and been in countryside all the way.

Married women were not allowed to teach in schools, so Mabel stayed at home, and when Fred was a baby, she played the piano for hours every day. Hoyle never would take up an instrument himself: he later said that knowing, from an early age, how music ought to sound made it impossible for him to struggle through the trite little pieces beginners are taught to play.[9] Nevertheless, music would become a life-long passion.

While Ben was in France, Mabel received the serviceman's wife's allowance of a few pence a day; and it was not enough. The ban on married women teaching was relaxed as the young men were absorbed by the war, but Mrs Hoyle had a baby to look after, and she sought some other employment. It came along in 1916 in the form of silent films, which were shown in a cinema in Bingley. Mabel secured work as an accompanist, and she mustered the soundtrack for comedies, action films and romances by splicing together extracts from Beethoven sonatas. This was not the kind of music the cinema manager had had in mind, and he fired her, but when audiences at the cinema dwindled the manager asked her to come back: the cinema's former patrons had told him that they had only come for the music. So Mabel left young Fred tucked up in bed while she earned a living at the cinema in Bingley, and his earliest memory was of lying awake wondering what he would do if she did not come back.

Mabel always did come back. At home during the day, she turned school-teacher and while she could not coax Fred to the piano, she did give him a good start in other subjects, including mathematics: by the age of three or four he had worked out the multiplication tables. He was also fascinated by machines: at around this time, concerned neighbours called his mother to collect him after he had stood entranced in the street for two hours watching a steamroller fix the roads. He learned to read the clock by repeatedly asking his parents to tell him the time, and then seeing how their answer related to the position of the clock's hands. He did not, however, master reading until the age of seven: at the cinema, the subtitles on the silent films suddenly made perfect sense. Hoyle later attributed the late start – and persistent headaches – to a problem with his vision: he was extremely short-sighted.

Fred's apparent capacity to work out for himself the kinds of information other children were learning at school was sufficient evidence for his mother to believe that formal schooling was not an urgent requirement for her boy. However, when he was six, school became compulsory by law, and a boom in his father's business provided the fees for a small private school. His parents hoped that young Fred might be allowed to continue his exploration of interesting matters relatively unencumbered by the rote-learning that was standard in other schools. But within a year business slumped, and with

Mabel ill after the birth of a daughter the family rented out their house and went to stay as paying guests of an acquaintance in Essex in southern England. There, in the flat landscape of the Thames Estuary, Fred befriended a local boy who taught him the art of truanting, and he applied it extensively on his return to school in Yorkshire. This happened sooner than expected, because the tenant of 4 Milnerfield Villas had run up big bills and then disappeared. When Ben was back in business, Fred was sent again to the private school, where, ahead of the other pupils in many subjects and bored by the uselessness of others (the futility of learning Roman numerals stuck in his mind), he often contrived to be reported sick by another boy so that he could spend his days elsewhere. The reports of his illnesses spiralled to the point where concerned neighbours called at the house, thus alerting his parents to the fact that they were likely to be in serious trouble with the law. So they let their Fred choose a school in the hope that he might then go there more often, and he chose the one about a mile away that was attended by most of his friends among the local children.

While Fred performed creditably in class – and won prizes in essay competitions at the local Temperance Society on such diverse topics as 'Our Heritage', 'Temperate in all Things' and 'In Business, Serving'[10] – he still contrived to present himself infrequently at school, often faking illness to legitimize his absence. Then an incident brought the truanting to the attention of higher authorities: Fred abandoned school after a teacher conducting a nature class insisted that a certain local flower had five petals. The specimen Fred had collected clearly had six. He was pondering the teacher's neglect of the evidence when she delivered a sharp slap to the side of his head, with the aim of recovering his attention (a contributing factor, according to Hoyle, to his later deafness). He was not prepared to accept punishment when it was the teacher who had been wrong, and he walked out of school. Three weeks later his parents were summoned to bring their boy before the local education committee. Fred had kept his flower, and he produced it, wilted but still bearing six petals, and told the committee of the teacher's mistake. The case was dropped, but the incident served to confirm Fred in his disregard for formal education.

It was simple enough to leave home in the morning as if for school, but to head instead for the factories and mills, where Fred, with his chubby face and unruly dark wavy hair, would hang around watching the machines and asking the workmen questions. A favourite spot was Bingley's celebrated Five-Rise Lock, which for 150 years had been raising barges through five lock gates up a steep incline on the Leeds–Liverpool Canal. For a while, Fred's ambition was to be a bargee. Then he was off into the hills, exploring for birds and flowers, rocks and clouds, until tea-time, when the dishevelled lad would saunter home as if from school.

There were more delights at home. Ben Hoyle had left school at the age of eleven, but was an enthusiastic reader, particularly about science. One particular treasure for Fred was Ben's chemistry book. With some cast-off equipment and a few new pieces, saved for and bought in a chemist's shop in nearby Bradford, Hoyle repeated many of the experiments described in the book, particularly enjoying those that involved explosions or bad smells. The kitchen at 4 Milnerfield Villas bore the scars. In 1923, an Easter fair visited nearby Shipley Glen, and Fred and his father set off gleefully to scoop up prizes on the shooting and darts. Another attraction caught their eye: a small tent to which people were admitted one at a time at a charge of the substantial sum of sixpence. The growing queue hinted at wonders within, and the Hoyles joined it. When their turn came, Fred was left outside while his father went in to see what was causing such excitement. He was amply rewarded for his sixpence: in the little tent was a 'wireless' – it was one of the earliest radios. Ben Hoyle was astounded by this device, and father and son mustered the components and connections to build their own wireless, which they tuned in to the little radio stations that were springing up all over Britain.

There was church on Sundays, where Fred sang in the choir and giggled with his friends when the lay preachers ranted and fumed about the latest most damnable sin. He worried, though, that the gospels told different versions of the same stories, and, being interested in angels, he was disappointed to find that Matthew, Mark, Luke and John each gave different reports of the number of angels attending Christ's tomb. That was something, he felt, that 'even the most slow witted person in my native village would have been able to report correctly.'[11] Clearly the scriptures were unreliable, and Fred abandoned church, and Christianity, very soon thereafter.

When Fred was old enough to qualify for a ticket to the public library, it became a favourite haunt. Bingley Library was a splendid example of the great tradition of English public libraries. Built in 1864 as a Mechanics' Institute offering classes for working men, it presented a solid stone face to the High Street, but contained many treasures for its eager clientele.[12] Truanting schoolboys were not turned away. Fred's favourite authors were T.E. Lawrence, 'Lawrence of Arabia', for action and adventure in distant lands, and H.G. Wells, whose *Short History of the World*, with its journey from the physical origins of the Earth to post-war reconstruction, was far preferable, in Fred's opinion, to the 'succession of kings' variety of history favoured in schools.[13] He also admired Sir Arthur Eddington, the Cambridge astronomer made famous by his demonstration of general relativity in 1919, whose popular books were great favourites.

Fred's parents were determined that their unruly boy should win a scholarship to study at Bingley Grammar School from the age of eleven, and with little time left to make good the gaps in his education, they moved him to another new school in a nearby village. The headmaster there, Tom Murgatroyd, was a former colleague of his mother's, and he became their last hope. Murgatroyd took an idiosyncratic view of education: he had gardening at the core of his curriculum. Fred was not keen on gardening and opted to do arithmetic instead. Clearly a bright lad, he was entered for the grammar school scholarship examination in the spring of 1926, but with so little experience of formal education he was baffled by the exam paper. Despite some effort in the arithmetic section he was not successful, but it happened that very few scholarships were awarded on the basis of the exam that year, and some of the failed candidates were recalled for a second chance. Thus Hoyle was sent for an interview to Bingley Grammar School, where he met the headmaster Alan Smailes, a product of the fearsome Mathematical Tripos, the gruelling mathematics course at Cambridge University. Smailes listened to Hoyle chattering on about the books about stars in the library and his home-made chemistry set, and then sent him off to see the chemistry master. Shortly afterwards, Murgatroyd received word that his errant pupil had been selected for a place at Bingley Grammar.

Down the hill, along the canal, past the railway line and along the main road to the grammar school: Fred walked this two-mile route through icy Yorkshire winters four times a day – there were no school lunches – hoping to catch a glimpse of the boat train on its way to the Southampton Docks to spur his dreams of sailing to America. School uniform was not obligatory because some pupils could not afford it, but the school cap was: and it marked Fred out from his village friends as he made his way to the grammar school. He was distracted from their games now, caught by an observation or a thought, gazing at the night sky instead of larking about. A friend of the Hoyles had an astronomical telescope, and some evenings father and son would walk the eight miles to his house and spend the night watching the broad bright moorland skies, plodding home at dawn, happy and exhausted.[14] They did not always agree about science: Ben thought Darwin's theory of evolution was 'a wonderful thing, the greatest thing in science,' Hoyle recalled. Fred would counter that Newton was the greatest thing in science. But his father:

> . . . would try to explain to me Darwin's idea because he was an uneducated man – self-educated I should say. Maybe he didn't do too good a job. And so, I started by thinking this is a lot of bunkum, you see, at the age of 12 or so. I just didn't believe it. I had been brought up in the country and I knew all the flowers and all this sort of stuff. And so I just did not believe it. I had that feeling that these were 'Just So Stories.'[15]

In 1927, Sir Arthur Eddington's *Stars and Atoms*, a little book of lectures, came to Bingley Library, and Fred felt his destiny was decided: working on a canal barge might be a good life, but nothing could beat the great excitement of science. He was going to become a scientist.

The library still seemed more interesting than schoolwork, and it took Fred some years to realize that there was advantage to be gained from playing the educational game if he wanted to reach his longer-term goals. Eventually he would decide to keep a little time for his favourite books, his radio set and new excitements such as the works of Sir James Jeans, and apply himself to schoolwork. He raced though the syllabus and gained his School Leaving Certificate a year early, at the age of 15.

Sixth-form study at Bingley Grammar was a rather *ad hoc* affair, as few children undertook the Higher Certificate: for most pupils, their School Leaving Certificate was, as its name implies, their ticket out into the world. The sixth-form boys studying science set up their own laboratory and were left to solve problems set by the teachers. A severe economic depression was setting in, and sixth-form pupils' parents were given £15 per year by the government in compensation for the income normally brought in by a teenage worker. Typically, Fred's parents gave him the money and let him spend it as he saw fit. One extravagance was a six-shilling season-ticket to a series of six concerts given in Bradford by the famous Hallé Orchestra, conducted by Thomas Beecham: each concert was, Fred thought, 'a good bob's worth.'[16] In 1932, at the age of 17, he did well enough in his examination for the Higher Certificate to be in line for a local scholarship, which suited him very well as he was hoping for a place to study chemistry at Leeds University, a short train ride from Bingley. The Depression had only hardened his resolve to become a scientist: if one's fortunes could be as vulnerable as those of the hard-working men of Bingley, one might as well forget about trying to make money and do something interesting instead.[17] But the Depression was to handicap him too: cuts in education spending brought the qualifying mark for the local scholarships up into line with the rather more demanding national scholarships, and Hoyle was stuck at school for another year.

Hoyle was sure that a repeat of the previous year would dull his senses, so he asked Smailes, his headmaster, if there was perhaps something more interesting he might do. Smailes, remembering his own studies for the Tripos, gave Fred and his friend Fred Jackson the scholarship exam papers in physical sciences and mathematics for Cambridge University. The chemistry master bought the boys advanced textbooks with his own money, and the physics master encouraged them along. The maths master was Smailes himself, but during the day his duties as headmaster left no time for extra classes, so he tutored the boys in the evenings at his home. The two Freds set off for Cambridge to sit their exam in December 1932, Jackson for St John's College and Hoyle for Emmanuel College. Hoyle was astounded by the grandeur of the dining hall at Emmanuel, and by the way the undergraduates extricated themselves from the long tables by jumping on to them and walking down their length. Having invented his own laboratory procedures with makeshift equipment at home, he was awed and bewildered by the Cavendish Laboratory where he took a practical exam. Shortly after their return to Bingley, Jackson heard that he was admitted to St John's, but Hoyle had a letter from Emmanuel College telling him he had failed in some crucial aspects of his exam, mathematics being the most important of these.

By now though, Leeds University was forgotten: Hoyle was determined to go to Cambridge. The Cavendish Laboratory had become the world centre for physics under Ernest Rutherford, the charming New Zealander whose practical skills and ingenuity had brought great advances in atomic physics there. Rutherford and his team had identified the nucleus as a tiny dense dot in atoms full of space; and, just as Hoyle was struggling with his exams, James Chadwick had found the solution to a baffling problem in atomic structure: he discovered a new particle, the neutron. James Jeans had trained at Cambridge as a scholarship boy; Arthur Eddington was Plumian Professor there; there could be nowhere else for Hoyle. So Smailes set him to work for the next round of entrance examinations that would be held in a few months' time. Again, a near miss: he achieved entry standard but not a Cambridge scholarship – he would be admitted to study, but only if he could find the money to pay for it. So Hoyle then had to do his utmost to earn a national scholarship in his second attempt at his Higher Certificate exams in June. It was late summer before he knew that he was off to Cambridge.

When Hoyle got off the train in Cambridge in the autumn of 1933, it did not cross his mind to hire a taxi, and he dragged his suitcase for half a mile to his lodgings. Other boys were more used to such small luxuries as taxis: the student body was dominated by the offspring of the wealthy. Hoyle noticed the young men in jodhpurs and small flat caps – the uniform of the county squire – who called to each other in over-loud voices. But there were other scholarship students like Hoyle, and also students from even further afield than Yorkshire: they came to Cambridge from across the Commonwealth, and the tattered Empire could still promise Cambridge men jobs almost anywhere in the world. There were fewer women: they could gain a Cambridge education, though not a university degree, at Newnham College.

Hoyle's college was Emmanuel, like his headmaster Smailes. He was required to pay for five dinners per week in college halls, and 'having paid for them we ate them, of course.'[18] Emmanuel Hall had seemed so grand just a few months before, but now it was dingy, and the food poor (throughout his life, Hoyle would record his opinions of meals). However, the dinners were a good opportunity to meet fellow students, as their lodgings, licensed by the university, were scattered across the residential streets of Cambridge. The landlord or landlady of these 'digs' was required to note whether any student was out late in the evenings, and too many late nights or other indecorous behaviour would be reported to the University and could result in 'rustication', a temporary suspension from college, or even in 'sending down', a banishment from which one did not return. Hoyle found the university discipline as irritating as he had the rules of schooling, but he stayed out of trouble except for one occasion when he was seen in the street without his academic gown. His experience of Yorkshire weather stood him in good stead, for Cambridge was usually cold, indoors and out. Lectures could be anywhere in the dispersed colleges of the university, and the ubiquitous bicycle was employed by many students simply because on foot, they could not get from one lecture to the next in time.

Rather than enter straight away for the Natural Sciences degree, Hoyle's tutor suggested he take Part I of the Mathematical Tripos. His chemistry and physics were already quite advanced, so he could pick those up again in his second year, but the Tripos would bring his maths up to scratch.

Although he did not look set to distinguish himself in mathematics, Hoyle decided that many of his scientific heroes, Newton and Eddington among them, had become great scientists by first becoming good mathematicians, and so he would stick with mathematics for now and become a scientist later. His mathematics steadily improved, and Hoyle abandoned chemistry and physics classes to stay with the Tripos for the full three years. His applied mathematics classes took him into problems in theoretical physics, and in his final year he attended a lecture course given by Eddington on relativity. Hoyle found it difficult to follow his hero's lectures: Eddington was a slight, shy man with a high-pitched voice, and he often failed to reach the end of his sentences.

Outside of formal teaching, Cambridge offered many opportunities for sharing thoughts about the state of the world. The arrival of academics from Europe who were fleeing the spread of Nazism brought political concerns sharply to the fore. When the Germans occupied the Rhineland in 1936, Hoyle and his friends thought that this transgression would be swiftly punished, putting thoughts of further expansion out of Hitler's mind; and when that did not happen, they could not decide whether the human race was too sensible to allow another world war, or whether war was inevitable. Disillusionment with the Government was furthered among the students by its lack of support for the democratically elected government in Spain during the civil war there. The communists offered nothing to interest Hoyle, and despite ample evidence of what he called 'Soviet propaganda' in Cambridge, he saw little indication of the next generation of spies that was to emerge there after the war. Hoyle enjoyed the communist newspaper the *Daily Worker*, though, because it carried science articles by the biologist J.B.S. Haldane, one of the great science popularizers of his generation. When Haldane came to speak at Cambridge, Hoyle eagerly joined the audience, but found the lecture 'so full of Marxist jargon as to be incomprehensible. . . . I was amazed that a man who could write such beautiful science articles could talk such codswallop.'[19] (Despite this poor first impression, Hoyle would correspond with Haldane in the 1950s.)

Hoyle was fascinated by a visit in 1938 by the mathematician and philosopher Bertrand Russell, whose patrician profile, so unlike his own pudgy face, Hoyle admired; Russell's lecture, however, in which he proposed that modern weaponry was sufficiently terrible to act as a deterrent to war, was less impressive. Russell was of aristocratic lineage – he was the third Earl Russell of Kingston Russell and Viscount Amberley of Amberley and of Ardsalla – and gave Hoyle his first glimpse of what he described as 'the divine right of class':[20] social position that allowed one to get things done by pulling strings invisible to ordinary men like himself. Generally, though, Hoyle was not interested in politics, and he resisted the efforts of student political societies to recruit him: he had been fired up by the consequences of the Depression for a working town like Bingley while at school, but apart from keeping a weather eye on events in Europe, he now had other things on his mind.

Hoyle had joined the students' chess club – chess was to become a lifelong passion – and was soon in the university team; but his social life in Cambridge revolved around outdoor pursuits, in which he found a great release from the pressures of studying. He spent his holidays walking the hills, first with schoolfriends and then with the Cambridge Rambling Club, which also made local expeditions at weekends, sometimes of 30 miles in a day. Through this club he met research students, including George Carson, who had come from Northern Ireland to do a PhD in botany. Carson interested Hoyle in ideas from biology, as well as in canoeing, to which they applied themselves with fierce dedication on the River Cam. Hoyle became committed to physical fitness, and was proud of his growing muscles as walking and canoeing crowded his schedule. He noted with pleasure that his 115 pounds on arrival at Cambridge had become 150 pounds by the time he graduated.

Hoyle also attended classes with the research students, listening to lectures in quantum field theory by the émigré physicists Rudolph Peierls and Max Born, who had brought a glimpse of Continental theoretical sophistication to the more practically oriented Cavendish Laboratory. Hoyle asked Peierls if he would supervise him as a research student, and, as his three years as an undergraduate were coming to an end, he applied to the education authorities in Yorkshire to extend his scholarship for a year. Then he set out on a hitchhiking trip that was to be the first of many holidays in Cornwall, a county in the south-west of England with warm summer weather and a long, dramatic coastline. While there, he received a telegram from George Carson telling him that he had achieved honours with distinction in the Tripos, and a share of the Mayhew Prize for applied mathematics – the area of mathematics that overlaps with theoretical physics. His success was reported in the local paper back in Bingley, and he would keep the clipping for life.[21]

Hoyle had become a good mathematician. Now he could start on the second phase of his plan, and become a scientist.

Having graduated, Hoyle was free of some of the restrictions of student life. He was allowed to move out of licensed lodgings and find his own accommodation – a move that saved him £75 a year, a sum that, in the light of earlier hardships, 'had a touch of real wealth about it.'[22] He shared cheap rooms with friends from the Rambling Club. With the Mayhew Prize and an excellent degree to his credit, Hoyle once again approached Rudolph Peierls and asked to be his research student. Peierls took him on, and gave him some interesting problems in nuclear physics to think about. His work on these problems won Hoyle the Smith's Prize, an award of money for achievement by a graduate student in research in applied mathematics or theoretical physics (previous winners included Jeans and Eddington). Shortly afterwards Hoyle also successfully applied to the Goldsmiths' Company for a scholarship which amounted to £350 a year – enough for jodhpurs and a small flat cap, 'if I'd been of a mind', Hoyle noted.[23] His reliance on his home county of Yorkshire was at an end, and he expressed his gratitude for its generosity in a letter to chairman of the Education Committee there. Hoyle published his first scientific paper in 1937, at the rather young age of 21, on, at Peierls' suggestion, the Fermi interaction, a process in which a neutron breaks down into a three different particles.[24]

But what to do next? In 1937 Peierls moved to a post at the University of Birmingham – a move that would soon lead to his crucial involvement in the atomic bomb programme. Hoyle visited him there with a view to joining him, but Cambridge drew him back. He was working on a problem in the decay of nuclear particles which also interested his friend the young theoretical physicist Maurice Pryce, who, despite being just two years older than Hoyle, was in a position to stand as his PhD supervisor. The research degree for which a PhD is awarded was not a requirement then for a university career, especially in mathematics where the undergraduate degree was considered a sufficient measure of a student's competence; and neither Hoyle nor Pryce was keen on it as a start in research because they felt it constrained the researcher during what ought to be a creative period. But registering as a PhD student gave Hoyle some tax advantages, so a PhD student he became, though he never completed the degree.[25]

There were few research students then, and they spent time together irrespective of their subject. 'It was one of the virtues of small numbers that we were on arguing terms with each other, not just in our own experience as it happened to be, but over the sciences at large, and even in such apparently distant matters as history and the classics,' Hoyle recalled.[26] They were not shy about going against the mainstream: for example, one of the great controversies of the time was over the theory of continental drift as proposed by Alfred Wegener, which was not considered respectable in Cambridge.[27] Later, Hoyle would feel lucky that his friend Joe Jennings had spoken up for Wegener: the theory was eventually widely accepted. Carson was sceptical about Darwin's theory of evolution, and thought that if he worked through it diligently, 'if he did the right mathematics',[28] he would find its fatal flaw. Carson became Hoyle's roommate in 1938, and introduced him to the distinguished botanist and geneticist Cyril Darlington, a meeting that, Hoyle later noted, 'would exert a considerable influence on me down the years.'[29] Darlington saw human culture as a result of genetics, and he wrote a kind of universal biological history of the human race rather in the spirit of Wells's *Short History of the World* that had so impressed Hoyle as a boy.

Hoyle was repeatedly denied membership of the exclusive research club Delta-Squared V ($\Delta^2 V$ is a famous function in applied mathematics) until Pryce moved from Secretary to President and someone was needed to fulfil the secretarial duties. Hoyle was surprised and pleased to be elected a member at last, and then he realized what was afoot when he was swiftly elevated to the post of Secretary. His job involved inviting guest speakers, and one of his first triumphs was to secure a talk from the great physical theorist Paul Dirac, who was a Fellow of St John's and a member of the Faculty of Mathematics. When Pryce took up a lectureship at Liverpool University, Hoyle asked Dirac if he would become his supervisor despite the fact that he had no intention of doing a PhD. Dirac did not usually take on research students, but since Hoyle did not want to be supervised, the arrangement suited them both, and Dirac accepted the job. He was to become an important ally.

Another prospective speaker for Delta-Squared V, the St John's mathematician Ray Lyttleton, turned Hoyle down. Hoyle responded with the rather ungracious remark that he hoped he would be able to do as much for Lyttleton one day, which amused Lyttleton immensely. They became firm friends. Lyttleton was five years older than Hoyle, and had followed his Cambridge education with a spell at Princeton University in the USA, where he had worked with the celebrated astronomer Henry Norris Russell on the theory of the formation of the planets. He had returned to Cambridge in 1937 and established a research school for theoretical astronomy. Lyttleton had turned down the invitation to speak at the Delta-Squared V club because he was deep in a problem in what became known as accretion: the process by which a large body in space, such as a star, draws to itself, by virtue of its own gravitational field, matter from the surrounding medium.

This is an important process in the evolution of stars. Lyttleton was not convinced by Eddington's treatment of accretion in his book *The Internal Constitution of the Stars*, which gave an accretion rate that Lyttleton thought was too slow. Lyttleton explained this to Hoyle, who, as a research student unconstrained by the demands of a PhD, was free to consider whatever interesting problems came along. Hoyle did consider this problem, and found that he agreed with Lyttleton. They concluded that Eddington's calculation was correct if large chunks of matter were being pulled in by the gravitational field of a body, but if the surrounding medium were a gas, comprised of tiny particles interacting with each other, then Eddington's accretion rate was too low. According to Hoyle and Lyttleton, when the surrounding medium was a gas, these large bodies in space were behaving as though they were bigger than their physical size. Hoyle found that this apparent size could be determined from a calculation that took into account the mass of the body, the gravitational constant, and the velocities of the particles in the gaseous medium. He and Lyttleton thought about the problem for some weeks, decided they were right, and concluded that therefore accretion was 'far stronger and more important in astronomy than had previously been supposed.'[30] Lyttleton thought about possible applications of their idea, one of which is still used to account for the evolution of binary systems – pairs of stars, where each star orbits the other. The two men wrote up their work and submitted the paper to the highly regarded journal the *Monthly Notices of the Royal Astronomical Society*, in the hope that it would be accepted for publication.

Hoyle's studentship would not last long, and he still needed to secure a position in Cambridge. The future in physics there was uncertain: Rutherford had died in 1937, and the Cavendish Laboratory had been placed under the direction of a committee led by the physicist John Cockcroft until Rutherford's successor could be appointed. Fundamental experimental physics was in decline there, and the big names of the period, including James Chadwick, discoverer of the neutron only five years previously, were finding jobs elsewhere. It was hard to predict where the lab would head under new direction. Hoyle's future would be determined by his next source of funding, and he had already been turned down for Fellowships at St John's and Emmanuel. However, true to form, he tried again, and with support from Cockcroft and Dirac he was appointed to a three-year Fellowship at St John's from May 1939. A few weeks later he also won an award from the Commission for the Exhibition of 1851, bringing his annual income to an unprecedented £850 a year – at a time when craftsmen among his former schoolfriends would have been lucky to earn £200 per year.

The money made the immediate future less uncertain. But the news from the Royal Astronomical Society was less encouraging. The Society 'did not spar around ... for very long' with Hoyle and Lyttleton's paper on accretion.[31] They were shocked when their paper was rejected and returned. Scientific papers submitted to journals are first read, in confidence, by a small number of other scientists with relevant knowledge, known for this purpose as referees, who are nominated by the journal to advise on the paper's competence, originality, and likely interest to the scientific community. Hoyle and Lyttleton were told that, on the advice of two referees, the Society had decided that their paper was not suitable for publication in the *Monthly Notices*. They were not offered a chance to improve it, and they could not find much of substance in the referees' negative reports. Hoyle was offended and baffled:

> Both Lyttleton and I had scored well in the Cambridge Mathematical Tripos, and we had both done well enough in research to this point to have been awarded an open Fellowship to St John's. Yet here we had a paper, about which we had thought intensely, rejected pretty well at sight. Why?[32]

They could not let the matter rest. Hoyle and Lyttleton took a train to Oxford to see Harry Plaskett, Professor of Astronomy and Secretary of the Royal Astronomical Society. Plaskett was courteous but adamant about the Society's decision, and Hoyle and Lyttleton returned to Cambridge with nothing to show for their trip.

Hoyle and Lyttleton subsequently found some flaws in their paper, and they thought they might have been too ambitious in their claims for their idea, rather than introducing it in a gentler fashion. Hoyle later reflected that he had perhaps been ill-prepared for the more conservative community of astronomers by his start in the adventurous world of nuclear physics. There, 'in the wake of the discovery of quantum mechanics ... everybody was avidly searching for new ideas.'[33] The physicists were happy to run with an idea for a while, even if it was not yet fully worked out, but astronomers seemed to be more cautious. For Hoyle, this brush with the referees of the *Monthly Notices* was the beginning of a life-long disdain for the refereeing process. It was another two years before the Royal Astronomical Society published any of Hoyle and Lyttleton's work on accretion.

Still smarting from that rebuff, Hoyle and Lyttleton submitted their work to the journal the *Proceedings of the Cambridge Philosophical Society*, and after a nervous wait they found that their paper had been accepted. They later heard that Lyttleton's friend at Princeton, Henry Norris Russell, had been an influential referee.[34] In all, they published eight papers together between 1939 and 1942, including one entitled 'The effect of interstellar matter on climatic variation' – a subject to which Hoyle was to return many times in his long career[35] – and one in *Nature* on the evolution of the stars.[36] This paper disputed the idea that composition of stars was uniform: it was thought to be so because, as Eddington had demonstrated with the gaseous, relatively cool stars known as red giants, the rotating motion of the star itself swirls the gases of which they are made into a homogeneous mixture. This mixing also meant that the temperature was always evening out across the star: heat energy generated from the nuclear reactions in the star would be swirled around too. Hoyle and Lyttleton proposed instead that while red giants consist mainly of swirling gas, they would develop, at their centre, a very hot, very dense core of matter in which particles and atoms would be smashed together in very high temperatures and pressures. Maybe, then, the heart of stars held the answer to a question that was exercising many astronomers: how are the elements made?[37]

The work on accretion laid the basis for a fruitful future programme, for both Hoyle and a wider community. It also marked a turning point in Hoyle's interests: he would take what he knew about particle physics and quantum mechanics and devote his energies to problems in astronomy.

Hoyle reached another turning point in his life in 1939, during a tea engagement in May at the Dorothy Café in Cambridge, with an old friend, an Emmanuel mathematician who was now a schoolteacher at Scunthorpe Grammar School, in north Lincolnshire, just across the border from Yorkshire. With the teacher were two of his former pupils, sisters Jeanne and Barbara Clark. Jeanne was studying at Homerton College, which specialized in training teachers, and Barbara, still in school-girl plaits, had secured an interview for admission to Girton College, a college for women that would shortly become part of the University. Hoyle was instantly convinced that Barbara was to be his wife, and not long after that first meeting, in a newly acquired car and with the ink still wet on his driving licence, Hoyle drove north to his friend's school with the intention of finding her. This he did, and while she fussed over the holes in his shoes he persuaded her to come on holiday with him in August. This triumph achieved, he drove back to Cambridge, and then realized he still did not have her address. He wrote to her care of his friend's school, telling her: 'Don't be surprised at this: I never do anything right.'[38] Barbara was not deterred: she was an energetic young woman, and Hoyle's organizational failings presented her with a challenge she could meet. Their holiday, a tour of the Lake District in north-west England, was a great success. News from Europe, however, was very bad, and, convinced that war was imminent, Hoyle and Barbara decided to marry as soon as they could. Barbara's parents came round to the idea, and the ceremony was held at the end of December. Barbara chose a church wedding, and Hoyle felt obliged to confess his atheism to the vicar, who agreed to keep the service as short as decently possible.[39] The marriage, after such a swift courtship, was to endure for more than 60 years. But in that first winter of the war, they returned from a brief honeymoon to a fragmented Cambridge, with Hoyle's Fellowship and the 1851 award suspended for the duration, and with no idea of what would happen next.

2
Hut no. 2

War posting; radar; meeting Bondi and Gold; radar
problems; a trip to the USA; the origin of the elements;
return to Cambridge

I n January 1940, when Hoyle and his new wife Barbara returned from their honeymoon, they found Cambridge transformed by the war. Many of the university's scientific staff were gone, already set to work on war projects in secret defence research stations. The nuclear physicists had formed their own private group within the Cavendish Laboratory, concentrating on nuclear fission and the possibility of an atomic bomb. Former staff had come out of retirement to continue the teaching, and the place livened up when the University of London evacuated its students away from the bombing to the rather safer Cambridge.[1] What Hoyle recalled as a 'blizzard of regulations' was imposed upon the nation, Cambridge included: all men of working age were to be registered, children in cities sent to the country, all windows blacked out after nightfall, and food rationed. At least Lyttleton was still in Cambridge, and while they waited for their war postings he and Hoyle developed their accretion work, studying in particular the clouds of hydrogen in space that they thought might be sucked into the processes at work inside red giant stars.

In the autumn of 1940 Hoyle was summoned to the Admiralty Signals School, which became better known under a name it acquired in 1942: the Admiralty Signals Establishment, or ASE. The Signals School was a research station near Chichester in Sussex on the south coast of England, and its task was to develop radar systems. Radar was to become the biggest scientific project of World War II, far bigger than the atomic bomb project, and it would absorb physicists, engineers and mathematicians in their thousands.

The purpose of radar was to detect enemy aircraft. The problem of attacks from the air had been anticipated long before the war: in 1934 an extensive air exercise, including mock raids on London, was held to test air defences, and more than half the bombers had reached their targets unchallenged, despite the fact that their routes and destinations were known in advance to the pilots who were supposed to intercept them.[2] The Army began to experiment with listening devices that could pick up the sound of aircraft engines, and they built huge concrete dishes that focused sound waves onto a microphone. These worked provided the surroundings were quiet, but during a crucial visit to the project by some senior civil servants from the Air Ministry, a milkman clattered past with his horse and cart and ruined the demonstration.[3] The Ministry was appalled, and set up a committee of advisors, under Henry Tizard, to review the situation and suggest possible solutions. Tizard was Rector of Imperial College in London, a leading centre for physics and engineering research, and he had many years' experience in aeronautics, including during World War I.[4]

Tizard was familiar with a technique being used to measure atmospheric phenomena by Professor E. V. Appleton (later Sir Edward) at King's College London, which involved analysing the reflections from radio waves shone into the sky. And since the earliest days of radio transmission, reflections of radio waves had been tried as a means of detecting objects: in 1904, for example, Christian Hülsmeyer, a German engineer, had filed patents for a radio device for helping ships to avoid collisions. At the first meeting of Tizard's committee, in January 1935, it quickly reached a consensus: it ought to be possible to detect enemy aircraft by collecting the reflections from them of radio waves beamed into the sky.[5]

The idea of using beams of radiation as weapons was popular in science fiction in the 1930s, and some calculations were undertaken to determine whether the radio waves might melt the aircraft, or injure the pilot. The results showed that the aircraft and pilot would be safe from the beams, but that enough radiation would be reflected for it to be detectable.[6] Within six months, test equipment was detecting planes 15 miles away, and the distance increased with every test. The technology became known as radar, short for radio detection and ranging. Immediately following that successful experiment in the summer of 1935, a programme of works began to build a series of radar stations, the Chain Home system, along the south and east coasts of England, and their data were reported by telephone for analysis at a central operations room.[7]

In the early trials, results looked good should there be mass bombing raids by day, but were not so good for a lone attacker at night. Tizard urged the Air Ministry to allow radar sets on aeroplanes to complement those on the ground. One of Appleton's former PhD students, the Welsh physicist Edward G. Bowen, who had joined the radar project from University College Swansea, was given the task of developing new equipment that could be accommodated within the confines of a fighter plane. By the autumn of 1937, Bowen had some rough-and-ready equipment that could detect ships and aircraft even in very bad weather. Bowen's project took a step forward in 1939 when the physicists Henry Boot and John Randall at Birmingham University invented an electron tube called the resonant-cavity magnetron. It generated very powerful high-frequency radio beams with short wavelengths, which meant the aerials could be shorter, and the more intense beams survived over long distances and were much easier to detect.[8] With the magnetron installed, the airborne sets, with their short aerials and intense beams, worked very well.

The numbers of staff working on radar grew with the onset of war, and it was a very large operation by the time Hoyle joined in 1940. In Chichester he was reunited with his old Cambridge friend Maurice Pryce, though Pryce was eventually to be posted elsewhere on a secret mission. The staff there were not sure what to do with Hoyle: it was difficult to see what job would be suitable for a mathematician with experience in problems in theoretical physics and astronomy. Indeed, according to one colleague at ASE, many people there were not convinced that theoretical people had anything at all to contribute[9] – radar was seen as a practical matter, and theory was beside the point: after all, the magnetron was working very well despite the fact that no-one understood how. Eventually Hoyle was assigned to help out with aerials and propagation – that is, sending the radio waves from one place to another. The aerials division consisted of some fields and primitive huts known as Nutbourne, in the countryside north-east of Portsmouth on the south coast. The aerials were housed in horse-boxes so that they could be wheeled into different positions. Hoyle moved into Hut No. 2, where he read his way through a huge pile of literature about radar. He had given up his car, and now got about on an ancient bicycle – transport for which his wardrobe was ill-suited, especially in winter. Civilians on war service were not entitled to any sort of uniform or special kit, and clothes were in short supply, so Hoyle found a second-hand hat and a cape made from a tarpaulin, and, looking like something out of a melodrama according to Pryce, he pedalled about his business, developing a deep loathing of bicycles that was to last a lifetime.

Hoyle was joined in Hut No. 2 by a recent Cambridge mathematics graduate, 20-year-old Cyril Domb, who recalls spending his formative years under Hoyle's supervision.[10] Domb was being trained in radar technology, but when he proved less than adept at practical problems, he was sent to join Hoyle – a situation Domb found much more congenial. Hoyle announced that he had no great inclination towards mathematics, although he was committed to it as a tool for scientific exploration, and he suggested that Domb take on all the mathematical work. Hoyle also told Domb that while he thought it important to devote time and thought to the war effort, they should not stop thinking about fundamental problems in physics – indeed, that could only be detrimental to the national interest. Domb was surprised to see that Hoyle spent his free time devouring cheap science fiction.

Hoyle thought that Domb, as a recent graduate, should be looking for a field in which to do basic research. Hoyle described his own fruitful move from nuclear physics to astronomy, and thought Domb might do well in astronomy too. Hoyle's admiration for his childhood heroes had clearly waned, for he advised Domb:

> The leading British authorities in the field are Eddington, Jeans and Milne [Eddington's counterpart at Oxford University]. Of these three, Jeans and Milne have done nothing really of any significance, and Eddington may have done one or two significant things but he is still pretty mediocre. In a field in which the top researchers are so mediocre, you should have no difficulty in making your mark very quickly.[11]

One morning shortly after offering this advice, Hoyle came into the Hut and showed Domb the latest edition of the journal the *Monthly Notices of the Royal Astronomical Society*. He told Domb that there was an opportunity for him: Eddington had published a paper that contained an elementary mistake which Domb would easily be able to spot. Hoyle encouraged Domb to write a letter to the Royal Astronomical Society, and thereby launch his research career. Domb felt unable to follow the advice: 'Since, only a few weeks before, I had taken Part III [of the Tripos] and been examined by Eddington, it didn't seem very appropriate to start my career in this manner.'[12]

In the spring of 1942, Pryce told Hoyle he had an 'astonishingly fast' mathematician for him, and introduced another Cambridge graduate to Hut No. 2: Hermann Bondi.[13] Bondi was Austrian, a few years younger than Hoyle, and had as a schoolboy decided on Cambridge when he met Eddington in Vienna (who was impressive despite being 'silent to a fault'[14]).

He began his degree as a foreign student of mathematics at Trinity College in 1937, determined to make a career in England. After a weak start Bondi quickly overtook his classmates to impress his tutors with his skill and originality. But in May 1940, before he could complete his degree, he was interned as a citizen of an enemy nation – Austria was now under German rule – and sent to a camp in Quebec. Despite his detention he was free from the wartime hardships of Britain, and he enjoyed the landscape and the physics classes he offered to other internees. Bondi's family was Jewish, and his parents, who had left Vienna at his urging just hours before Nazi troops marched into Austria,[15] were by now settled in New York. While he was in Quebec they tried to organize his immigration into the USA; but even with Einstein's intervention the bureaucratic hurdles could not be overcome, and when Bondi was released after 15 months of internment he returned to England to complete his studies.[16] In April 1942 he was sent to work in the Radar Research Group Headquarters at Portsmouth. There he heard talk of 'a very interesting scientist called Fred Hoyle – who was a little bit of a wild man but very original.'[17] Pryce made the introductions soon afterwards, and Hoyle and Bondi got on well. Bondi recalled an occasion early in their friendship when Hoyle's unusual logic impressed him, when they were walking together late one night and a bombing raid began on Portsmouth:

> I shook in my shoes, but Fred walked on, totally unconcerned. And then the anti-aircraft guns started up, and shrapnel started falling. Fred immediately hid in the nearest doorway, and I asked him what that was all about. He said that he never thought the Germans would be clever enough to kill him, but the British Army might be.[18]

Suddenly in the summer of 1942, the operation moved north to occupy a school in Witley in Surrey. At the time the staff thought that they were being moved out of range of the bombing raids on Portsmouth, but they later found that British troops had returned from a raid into Germany with a German radar set, and they were being moved away from coast to put them out of reach of German troops on a similar mission.[19] The staff were reorganized, and a new theoretical group was constituted of ten or so people, including Domb, with Hoyle as head and Bondi as deputy. Bondi later recalled that the quality of scientists there must have been high, because of those ten, five went on to become Fellows of the Royal Society, the UK's most prestigious scientific society. Bondi was very impressed with the confidence Hoyle had in his team. He seemed to have a keen eye for a good person and a good idea, and his support would then be unqualified.

Hoyle's leadership consisted largely of allowing his team to follow their own ideas and then backing everyone vigorously.[20]

Hoyle's unqualified confidence was also important in the career of another Austrian, Thomas Gold. Gold, who was known to his friends as Tommy, was a bright boy from a wealthy Viennese family who had spent his teens idling and getting into trouble. At school in Germany and then Switzerland, he showed an aptitude for mathematics, and he decided to become a scientist. Like Hoyle and Bondi, he was captivated by popular science books, including those of Jeans and Eddington.[21] His parents, who were now living in England, advised against his choice of physics for his university degree and he settled for engineering instead; he entered Cambridge University in 1937. His parents thought he would calm down and apply himself there, but Gold did not pay much attention to his classes. Cambridge was an ideal playground. He later told a colleague that it was much more interesting to climb King's College Chapel than to go to engineering lectures, and that spending one's life thinking about how to avoid the people who do not like King's College Chapel being climbed had not been conducive to good examination results.[22] Gold had therefore not learnt much about engineering by the time he was interned in May 1940. He was sent to the camp in Quebec on the same ship as Bondi, and they became friends – they found that their parents had known each other back in Vienna. Gold was one of Bondi's keenest students in the informal physics classes in the internment camp.

Gold spent an uncomfortable year at the camp, and was relieved to get back to Cambridge and finish his studies. He graduated with a fourth class degree, which made him a poor candidate for a research career. But Bondi had a high regard for Gold's intellectual ability, and suggested to Hoyle that he would be an ideal man to join them. Hoyle went into battle with the bureaucrats, who could not understand why a fourth-class engineering student could be useful in the theoretical group. Hoyle's obstinacy and Bondi's diplomacy eventually won the day, and Gold joined the theoretical group in November 1943. He brought with him not only vital technical skills but also fine theoretical insight, and despite his poor qualifications on paper, he nevertheless quickly earned a very positive reputation for the high quality of his work.

Gold's playboy lifestyle had left him ill-equipped to tolerate the rather basic living accommodation provided for them by the Admiralty. He and Bondi were taking refuge from the spartan conditions by visiting Gold's parents in London at the weekends. Since in Witley they were surrounded by the commodious homes of rural Surrey, Gold suggested to Bondi that the two of them rent a house, which they did in the nearby village of Dunsfold. The former farmhouse was quiet and surrounded only by fields until the Canadian Air Force built a runway nearby from which to launch their fighter bombers, and the occasional crash on take-off made the two men fear allied planes more than enemy ones.[23] Gold hired a cleaning lady, and shared his cooking skills with Bondi. Food was in short supply and closely rationed – shoppers could buy their groceries from designated shops on presentation of coupons from their ration book – but the local farmers were generous with meat, and a chicken would find its way to Bondi and Gold's kitchen most Sundays, where a device invented by Gold and consisting of an electric switch, a length of fusewire, a spanner and a muffled alarm clock would ensure that the oven would light itself without either of them having to get out bed.[24] Hoyle was still living, with Barbara and their new-born son Geoffrey, near the aerial establishment in Chichester, and his daily commute to and from Witley that winter was a chore. He asked Bondi and Gold if he might stay with them a few nights a week, and they were happy to agree, not least because Hoyle's newly acquired car – a 1928 two-seater Singer, with a dickey seat, that he had bought for £5 – now made their own journey to work much more comfortable. Bondi rode in the more exposed dickey seat because he owned a balaclava helmet. Eventually Bondi and Gold extended the house so that Barbara, Geoffrey and the Hoyles' second child Elizabeth could stay there too.

Hoyle's team tackled a variety of problems for the Admiralty. Some of their solutions to war problems were short and neat. Later, during the war against Japan, the need arose for a safe code that could be deciphered quickly. The codes used in the European War were extremely secure, but took time to decipher; what was needed was an encryption and decryption machine that was fast. Hoyle suggested that such a device might consist of two Welshmen, and their code could be the Welsh language. He doubted whether any Japanese knew Welsh, so the code would be secure, and decryption would be instantaneous.[25]

But most of their work was scientific as well as ingenious. Domb identified three areas in which the theoretical group made important contributions: radar data concerning the flying height of aircraft as well as their distance away; the identification of the foil strips called 'Window' ('chaff' to the Americans) that were released from aircraft to blur the radar reflections; and unusual and noisy radar signals over the sea.

One of their first challenges was the height problem. The radar system was efficient at measuring how far away aircraft were, but could not detect the height at which they were flying, which made intercepting them rather difficult. Hoyle decided that this was actually a rather simple problem, as many of the complicating factors could be ignored given the scale of the situation under study. He collected the relevant data from his experimental colleagues, assumed the errors in the data to apply randomly, and by some simple geometry he worked out a method for determining height from the interference in the outbound and returning radar beams. In essence, the higher a plane was flying towards a detector, the sooner it would be detected – that is, it would be detected while it was still a long way off – while low-flying aircraft would get in much closer before showing up on the detector screens. Hoyle thought that a graph of distance at detection versus height could be the answer, and set out to calibrate it. But what looked like a simple solution to an important problem turned out to be useless in practice: it simply did not work for real aircraft, and the height estimates were invariably wrong. Hoyle had to think again, and he took long walks through the countryside to ponder his mistake. He had assumed a degree of error in every part of the system, which he was quite sure was the sensible thing to do. However, in line with the teaching he had received from Eddington he had assumed the errors to be random, so that an over-estimate of one measurement might cancel out an under-estimate of another, and the error overall would not be too different from any one typical error. But instead it appeared that the errors were systematic: they all stretched the measurements in the same direction, with the result that the errors piled on top of each other instead of cancelling out. It taught Hoyle a sharp lesson:

> Why were the errors not random? Because each person had included a measure of hope in his estimate, and the estimates, therefore, were all in the same optimistic direction. It was the first time I had run into human factors so strongly affecting a scientific calculation. It was not to be the last time, for this was to be an ever-recurring feature of controversial issues in later years.[26]

Hoyle decided there was no point in trying to persuade the various staff to moderate their measure of hope, so instead he drew up a collection of graphs and issued them to the radar operators. They would then test the graphs against a friendly plane flying at a known height, and see which graph best reflected their particular set-up. The operators could then use their chosen graph reliably to determine the heights of enemy planes. This technique was used with great success until the end of the war.[27]

Scientists engaged in war work were not allowed to take out patents on their inventions and thereby earn money from them, but after the war they were entitled to apply instead to a special fund that would compensate them for this lost opportunity. The inventions needed to be of 'exceptional brilliance or value to the service.' So in July 1947, Hoyle wrote to the Admiralty about his height–distance graphs.

> I fully appreciate that the present invention was made in accordance with my duties as an employee of the Admiralty. Nevertheless, I feel justified in making this claim because I accepted, as a matter of urgency, the very low salary of £350 per annum when I joined the Admiralty. This represents about half of the income I was receiving at Cambridge before the war.[28]

Hoyle waited five months to hear that his work was not considered of exceptional brilliance or value, and he did not receive any further payment for it.

The next task for the theoreticians was to do with the foil strips called Window. Anything connected with Window had been classified as top secret, and it had never been used as the British felt that they would lose any advantage as soon as the Germans copied it. As it turned out, the Germans had thought of it too, but were not using it for the same reason. But the British decided to use Window in some crucial attacks in the summer of 1943, and in anticipation of the Germans responding in kind, Hoyle was asked if there was any method of discriminating between the radar reflections from an aircraft and from Window. Hoyle decided that they should first simplify the problem by investigating the detailed character of reflections from aircraft of a single radar pulse. For this, they needed a continuous supply of flying aircraft. Hoyle had recently been on holiday to the Drummond Steps on the North Cornwall coast, an area he knew well, and where Liberator bombers flew round the clock from the RAF station at St Eval. He thought they could set up a base at the coastal command station at St Merryn, along the coast from St Eval, and position the radar set near the Drummond Steps. So they hitched up their horsebox and headed for Cornwall. To Hoyle's great delight, local produce made the food shortages much less severe there, and the team looked set to enjoy the change of scene.

The radar signals were in pulses produced at a rate of 500 per second – far too fast to see any individual pulse by eye – so Gold built a camera in which photographic paper whizzed past the aperture fast enough to catch the pulses as individual dots. In between the severe electric shocks inflicted by the ramshackle radar equipment, they found that there was a great variety of dots on the films. They reasoned that an aircraft moving at great speed in one direction would make a changing pattern of reflections, whereas an object such as a foil strip of Window wafting on the breeze would give a much steadier pattern. Adjacent dots would be different if they were the trace of a fast-moving plane, but the same if they were from Window. Capturing this difference between dots, when it occurred, was the next challenge. Gold decided that instead of turning the reflections into visible dots, he could turn them into sound, and use mercury delay lines to compare one pulse with the next electronically. The even pulses from Window would make the same sound, and subtracting one sound from the next electronically would leave little trace. But different pulses would make different sounds, and this difference could be captured to reveal the approach of an aircraft. The idea was not ready for practical use during the war, but was used for many years thereafter.

The theoretical group's third problem was to do with a phenomenon called anomalous propagation. The shorter-wavelength radiation produced by the magnetron was behaving oddly: aircraft could sometimes be detected at three or four times the usual distance. This was useful when detecting enemy aircraft, but it also left allied forces more vulnerable to detection themselves. The best guess was that this anomalous propagation was due to particular atmospheric conditions, and a Radio Meteorological Panel was established to investigate it. Hoyle joined the Panel. There, a meteorologist claimed to be able to predict the onset of anomalous propagation, and published results which were accurate in 80 per cent of cases. But a biologist in the theoretical group, Jerry Pumphrey, noticed that anomalous propagation tended to last for a week or two, so that it was a reasonably safe bet to assume that on any given day, tomorrow would be like today. Using this method, he was able to predict anomalous propagation accurately in 85 per cent of cases. Then Hoyle pointed out that for practical purposes they needed to be able to predict the start and end of periods of anomalous propagation, and not whether it would be there or not on any particular day. For that, Pumphrey's criterion was no good at all.

This problem occupied the group alongside another problem: sea clutter. This was the noise introduced into the radar signal as it bounced off the ever-changing waves of the sea. Under the auspices of the Radio Meteorological Panel, anomalous propagation and sea clutter were to be the focus of a continuous research programme, rather than intermittent experiments as and when ships and aircraft happened to be available. Hoyle was charged with setting up a radar transmitter on the top of Mount Snowdon in North Wales, which, at 3,500 feet high – it is the highest British mountain outside Scotland – is at about the height of a low-flying aircraft.

The receiver was to be in Northern Ireland, across the Irish Sea. But Hoyle's team had their own plan for a receiver in Wales, near Fishguard, across Cardigan Bay from the mountain.

Snowdon had been a tourist attraction before the war, and there was a hotel near the peak that was opened up for the research. There was also a mountain railway that could be used to carry the equipment and supplies. Once the equipment was up and running, all members of the theoretical group took their turn on the mountain. Domb found it exhilarating. Hoyle accompanied him on his first trip to teach him what to do, and the first problem they faced was that they had just missed the mountain train. Although Hoyle felt rather out of shape compared to his pre-war fitness, he told a horrified Domb – a Londoner not used to the great outdoors – that they could walk up Snowdon: 3,000 feet at 1,000 feet per hour would only take them three hours. To Domb's great surprise, they accomplished the climb in two and a half hours.[29]

The transmissions over Cardigan Bay did not reveal any anomalous propagation, which suggested that it was a phenomenon of the lower layers of the atmosphere. Bondi made some progress with sea clutter. But Hoyle was out of the country when the formal report was due at the end of the project late in 1944, and he never wrote it. He gave verbal reports to various meetings and committees, but was back in Cambridge by the time the rather large bills came in the following year. To their great embarrassment, no member of the permanent staff of the Admiralty could explain what the project had been about.

Hoyle had told Domb that the war was no reason to forget about the big problems in science, and he practiced what he preached. In the evenings, after a day spent with radar equipment, Hoyle, the mathematician turned physicist turned astronomer, Gold the engineer and Bondi the mathematician talked about astrophysics. According to Gold, Hoyle would badger them relentlessly with questions, forcing them into discussion, and pressing Bondi into endless calculations. Bondi was acknowledged to be the most accomplished mathematician of the three; and he thought that Hoyle sometimes used 'somewhat shaky' means to reach his goals, even though the goals themselves were invariably sound:

> I remember many occasions when I read one of his papers, and was totally unconvinced by the mathematical derivations of the final result. I decided to

work through properly and found to my utter fury that his conclusion was quite right. His tremendous physical insight meant that his mathematics was just a way of hiding or making respectable his derivations, in the hope that nobody else would detect the holes.[30]

It would become one of Hoyle's trademarks that he could often get the right answer even when his means for demonstrating it were simply wrong.

Like Hoyle, Bondi was working on accretion. He enjoyed Hoyle and Lyttleton's idea of the accretion of interstellar matter to stars as an on-going process in a changing universe – he preferred that to the more traditional idea of empty space across which old and 'finished' stars were scattered.[31] Bondi's work on on accretion earned him a Fellowship of Trinity College Cambridge in 1943 – a source of great satisfaction and a secure prospect for after the war. He found that their war research gave him some useful leads into the problems of accretion: the radar work revolved around the poorly understood but very effective magnetron, and Bondi was one of many who set himself the task of deciphering it. Domb remembers Bondi despairing of the flood of papers about the magnetron, of which he could be sure that 'when the nth paper comes along, everything up to $n-1$ is wrong.'[32] Pondering the interaction of its two electron beams on crossing paths, Bondi compared that situation with the streams of particles that are drawn in to form the accretion core of a star, and turned to his two friends in the evenings for criticism, argument and speculation.

Another accretion problem that interested them was the behaviour of the Sun's corona, on which Bondi and Hoyle worked for quite some time. They could not make the particle physics fit with the laws of thermodynamics. Bondi thought that they were in an inescapable hole, but Hoyle saw a way out, and wrote up the work under both their names. Bondi protested that his only contribution had been to be wrong, and therefore he did not merit the credit. He asked Hoyle to remove his name, and Hoyle replied that he was quite prepared to leave Bondi's name off 'as long as you agree that I can write that I am indebted to you for leading me by the nose for two years.' Bondi accepted the lesser embarrassment of joint authorship of the paper.[33]

Another interest of the three men during their evenings in Witley was cosmology. One of Hoyle's obsessions was the work of the distinguished American observational astronomer Edwin Hubble, who in 1929 had found startling evidence that the universe is expanding, thus favouring cosmo-logical theories that proposed an evolving universe over those that presented a more static picture of its history. For many in astrophysics, Hubble's work sent a very clear message about the history and character of the universe. But Hoyle was not so sure: he was convinced there were other stories to be told about the universe, and he nagged at his friends to help him work out what those stories could be.[34]

At the end of 1944, Hoyle escaped writing the formal report on anomalous propagation, and fulfilled his childhood dream of sailing to America: he was sent to Washington as a representative of the Radio Meteorological Panel, to attend a conference on radar. He crossed the Atlantic on an American ship taking troops home for Christmas. With him on the trip was an old friend, another Panel member Frank Westwater, who had a been a prize-winner in both the Tripos and in astronomy at Cambridge a year ahead of Hoyle, and was now a captain in the British Navy. The ship, which was impressively stoked with American provisions, arrived in New York in good time, and Hoyle and Westwater had three days to spare before their duties started. They took the train to Princeton New Jersey, where they visited the university to see Lyttleton's old mentor, Henry Norris Russell, a pioneer of the study of the structure and evolution of stars. As Hoyle and Westwater's formal itinerary was to take them from the conference in Washington to the naval base in San Diego, California, Russell arranged for them to visit Mount Wilson Observatory at the California Institute of Technology in Pasadena. Hoyle was smitten with California: he was impressed by the hospitality, the weather, the beaches and mountains, and the light in the clear air, and by the food that was good and plentiful at every meal. Westwater however created a tense moment by offering some bottles of wine during a convivial dinner at the Observatory's staff residence: it was known locally as the Monastery because women and alcohol were banned. There Hoyle met the influential German astronomer Walter Baade, who brought him up to date with his papers that Hoyle had missed during his war work, about the extremely high temperatures in the exploding stars known as supernovae.

Rather than return from the USA by sea (boats were too slow, and he suffered from seasickness), Hoyle had arranged to fly back from Montreal after visiting a radar station in Ottawa. His clothes were woefully inadequate for the freezing Canadian winter, but when the weather delayed his flight home, he took the opportunity to visit Maurice Pryce, whose secret mission was by now revealed: he had joined the Manhattan Project, which had swept up thousands of scientists and engineers from across North

America and Europe to build an atomic bomb. Pryce was working in a British and Canadian team based at Chalk River just outside Montreal. The particular combination of expertise at Chalk River gave Hoyle an idea of the problem they were attempting to solve. The challenge was to trigger the nuclear reactions that would lead to a nuclear explosion, and the solution was implosion. If a spherical shell made of the heavy radio-active element plutonium could be made to collapse in on itself with great enough force, then a chain reaction could start, with the plutonium nuclei disintegrating into lighter elements and blasting the left-over particles into other nuclei, which would disintegrate in turn, blasting out more particles. If the intensity of these nuclear reactions could be sustained for long enough, a nuclear explosion would occur. The great heat and pressure of the nuclear explosion was, thought Hoyle, not unlike the situation inside a star. Accretion theory showed that the gravitational field around a star would draw in material from space, and as the core of the star grew the force of attraction would increase, drawing in more material and increas-ing the pressure inside the star, intensifying reactions and raising the tem-perature there until it is relieved by an explosion. Baade's exploding supernovae showed this intense heat. If an imploding plutonium shell could create the temperatures and pressures that would transform elem-ents and produce a nuclear explosion, then what could the great pressures and temperatures in stars do to the material they engulfed? With exploding supernovae and the atomic bomb in mind, Hoyle set to work calculating the range of possible nuclear reactions that could take place at the tem-peratures and pressures found in stars. He found that the physical con-ditions there could indeed account for the formation of a wide range of elements, each one turning into another in nuclear reactions, and pro-duced in roughly the proportions in which they are found on Earth. Hoyle felt that he had glimpsed a solution to one of the great mysteries of physics: the origin of the elements. Suddenly he felt very optimistic about his career:

> [I]t was with some confidence that I contemplated the future. I felt I had a kind of passport to success in my pocket ... it was the key to the origin of the chemical elements that I had luckily stumbled upon.[35]

This project would occupy Hoyle for several years.

When Hoyle got back to England, he found that he was in deep trouble for the Mount Wilson detour: the Foreign Office had told the Admiralty that he had abandoned radar business to meet the Caltech astronomers. Hoyle talked his way out of trouble: he invented a story about atmospheric conditions in California being interesting from the point of view of short-wave propagation, and that he had been asking the astronomers' advice. Privately he wondered whether Baade being German might have been the reason for the reprimand. But in many ways it had been a fruitful trip. Apart from the invaluable scientific intelligence gained, Hoyle would remember Frank Westwater twenty years later, and involve him in another adventure.

The European war ended in May 1945, and it seemed likely to the scientists at Witley that the British would be much less involved in the Japanese war, and that there was a good chance that manpower restrictions would be eased. Indeed, scientists with teaching appointments were released very quickly, and Bondi, who had held a Cambridge lectureship since 1942 and a Fellowship at Trinity since 1943, left immediately for the university. Hoyle's St John's Fellowship was all very well, but the £250 it paid per year was not sufficient to keep a family. He had been looking for a lectureship, but had not seen any advertised that he thought he might get.[36] Barbara was worried about how they would make a living, but Hoyle was confident that it would all come right in the end: after all, he had the key to origin of the elements, and he felt 'like the young Richard Wagner [as he] toured the courts of Europe with the score of the overture to Rienzi in his pocket.'

They did not have to wait long: many Cambridge staff had taken appointments elsewhere during the war, and in the summer of 1945 St John's filled its vacancies with a number of temporary junior lectureships at £200 per year, one of which went to Hoyle. But since his teaching commitments were not to start immediately, he had to stay on with the Admiralty. He was sent to Germany to see the radar equipment there, but found few opportunities to pick useful information out of the devastation. Back at Witley, he let it be known that he had a new idea for explaining the abundance of the elements in the universe, which he regarded as far more important than anything he had done before. As the weeks rolled by, he became increasingly frustrated until one morning his new idea could wait no longer: he was leaving ASE to go to Cambridge. He explained to the senior administrative officer that he had no objections to participating in the war against Japan, but he knew from previous experience that any contribution that he made would take 18 months before it was put into practical use, and he was quite sure that the War would be over by then. But in the mean time he had an idea of supreme scientific importance which he urgently needed to develop, and the literature he needed to consult was available at Cambridge. To everyone's surprise, the officer was very sympathetic to this (a civilian, they suspected, would not have been). He suggested that perhaps Hoyle could spend some of his time at Cambridge thinking about Admiralty problems, so that he could legitimately classify Hoyle as 'away on duty.' Hoyle returned to Cambridge but felt unhappy with the less than proper arrangement, and so shortly thereafter the Admiralty came up with a formula by which he could be officially released. A year or so later, among a series of papers on the structure of nebulae,[37] he published his first paper on the origin of the elements.

When they returned to Cambridge, Bondi and Gold, as single men, could live in College, but Hoyle needed a house for his young family. House prices in Cambridge were prohibitively high, and the nearest house that Hoyle could afford was in the village of Quendon, 25 miles to the south. This left him without a base in Cambridge, so he used Bondi's room in Trinity College. When Bondi married the astronomer Christine Stockman (a mathematics student of Hoyle's) in November 1947, they moved into a little flat in Trinity Street, and that then became Hoyle's Cambridge headquarters. He would turn up there most days, and engage Bondi in a discussion that lasted all day. Often Gold came too. Bondi recalls Hoyle's characterization of these discussions: they were usually of the kind, Hoyle said, where they referred to some problem in physics, and started to argue about it in the morning. They would still be arguing about it when Hoyle went home in the evening, but the two men would be as far apart as ever, because each would have convinced the other of his own starting point of view but had lost faith in it himself. Hoyle and Bondi also collaborated in teaching: Hoyle, being a few years older than Bondi and Gold, had reached the point soonest where he was required to coach students in the Mathematical Tripos. Bondi found coaching rather easier than Hoyle did, and helped him with it – Hoyle called Bondi 'the supervisor's supervisor.'

Gold's time was largely taken up in the analysis of the sense of hearing, for which he was funded by the Medical Research Council: he used his wartime experience in signal processing to determine the physics of the inner ear. The work would turn out to be very important, but at the time the doctors and physiologists and physicists were in different worlds, and communication between them was very poor. The post-war subject of 'biophysics' had yet to find its feet.

The three friends also spent time together keeping up their war-time habit of discussing problems in physics and astronomy. Out of these discussions would come an idea in cosmology that was to make their names, and then haunt them for decades.

3
Into the limelight

A new cosmology; post-war gloom; steady-state theory in the spotlight; The Nature of the Universe; *the Massey conference; confronting Ryle*

One evening in 1946, Hoyle, Bondi and Gold strolled across Cambridge to the cinema. The film they were going to see was *Dead of Night*, the first post-war release from the famous Ealing Studios and, unusually for Ealing, a horror story. It consisted of a number of short films by different directors, all linked into an overall plot, and it showcased the best of Ealing talent. The tension arises in the film when guests at a country house weekend find that they have all appeared to one of their number, Walter Craig, in a nightmare. Craig knows that at the end of his nightmare, he kills one of the guests, and he decides he must leave to prevent this awful act. But the guests encourage him to stay, and Craig warns them that they can test his claim because, according to his dream, a dark-haired woman will come asking for money. The guests are fascinated – all except for Dr Van Straaten, a psychiatrist, who is sure that he can find a rational explanation for Craig's dream. As the guests each contribute their own supernatural stories, the film becomes an anthology of chilling tales, among them the now famous sequence in which a ventriloquist, played by Michael Redgrave, is driven to madness by the taunts of his own dummy. The suspense mounts when another guest arrives – a dark-haired woman who asks for money to pay her taxi fare. After the guests have told their stories, Craig is driven to strangle Van Straaten, and then he falls into a troubled sleep where he once again relives the guests' stories. He is relieved to wake and find himself at home in his own bed. When the telephone rings and an old friend invites him to spend the weekend at his country house, Craig gladly accepts; but when he arrives at the house, it suddenly looks terrifyingly familiar.

As Hoyle, Bondi and Gold walked back from the cinema to Bondi's rooms in Trinity, Gold suggested that films should be made in a loop with a circular plot and shown continuously in the cinemas, so that the audience could turn up at whatever time they chose.[1] He then asked his friends: 'What if the universe is like that?'[2]

The idea of a universe that continually circled back on its own beginning, always changing in detail yet still the same, fascinated the three men. Although it was not a new idea, it was at odds with the trend in cosmology, which, the wake of Edwin Hubble's great discovery of the expanding universe, was towards a linear history of a universe developing from a specific beginning in time – an idea known as evolutionary cosmology. It was this cosmology that had been subject to Gold, Bondi and Hoyle's critical scrutiny during their long discussions at Witley. But the field was wide open for new challengers: recent science had given cosmology a new lease of life, and there was plenty of scope to guide its new career.

The new scientific cosmology had emerged from a combination of technical and conceptual developments in the early decades of the twentieth century. Telescopes were seeing further and more clearly, making the cosmos a bigger and more complex place. The localized, specific cosmologies of the immediate, known universe were no longer enough, and grand general theories of an increasingly mysterious universe were sought instead. Einstein's general theory of relativity, published in 1915, provided a framework, and by the mid-1920s scientists were fitting new observations into a relativistic universe. As the new quantum mechanics also took hold, scientists saw not the perfect predictability of a classical cosmos but the apparent haphazardness of a quantum system. The universe was, as the Belgian mathematician Georges Lemaître put it, a place where 'something really happens.'[3] In 1926, in recognition of a cosmological, quantum-mechanical free-for-all in keeping with the spirit of the time, Eddington wrote in *Stars and Atoms* that: 'The music of the spheres has almost a suggestion of – jazz.'[4]

Many cosmological theories competed for attention in the 1920s, but those that presented a universe that evolved over time gained a lead from Hubble's work at the end of the decade. Hubble demonstrated that there were other galaxies like our own scattered throughout the cosmos, then showed that they were rushing apart in a way that implied that the universe was expanding. The change in wavelength of light coming from distant galaxies, a phenomenon known as the redshift, indicated the speed at which they were moving, and this provided a measure of the rate of expansion. To some scientists, the rate of this expansion gave important clues to the age of the universe.

Another step for evolutionary approaches to cosmology was taken in 1931, when a short paper from Lemaître occupied barely a column in the research journal *Nature*.[5] The paper's title, 'The beginning of the world from the point of view of quantum theory,' neatly joined the very big to the very small. Lemaître imagined in reverse the physics of the expanding universe, and posited a return to an original 'single quantum,' at which point 'notions of space and time would altogether fail to have any meaning. . . .' It was this moment – the moment immediately before space and time began to have meaning – that Lemaître suggested as 'the beginning of the world.' The interest provoked by Lemaître's idea was tinged with some scepticism: many astronomers believed that the expansion of the universe from a single quantum would leave measurable traces of radiation, and they were not convinced by Lemaître's subsequent suggestion that these traces could be the as-yet unexplained cosmic rays.

During the war, the nuclear physicist George Gamow had been considering this outburst from a single state as an explosive beginning to the cosmos. When the USA entered the war in 1942, Gamow, a Russian, had been in Washington for eight years; but unlike most of his colleagues he was not granted the necessary security clearance for the Manhattan Project, so he stayed on in his university post, and served as an occasional consultant on high explosives to the US Navy.[6] Thus Gamow was unusual among atomic scientists in being able to spend the war years working on a subject of his own choosing, which was, as he wrote in 1945, 'problems on the borderline between nuclear physics and cosmology.'[7] In 1946, he published a paper in which he used relativistic and nuclear physics to describe how all the matter that comprises the universe was produced during the initial very rapid cooling brought about by a sudden, very rapid expansion of a primeval entity.[8] A mathematical formulation was published a year later.[9] Thus the evolutionary cosmos became one with a specific, explosive beginning that coincided with the beginning of time, and which then expanded through time, and to give the cosmos of which our observations today are a snapshot.

This was not a perfect theory: among other problems, it produced a universe that was younger than much of its contents, and that contained no elements of atomic number greater than four – that is, only hydrogen, helium, beryllium and lithium. Gamow set his PhD student Ralph Alpher to work on the elements problem, and by 1948 they were able to publish work that plugged some of the gaps. It was a theory that appealed to nuclear physicists, but there were few in astronomy who took notice of it, and few physicists who had the skills in both mathematics and particle physics that were needed if they were to think about the cosmology of a relativistic universe generating elements. But this latest formulation helped to keep evolutionary ideas at the forefront of thinking, and the evolutionary cosmology, while not entirely persuasive, held sway.

During the long evenings at Witley, Hoyle, Bondi and Gold had already discussed their scepticism of the value set for the age of the universe derived from Hubble's work, 1.8 billion years. Hubble was held in great regard – he was, according to Bondi, 'more than life sized in the 30s and 40s, and . . . rightly so'[10] – but the three men knew that this age did not fit with the accepted value for the age of the Earth, which was, according to the geologists and physicists, 3 billion years. How could the Earth be older than the universe? Bondi called this 'a fearful contradiction.' Now Gamow's work was presenting the same problem. Nor were they happy with the universe having a finite beginning in time: how could a one-off event at the beginning of time ever be explained by the eternal laws of physics? And to assume, as Gamow did, that in the very different conditions of the early universe, present-day physics applied was, noted Bondi, 'daring, to put it politely.'[11] They were also uncomfortable with the evolutionary cosmology on a more visceral level: Hoyle would later compare the universe bursting out of nowhere to a 'party girl' jumping out of a birthday cake – 'it just wasn't dignified or elegant.'[12]

It was in this context that Hoyle, Bondi and Gold considered their never-ending universe – the cosmic film loop. 'These were the days,' Hoyle later wrote, '. . . when anything that anybody suggested was avidly discussed.'[13] But they soon hit a stumbling block: if the universe is expanding, how could it stay the same? Gold wondered whether a universe spreading itself ever more thinly could be restocked if matter were created within it,[14] and delighted by his idea, he tried to persuade his friends; but Bondi and Hoyle were not so confident: Hoyle thought they could 'disprove this before dinner.'[15] Dinner, recalled Bondi, was a little late that night, and eventually they all agreed that the creation of matter was indeed a possible solution.[16]

While Bondi and Gold enjoyed pondering the philosophical implications of a never-ending universe, Hoyle was dubious of their strategy,[17] and more interested in discussing his war-time idea about the origin of the elements. In November 1947 he gave a talk at Birmingham University where he discussed the building-up of elements in stars from a starting material of hydrogen. The audience was sceptical, and after the talk his former supervisor Rudolph Peierls, who had left Cambridge for Birmingham just before the war, suggested to Hoyle that if he was going to claim that the elements were made from hydrogen, he should also consider where the hydrogen came from.[18] This was a very difficult problem. Hoyle drove back to Cambridge from Birmingham through the snow in his little Singer car, muttering to himself, asking over and over again where the hydrogen came from. Soon after that trip, a combination of field equations and gravity gave him an answer: matter could indeed be created to restock the expanding universe.

Although the field equations Hoyle was working on implied the creation of matter, they did not provide a physical mechanism for it. Eventually he decided to let the mathematics suggest that matter is continuously created throughout space and time, and not worry too much about finding a physical explanation. According to Hoyle, 'there not being any immediate astrophysical objection to this concept, we accepted it as a working hypothesis.'[19] If new matter was being created everywhere all the time, the universe would have to expand to accommodate it, thus explaining Hubble's observation. So Hoyle worked out the mathematics of continuous creation, and was not worried that it violated the fundamental physical principle of the conservation of mass–energy – matter and energy are not supposed to appear out of nowhere. After all, the evolutionary cosmology was inadequate on this point too. Hoyle wrote up his work, and early in 1948 he submitted a paper, 'A new model for the expanding universe,' to the *Proceedings of the Physical Society*.

Bondi and Gold's never-ending universe, that looked the same whenever one turned up to look at it, was the product of their philosophical objection to evolutionary cosmologies. They concluded that 'if the universe is evolving and changing then there is no reason to trust what we call the laws of physics established by experiments here and now, to have permanent validity. This argument impressed us very strongly.'[20] One flaw they saw in evolutionary cosmologies was that seemingly arbitrary decisions had been made about which cosmic parameters were permanent and which changed over time. More reliable, they believed, was an approach that used what we know about the universe now to write its history. So Bondi and Gold worked from two principles. The first was the 'cosmological principle' which had emerged in cosmology during the 1930s: it states that the universe is homogeneous and isotropic – it looks the same from whatever vantage point and in all directions. The second was an innovation: Bondi and Gold's 'perfect cosmological principle' states that the cosmological principle holds over time. So the universe is by and large the same wherever, and whenever, you look. They wrote up their work under the title 'The steady-state theory of the expanding universe', and submitted the paper to the *Monthly Notices of the Royal Astronomical Society*.[21]

Hoyle was disappointed to receive a letter from the Physical Society rejecting his paper on continuous creation, 'in view of the acute shortage of paper which is forcing us to reject papers we would otherwise be glad to publish.' He was, however, advised that he should submit it to the *Monthly Notices of the Royal Astronomical Society*, to which, in terms of its content, it was in any case better suited.[22] Hoyle was offended by rejection on such unscientific grounds, and he sent his paper instead to the American journal *Physical Review*. There too it was rejected – it was too long. Only then did he take the Physical Society's advice and submit to the Royal Astronomical Society. The Secretary of the Society, William McCrea, Professor of Mathematics at Royal Holloway College of the University of London, was very taken with the idea, and the response was swift and positive:[23] Hoyle's paper on continuous creation appeared in the *Monthly Notices* shortly after Bondi and Gold's on steady-state theory, in the summer of 1948. Bondi and Gold did not know that Hoyle had been unsuccessful in his earlier bids for publication, and they had cited his paper, thinking it would be published before theirs.[24] Between them, the two papers presented the philosophy and mathematics of an unchanging universe, kept constant, despite its expansion, by the continuous creation of matter. The three men had produced the steady state cosmology, and after following their separate paths had been brought together again by the Royal Astronomical Society, under the scrutiny not of the physicists, but at the heart of the British astronomical establishment.

Although the *Proceedings of the Physical Society* for 1948 is not a particularly slender volume, a shortage of paper was a common problem for publishers, just as a shortage of most things was a problem for most people in post-war Britain. Resources had been devoured during the war, and manufacturing and agriculture were severely disrupted. As part of its five-year plan for economic recovery, the government imposed strict controls on production and consumption. Mostly, people bore the restrictions with stoicism, even though they eventually lasted until 1954. Furniture was available only to newly-weds; parents bought redundant army kit as Christmas presents for their children. Fuel, clothes and food were still closely rationed, which was rather tiresome, especially for people like Hoyle who took great interest in their food. In 1945, one could expect no more than two ounces of cheese each week, and one egg per fortnight. The rather unpopular brown bread replaced the favoured white loaf, because brown bread used more of the grain. New foods such as whalemeat were rarely popular, and horse was eaten by people happy to have 'steak' and never mind where it came from. Supplies in the shops were sporadic, and the queues could be so long that family members would take turns standing in line for their rations. When the first shipment of bananas in five years arrived in the port of Bristol from the Caribbean, it was greeted by a civic reception.[25]

The privileges of Cambridge University did not protect its members from this austere life: food and fuel were in short supply there too. The winter of 1947–1948 was bitterly cold, with temperatures plunging to −20°C at night. Hoyle's rooms in New Court at St John's, where he taught his mathematics students, were extremely cold: 'by common consent [New Court] was the worst place in College to be,' he noted.[26] The rooms, built in the nineteenth century, were spacious and had big fireplaces that would originally have been tended by servants – the fires made up and lit in the morning and stoked during the day. But in that harsh, austere winter, there were no servants, and worse: there was very little coal. Hoyle's ration was one bag per week. He envied his colleague across the landing whose room had been fitted with a gas fire.

Staff and students survived their hour-long tutorials in New Court by wearing as many clothes as possible. New clothes were also subject to rationing, so the challenge was to muster as many old clothes as could be worn at the same time, turning even the most elegant of students into shabby balls of wool and tweed. One student, Abdus Salam, the future Nobel Laureate, had arrived from India the previous summer and had expected England to be cold; but he had not anticipated the shortage of food. His ration book was, as for all students, turned over to the College which undertook to feed him, and Salam trudged through the snow to scour the markets looking for local produce that the farmers around Cambridge could sell outside of the rationing system. He found consolation in a supply of apples, which he ate at every opportunity. Hoyle, who was not very comfortable with one-to-one teaching, nevertheless enjoyed tutorials with the talented Salam, but in that terrible winter, he 'would be anticipating the end of the hour, when it would be possible to rush to the Combination Room, where an austerity fire would be burning, and Abdus would no doubt be anticipating his next apple.'[27]

The privations did not deflect Hoyle from pursuing a cause dear to his heart: what he saw as the travesty of the refereeing process that prevented good papers from being published in reputable journals. Along with Lyttleton and Bondi, he decided that the Royal Astronomical Society was being far harder on theoretical papers than it was on observational papers. Bondi undertook a survey of published papers in astronomy and compiled a list of those that had subsequently turned out to contain mistakes. He found that observational mistakes were far more common than theoretical mistakes, suggesting that the refereeing process was less good at weeding out poor observations than it was at detecting poor theory. Bondi wrote a paper on his findings and submitted it to the Royal Astronomical Society, and it was duly refereed: the verdict returned to Bondi was that his conclusions were wrong so the paper could not be published. However, Hoyle was to find that the real reason why the Society would not publish the paper was that the majority of observational mistakes Bondi had detected had been made by astronomers who were not British, and it did not want to cause offence in the international community.[28] However, rather than say that the paper was potentially politically embarrassing, the referees had declared the work incompetent. Hoyle was to interpret reactions to his work in the light of this experience for the rest of his career.

In 1948, the steady-state cosmology – or the 'new cosmology', as it was at first called – brought its three young authors to the attention of the wider astronomical community. Hoyle, Bondi and Gold all gave talks about it at seminars and professional meetings. Colleagues were interested, but usually critical. Bondi recalled that the number of people who found the theory attractive was very small; 'however, the number of people who took it quite seriously was rather large, though generally they did not like it.'[29] Particularly irksome to some was the creation of matter: in 1949, Herbert Dingle, astrophysicist and now Professor of History of Science at University College London – the same reviewer who, in *Nature* in 1930, had cautioned Sir James Jeans for his adventurous speculations in *The Mysterious Universe* – expressed his disdain at the idea of matter popping into existence out of nowhere to fill the gaps in the expanding universe. At the annual meeting of the British Association for the Advancement of Science, he said:

> Such a magic transformation in the observable world has not been admitted since the alchemists proposed to change lead into gold by means of the philosopher's stone. The new hypothesis is, in fact, precisely similar to that, except that the philosopher's stone has been forgotten and . . . the lead as well.[30]

Such strident comments only served to draw attention to the theory, and Hoyle, Bondi and Gold became known around Cambridge as the 'Cambridge Circus.'[31] But they were no longer close; at least, Hoyle and Bondi now had little contact. Their collaboration was drawing to an end just as the steady-state cosmology was coming to prominence. When Bondi had married and moved out of College, Hoyle had followed him to the flat in Trinity Street to continue their discussions; but when the Bondis' first child was born and they bought a house about a mile and a half away, Hoyle lost his Cambridge base. According to Bondi, 'Fred clearly thought this was high treason.'[32] Hoyle did not visit the Bondis for two years, until Lyttleton took him. While their occasional meetings were friendly, they drifted apart, and never collaborated in scientific work again.

Hoyle and Bondi had planned very different routes to achieve their aims. They both began their careers as outsiders, unconnected to Cambridge or London, professionally or socially; but Bondi's aim was to join and contribute to life at the heart of his adopted country – he took British citizenship in 1947.[33] His new wife was the daughter of a senior civil servant who had risen through the ranks to become Deputy Secretary – the equivalent of an admiral or general in the military service – and their marriage widened his social sphere and gave him an insight into London politics. Hoyle, on the other hand, preferred to stand apart: he mocked the honours system and many other aspects of what he called 'the establishment'; and he enjoyed telling stories of senior colleagues at Cambridge being deflated and ridiculed by the younger men.[34] In many ways, he remained the Yorkshire lad dragging a suitcase through the streets while the young gentlemen glided by in taxis.

Nevertheless, the steady-state theory still bound Hoyle, Bondi and Gold, and the minor celebrity it conferred upon them was something Hoyle could exploit. In the chilly austerity of the late 1940s, it allowed him to tread a new path: he decided he would try to make some money for his young family by exploring his potential as an author. His own earliest acquaintance with science had come through starry evenings on the Yorkshire hills, experiments in the kitchen, and trips to the public library for the latest little book of lectures by Eddington or Jeans; so he knew that science was not only for the academy. Nevertheless, his first book was a technical work: in 1949 Cambridge University Press published *Some Recent Researches in Solar Physics* in its prestigious Monograph series, and the large number of citations it achieved in colleagues' work was a measure of its success among astronomers. Hoyle also began writing journalistic pieces: among them were an article for the BBC's widely read magazine the *Listener*,[35] and another for the popular newspaper the *Star*, which asked 'Are there men on other planets?'[36] The newspaper article reflected the content of a discussion to which Hoyle had contributed on the BBC Radio's Third Programme a few weeks previously. This channel was a recent innovation: it had been launched in 1946 to broadcast literature and drama, classical music, lectures and serious discussion. The output of the early years was a cross between that of today's Radios 3 and 4, and the writer Edward Sackville-West declared it 'the greatest educative and civilising force England has known since the secularisation of the theatre in the sixteenth century' (novelist Evelyn Waugh, on the other hand, bought a radio specifically to listen to it, and said that 'nothing will confirm me more in my resolution to emigrate'[37]). Many of the Third Programme's talks were commissioned from academics, who were flattered by both the invitation and the fee. For his broadcast, Hoyle was joined by the geneticist Cyril Darlington, whom he had met through George Carson before the war, and they addressed the question 'Is there life elsewhere in the universe?'.

This interest in the possibility of extraterrestrial life was not unusual among scientists: for centuries, astronomers had been giving it serious thought.[38] As a teenager, Hoyle had read about life in space in popular astronomy books, including Jeans' *The Mysterious Universe*.[39] According to Jeans, we have stumbled into a universe that appears to be 'actively hostile to life like our own. For the most part, empty space is so cold that all life in it would be frozen; most of the matter in it is so hot as to make life on it impossible; space is traversed, and astronomical bodies continually bombarded, by radiation of a variety of kinds, much of which is probably inimical to, or even destructive of, life.'[40] And life itself is still a mystery:

> . . . is a living cell merely a group of ordinary atoms arranged in some ordinary way, or is it something more? Is it merely atoms, or is it atoms plus life? Or, to put it another way, could a sufficiently successful chemist create life out of the necessary atoms, as a boy can make a machine out of 'Meccano', *and then make it go?* We do not know the answer. When it comes, it will give us some indication whether other worlds in space are inhabited like ours. . . .[41]

Jeans was far from the only distinguished astronomer in the 1930s who discussed questions of biology; nor did biologists question his authority: in 1931, when H.G. Wells, Julian Huxley and G.P. Wells sought a source of opinion on extraterrestrial life for their monumental textbook *The Science of Life*, they could do no better than to reproduce an extensive quote from Jeans.[42]

As an undergraduate Hoyle had taken a keen interest in biology, and he had considered postgraduate study in biology, rather than in physics.[43] He still, years after the childhood discussions with his father, struggled with Darwinian theory, and part of his concern over the age of the universe was about how the natural world could have come to its present state in such a short time. So it was neither extraordinary, for Hoyle the astronomer, nor surprising, given his interest in biology, that he should have given some thought to questions of life in space, or that one of his earliest forays into popular writing concerned the origin of life. His article for the *Star* began with the easy familiarity that would become his trademark: he introduced a discussion of the very narrow range of physical conditions in which life on Earth can thrive by remarking that 'Lately, most of us seem to have found the nights a bit on the warm side,' and he continued by comparing the universe to a cricket pitch. Hoyle asked if life could arise on other planets, and said that 'biologists apparently feel fairly confident that it would' and that 'we can be pretty sure that creatures very similar to ourselves in mental capacity and appearance inhabit not only the earth but the universe as a whole.'

Among Hoyle's Cambridge colleagues was Peter Laslett, a research fellow in history at St John's who was also a producer for BBC Radio. They had often chatted over dinner in college. Laslett's job was to bring lecturers to the microphone on the Third Programme, and he had been working with the historian Herbert Butterfield to develop a series of talks for the spring of 1950. Butterfield was Professor of Modern History at Cambridge and a Fellow of Peterhouse College, and his book *The Origins of Modern Science* had been published in 1949 to great acclaim. The book argued that no understanding of the history of Western civilization could be complete without an appreciation of the history of science. It was published at a time when science was emerging as a strong force in British popular and political culture, and Butterfield hoped that the book would allow people from the humanities and the sciences to get better acquainted.[44]

The series was to be a significant one for the BBC, and Laslett was disappointed when Butterfield pulled out of the series just a few weeks before it was due on air. He looked around for someone who could fill in at short notice, and thought of Hoyle. Astronomy had been a popular subject on radio since the very earliest days of broadcasting, and Hoyle had the experience of his discussion about life on other planets. He had also given a lecture on sunspots, but it had not gone well – Hoyle felt it had been 'a disaster.'[45] As in his recent book on solar physics, Hoyle had been talking as if to colleagues, and not to the rather different audience for radio. But Laslett thought he could work with Hoyle to bring his broadcasting skills up to standard. So he asked Hoyle if he could come up with a series of lectures for Butterfield's slot. Hoyle found that he would get £50 for each of five talks – maybe more if there were to be an accompanying article in the *Listener* – so he could expect to earn in a few weeks a sum greater than the amount his Fellowship brought him each year. The offer was too good to refuse.[46] He and Laslett worked together on the scripts, finding ways to speak to the Third Programme's eclectic audience. According to Hoyle,

> [Laslett] knew me well enough to finally make it clear to me that the first thing one has to be aware of in preparing anything whether written or spoken is who you are writing for or who you are aiming to speak to – this isn't so obvious if it's radio because you don't see your audience, but it was absolutely vital to know . . . [Laslett] was very very clear – he was a historian by profession – and he absolutely insisted that everything was comprehensible to him to start with, and then he would do another higher order effect on the script to make sure. . . .[47]

The broadcasts, given live at 8pm on Saturdays of the snowbound January and February of 1950, were called *The Nature of the Universe*, and covered many topics in astronomy and cosmology. Because the preparation time had been so short, the talks were mostly written in the week before the broadcast – once one talk had been delivered, the next could be composed in between lectures during the busy Lent term, and batted back and forth between Hoyle and Laslett and finally to the BBC censors in time for Saturday night. Hoyle used up his family's petrol ration driving from Cambridge to Broadcasting House in the West End of London, which was invariably deserted, even on a Saturday night, because no-one had much petrol and the weather was so bad. He delivered his talks, which were packed full of homely analogies and familiar scenes, in his slow, warm Yorkshire voice. Regional accents were rare on the airwaves then, and the contrast with the typically plummy BBC announcer who introduced him each week was striking. The series started close to home with a talk on Earth and its environs, and moved on to the Sun and stars. In a talk on the expanding universe, Hoyle invented a nickname for the evolutionary cosmology:

> . . . the assumption [is] that the Universe started its life a finite time ago in a single huge explosion. On this supposition the present expansion is a relic of the violence of this explosion. This big bang idea seemed to me to be unsatisfactory even before detailed examination showed that it leads to serious difficulties.[48]

Some astronomers have since claimed that the phrase 'big bang' was meant to make the idea sound ridiculous, but Hoyle has insisted he had only tried to make it vivid.[49] In his lectures, Hoyle challenged the theory, offering as an alternative the steady-state cosmology with its continuous creation. He returned to Gold's metaphor, inspired by their trip to the cinema:

> Now let us suppose that the film is made from any space position in the Universe. To make the film, let a still picture be taken at each instant of time. This, by the

way, is what we are doing in our astronomical observations. We are actually taking the picture of the universe at one instant of time – the present. Next, let all the stills be run together so as to form a continuous film. What would the film look like? Galaxies would be observed to be continually condensing out of the background material. The general expansion of the whole system would be clear, but though the galaxies seemed to be moving away from us there would be a curious sameness about the film. It would be only in the details of each galaxy that changes would be seen. The overall picture would stay the same because of the compensation whereby the galaxies that were constantly disappearing through the expansion of the universe were replaced by newly forming galaxies. The casual observer who went to sleep during the showing of the film would find it difficult to see much change when he awoke. How long would a film show go on? It would go on for ever.[50]

The BBC used a Panel of nearly 4,000 listeners from different walks of life to provide an insight into audience reactions. Panel members reported enthusiastically on the broadcasts, awarding each of them an 'appreciation index' in the upper seventies compared to the average for Third Programme talks of 62. By the final programme, the index had risen to 87, 'the highest figure recorded for a reported talk on any Service since the beginning of 1946.'[51] Hoyle's descriptive style and familiar analogies were highly commended, with Panel members mostly reporting that they were pleased to find that they followed the technical content of the talks without effort. Hoyle's Yorkshire accent drew mixed reactions, however: some thought it disagreeable, and 'unsuitable in a scientific talk', while others found it homely and sincere. As the series progressed, some members reported that after initial reservations they had now grown accustomed to Hoyle's voice, while one, a chemical engineer, said 'when a man has something like this to say, what the devil does it matter about his voice and delivery.'[52] The Panel members knew they were hearing some revolutionary science: they noted a contrast with what they had read in Eddington and Jeans, and a baker reported 'a complete volte-face from our school day ideas.'[53]

The Nature of the Universe was a huge success with the Third Programme's 300,000 listeners, and, as the Panel members had recommended, the talks were soon repeated on the more popular Home Service radio channel. They were also printed verbatim in the *Listener*, which then had a circulation of 150,000, and then as a book, which, in its first six months, sold 77,000 copies, making it one of the biggest scientific best-sellers ever at the time.[54] The book was eventually published around the world, and in the USA it was serialized in *Harper's* magazine. Hoyle earned around £1,000 in royalties for 'quite a few years thereafter,'[55] money he saved. He was also 'inundated with all manner of invitations to dissipate my energies', including an offer of a job from the *Daily Express.*

In November 1950 the *Daily Graphic* announced: 'Fred Hoyle is First.'

> Who is Britain's most popular broadcaster of 1950? [Comedian Wilfred] Pickles? [Reporter] Richard Dimbleby? [Actress] Gladys Young? The answer – believe it or not – is Fred Hoyle, Esq., M.A., astronomer, mathematician, Fellow of St John's College, Cambridge, and University Lecturer in mathematics. He has been rated by radio listeners as a better broadcaster than anyone else whose popularity has been investigated by the BBC Audience Research Service. BBC pollsters found him more popular than Bertrand Russell, [philosopher] Dr Joad, [comedian] Tommy Handley or even Wilfred Pickles. . . . Hoyle's challenging ideas have been helped by his homely accent, his dogmatism, his confidence. But they have not avoided criticism. He has been attacked in pulpits up and down the country. . . .[56]

Hoyle's cosmology had indeed caused some disquiet in religious circles, particularly on the issue of continuous creation: like his childhood hero Jeans, he was now being denounced from the pulpit. In the final lecture, 'A personal view', Hoyle had given 'some consideration to contemporary religious beliefs.' The BBC censor had already moderated the script, because, according to Hoyle, 'I used the word "God". It wasn't that I said anything particularly bad about God. It was just that I wasn't licensed to use the word. If I had said "tonsils" it would have been the same, since I wasn't a licensed medical practitioner either.'[57] But even after the censor's attentions, words such as these from Hoyle still caused a stir in the rule-bound Britain of 1950:

> There is a good deal of cosmology in the Bible. My impression of it is that it is a remarkable conception, considering the time when it was written. But I think it can hardly be denied that the cosmology of the ancient Hebrews is only the merest daub compared with the sweeping grandeur of the picture revealed by modern science. This leads me to ask the question: is it in any way reasonable to suppose that it was given to the Hebrews to understand mysteries far deeper than anything we can comprehend, when it is quite clear that they were completely ignorant of many matters that seem commonplace to us? No, it seems to me that religion is but a desperate attempt to find an escape from the truly dreadful situation in which we find ourselves. Here we are in this wholly fantastic Universe with scarcely a clue as to whether our existence has any real significance. No wonder then that many people feel the need for some belief that gives them a sense of security, and no wonder that they become very angry with

people like me who say that the security is illusory. But I do not like the situation any better than they do. The difference is that I cannot see how the smallest advantage is to be gained from deceiving myself. We are in rather the situation of a man in a desperate difficult position on a steep mountain. A materialist is like a man who becomes crag fast and keeps on shouting: 'I'm safe, I'm safe!' because he doesn't fall off. The religious person is like a man who goes to the other extreme and rushes up the first route that shows the faintest hope of escape, and who is entirely reckless of the yawning precipices that lie below him.[58]

The Listener Research Panel had worried that such passages would arouse opposition; some members commented that Hoyle clearly did not know enough about Christianity to be entitled to criticize it. Others thought Hoyle had offered his remarks 'humbly and fairly,' and a hostel-keeper was 'delighted that the BBC has come off its pulpit for once.'[59] The commentator Malcolm Muggeridge took the former position: his review of the book in the *Daily Telegraph* said of Hoyle:

He . . . re-writes the Book of Genesis, not, truly, as poetically as the original, but with no less assurance – perhaps more. . . . There can be no reasonable doubt that Mr Hoyle knows more about the stars than Galileo did; but it is by no means certain that he knows more about Man and the circumstances of human life than, say, Shakespeare or the author of the Book of Job, or St Augustine, not to mention St Paul. Mr Hoyle, indeed, from time to time displays a certain intellectual arrogance, which is liable to characterise the contemporary scientist.[60]

The book was reviewed in *Nature* by Herbert Dingle, who had already vented his outrage at the idea of continuous creation. Dingle was still extremely sceptical of the cosmology outlined in the talks. He had also been crusading against 'unscientific' cosmology, or what he called 'cosmythology,'[61] from his review of Jeans in 1930 onwards; steady-state theory was merely his latest target.[62] According to a student studying history of science under Dingle in 1950, his class was required to listen to Hoyle's radio series, and write reviews of all the programmes.[63] In his *Nature* review Dingle criticized Hoyle for presenting a personal view of the universe, and suggested that he ought to have distinguished between his own ideas and the conventional position – a comment that was also made by the Listener Research Panel (though Hoyle appears to have been explicit on this point: after describing the evolutionary cosmology, he said 'My own view is very different.'[64]). Dingle also felt that Hoyle's imagination was 'insufficiently offset by critical discrimination.' This did not matter in the astronomical world, where 'others may be trusted to separate the sheep from the goats', but Dingle was uneasy about Hoyle sharing unconventional ideas with the public. He chastised Hoyle for his 'certainty that the conclusions now reached must be right; and . . . [his] conviction that there is not much more to learn.'[65] However, Dingle did note two of Hoyle's qualities which he believed accounted for the success of *The Nature of the Universe*:

> In the first place, he has imagination – not only the power to conceive new ideas, but also the ability to work out their consequences and visualize their implications. He can look out on the universe with vision unperverted by conventional expectations and can bring fresh ideas to bear on the problems. . . . His other quality is the ability to hold the reader's attention and carry him forward with understanding to the conclusion of the whole matter. He can . . . write sentences (not to choose the worst) like 'A long list of this sort of statement could be compiled,' and still leave the reader unwilling to exchange the volume for, say, *The Oxford Book of English Prose*. This is a great gift. It seems to arise from the natural ability to visualize clearly the type of mind of those addressed and to speak consistently in appropriate terms.[66]

Public reaction to Hoyle's broadcasts went beyond the fascination with the astronomy. His challenge to authority in both science and religion was astonishing in a country where people still waited dutifully in queues for their meagre rations. According to one young listener who later became a student of Hoyle's:

> . . . people were quite taken by the audaciousness of the idea of the steady-state theory, continuous creation, all of it. These days, if someone were to say that they didn't believe in the standard god of the Christian religion, it wouldn't merit a line on the back page. But in England at that time everyone was expected to hew to the absolute establishment line. The BBC was completely pro-King, pro-Church of England, pro-everything that held the establishment together.[67]

Yet for all his unorthodoxy, there was something cosy and familiar about Hoyle. His Yorkshire accent marked him out immediately from the usual BBC pundits. When he was invited to broadcast on cricket, the *Yorkshire Post*, the celebrated newspaper of Hoyle's home county, commented that '. . . listeners will be familiar with his lively style and warm Northern voice.'[68] The *Sunday Express* described Hoyle as 'that cheerful young man with a Northern accent and a familiar way of talking about stars,'[69] and the London *Standard* reported that 'As he walks about Cambridge, Hoyle is stopped by workmen who want to discuss his theories.'[70] But *The Nature of the Universe* was not only for laypeople. Physicist Charles Barnes, who was a research student at Cambridge in 1950, recalls:

...the famous set of [radio] lectures that Fred gave on astronomy. In the Cavendish lab virtually everything stopped every time he spoke ... this absolutely wonderful magnificent series of lectures that must have educated a whole country of physicists and astronomers.[71]

For Barnes, one of the key messages of the lectures was about the innovation in astrophysics that was to prove so fruitful in the post-war years: that links could be made between astronomy and nuclear physics.[72] Biologist J.B.S. Haldane, whose *Daily Worker* articles Hoyle had so enjoyed as a student, wrote to him, and they corresponded for many years; and so did the novelist J.B. Priestley, with whom Hoyle was to enjoy an enduring friendship.[73]

In 1950, preparations for the Festival of Britain were under way. This mix of exhibitions, events and dramatic modern architecture on the south bank of the Thames in central London aimed to recapture the spirit of the Great Exhibition of 1851 and to drag Britain out of the post-war gloom. It set great store by science as a positive factor in shaping the future of the nation. When the Festival opened in 1951, its millions of visitors found, in its Discovery Dome, that continuous creation was taking pride of place in the section on the nature of matter. The text, which drew heavily on *The Nature of the Universe*, explained how this new British idea helps us to understand the expanding universe. The universe would become empty as matter spread out, and we would not be able to see any galaxies other than our own. However, this is not the case, for 'it is thought that new galaxies are being formed as fast as others disappear, so the total number visible with a "perfect" telescope would always be about the same.'

Why then do we not run out of interstellar gas from which the galaxies are formed? We now think that the basic hydrogen is 'created' as fast as it is lost. So the universe is being continuously created and did not suddenly happen. If this is so, the universe we see goes on for all time – as hydrogen is used up, so more appears.[74]

Hoyle's public prominence was soon set in stone. Russian mosaic artist Boris Anrep had been commissioned, in 1926, to design a floor for the foyer of the National Gallery in London. The project would represent 'the intellectual life of the modern age,' and include portraits of a variety of figures including writer Edith Sitwell, painter Augustus John, Bertrand Russell and Ernest Rutherford. The last floor, *The Modern Virtues*, was completed in 1952; and among the virtues are 'defiance' ('Sir Winston Churchill stands before the white cliffs of Dover and defies an apocalyptic beast');

'delectation' ('Dame Margot Fonteyn listens to the Hon. Edward Sackville-West playing the harpsichord'); and 'leisure' ('T.S. Eliot contemplates Loch Ness and Einstein's formula'). The virtue 'pursuit' shows Hoyle as a steeple-jack, climbing to the stars.[75]

In 1954, Hoyle's status as a public figure was affirmed when he was the guest on 'Desert Island Discs', a BBC radio show then 12 years old, in which a well-known person discusses which recordings he would want with him were he to be shipwrecked on a desert island. Among Hoyle's choices were Clarke's *Trumpet Voluntary*; 'La ci darem' from Mozart's *Don Giovanni*; Laurence Olivier performing the rousing 'St Crispin's day' speech from Shakespeare's *Henry V*; Purcell's *Nymphs and Shepherds*, and, in an echo of his childhood, the Scherzo from Beethoven's Piano Sonata *Opus 106*. Hoyle talked about the international character of both music and science; he described space travel as a nuisance, and he urged observational astronomers to test cosmological theories. He thought he would be 'pretty bad' at coping on his desert island, though he could perhaps make a harpoon and catch fish; and he would be lonely: he had chosen the recording of Olivier so that he could have the comfort of human speech.[76]

In the mean time, the steady-state cosmology had taken its place among the many new ideas that were stirring up astronomers in the early 1950s. The flood of data from new telescopes, such as the 200-inch Hale Telescope at Mount Palomar in California, coupled with renewed scientific activity after the war, prompted conflicting interpretations and raised tempers. Now that there was plenty to fight over, astronomers were fighting to win, and sometimes the atmosphere was less than polite. One event that has stuck in the minds of astronomers for more than half a century took place in 1951 at University College London, during what became known as 'the Massey Conference', after its host, the Australian physicist Harrie Massey, who had recently come to UCL as Head of Physics and Astronomy. The conference, called 'The Dynamics of Ionized Media', looked at the behaviour of collections of charged particles – a topic of keen interest to astronomers.

It was just such an ionized medium, the ionosphere in the Earth's upper atmosphere, with its reflecting properties for radio-wave radiation, that had inspired the development of radar. Before the war, scientists developing radio technology had argued that stray signals interfering with their trans-missions were coming from space. The radar scientists had found the same radio signals coming from the Sun and elsewhere, and some of these scientists had applied themselves to exploring the signals when they returned to their laboratories after the war. The sources of the radio signals became known as 'radio stars', and exploring them was turning into a new branch of astronomy. The British and the Australians were investing heavily in radio-astronomy, partly to make good use of the expertise and equipment generated by the war, and partly, at least in the British case, because the radio-frequency observations were not interrupted by bad weather.[77] So many of the former radar scientists were keen to attend Massey's conference on ionized media, which became an important gathering for the new science of radio-astronomy.

Hoyle and Gold drove from Cambridge to the conference in London in Gold's unreliable Hillman car – a car that, despite costing ten times Hoyle's £5 Singer, was plagued by faults so esoteric that only someone of Gold's technical ingenuity could fix them, which he often had to do *en route*. A graduate student called Geoffrey Burbidge was studying astrophysics with Massey at the time of the conference, and was helping out by handing round and collecting in the slips of paper on which participants were writing comments and questions. Burbidge was new to astrophysics, and knew few of the people – he had, however, met Hoyle at a conference in Paris in 1948, an encounter memorable for Hoyle haggling over their bill in a restaurant: Hoyle knew no French, and the waitress, in his view, knew no arithmetic.[78] Burbidge would later acquire a reputation for outspokenness himself; but at the Massey Conference he was amazed by what he saw in one intemperate session, where the discussion concerned the nature and location of radio stars: the usual rules of academic decorum seemed to be abandoned. The spark for the change of mood was struck when one participant stood up and challenged the current idea that radio sources were companions to nearby stars. This man – Burbidge was later to find out that he was Tommy Gold – 'loudly and clearly stated the following proposition':

> As far as we can tell, these sources, these radio stars, whose origin is not known, are distributed fairly uniformly around us. This means that they are outside the solar system. We live on one side of a galaxy which is rotating, far from the centre, in other words we are not in the centre of the Milky Way. If they are distributed uniformly around us then they may well be very close to us, in which case the diameter of the volume in which they are contained is small compared to the thickness of the disk [the galaxy]. Or, they may be very very far away.[79]

The second alternative was the challenge: at the time, most people believed that radio stars were close by. Then, recalls Burbidge:

A very tall man with very bad teeth got up and essentially told [Gold] in no uncertain terms that he was stupid, he was a fool. And from the other side of the room another man got up and ... poured cold water on [Gold's] hypothesis.[80]

The second man was George McVittie, a cosmologist from the University of London. The first, very tall, man was Martin Ryle, a physicist from the Cavendish Laboratory at Cambridge. Ryle's own notes of his contribution to the exchange begin with 'I think the theoreticians have misunderstood the experimental data.'[81]

Hoyle came to Gold's defence: both he and Gold knew Ryle well, and they did not get along. Ryle was to become a powerful critic of the steady-state cosmology, and the tensions in their relationship were to complicate the scientific debate for the rest of their lives.

Gold thought that some of the friction between him and Ryle was due to his own poor relations with the Head of Radiophysics at the Cavendish Laboratory, Ryle's boss and mentor Jack Ratcliffe, who alongside his scientific work was also responsible for the administration of the laboratory. At the end of the war Gold had been awarded a grant by the Medical Research Council for work on ultrasound imaging, and had seen this as his passport out of the Admiralty and into the Cavendish, where the research was to have been conducted. However, the grant was not accepted by the Cavendish, a move for which Gold held Ratcliffe responsible. According to Gold, 'that made me very annoyed with Ratcliffe. And I suppose Ratcliffe knew that and then all his people somehow got to know that too.'[82] Gold nevertheless took up a Fellowship at Cambridge in 1947, and in 1948 he and Ryle were candidates for the same radio-astronomy post at the Cavendish. Ryle got the job, and tensions between him and Gold were such that Ratcliffe banned Gold from the premises. Overall, according to Gold, 'we weren't on good terms.'[83]

Ryle was just a few years younger than Hoyle, and he too had spent the war working on radar, at TRE, the Telecommunications Research Establishment in Worcestershire. But in many ways they were very different: Ryle was a middle-class, home-counties public schoolboy and Oxford physics graduate, and he was a newcomer in Cambridge in 1945 when he started his research on radio stars. He inspired great respect and affection in his friends, who remember him as deeply committed to all aspects of his work – he approached theoretical work and technical chores with the same vigour and commitment[84] – but others found him a rather difficult character.[85] One of Ryle's wartime colleagues was the physicist Bernard Lovell, from the University of Manchester, who went on to set up the radio-astronomy programme on the university's Jodrell Bank site after the war. Lovell had a good relationship with Ryle, but found him a very tense person with a violent temper.[86] Ryle had found the war-time situation very disturbing. His research had involved testing radar sets on bombers, and when a plane carrying an experiment crashed and his colleagues were killed, Ryle had found it deeply traumatic. According to another radar scientist and future colleague of Ryle's, the then Cambridge undergraduate Antony Hewish, there were also heated arguments over technical issues at TRE, and one story circulating there was that Ryle had thrown a bottle of ink at a senior officer:

> . . . when military planners were beginning, [Martin Ryle] thought, to be a bit casual about wasting people's lives in the air, . . . Martin would get very het-up about that. He had an extremely strong social conscience, and if anybody was risking lives unnecessarily he'd get very annoyed, naturally. . . . He was supposed to have hurled an ink-bottle at an air-vice-marshal . . . whether it hit I'm not sure. But that person never bore any grudge against Martin Ryle, and if Martin Ryle was inflamed about something and let off, the next day, if you were on good terms with him, he would be marvellous.[87]

Such intense outbursts were to be an ongoing feature: colleagues have described him as obsessive, fervent, passionate and religious (in the least complimentary sense – Ryle was well known to consider himself a Humanist[88]); and he has been credited with both genius and paranoia.[89] According to Gold he was 'a very egocentric, absolutely unbelievably intense person . . . he had too much intensity of beliefs that were not compatible with the science.'[90] He was a very quick thinker and did not suffer fools, as Hewish noticed:

> He was so incredibly bright, he would see things light years ahead of anybody else, and that tended to be an annoying feature sometimes – he just obviously knew the right answers all the time and people respected him enormously for that . . . and if he thought someone had been stupid he was inclined to be rather short with them.[91]

At the Massey Conference, it was clear whom Ryle identified as fools.

Hoyle and Gold were both in angry moods as Gold hammered the Hillman home to Cambridge, and they sought consolation along the way in a late-night fry-up at a transport café. Hoyle later told Burbidge that Gold was: '. . . absolutely beside himself with rage as they drove back to Cambridge, and the car broke down twice and that didn't make it any better.'[92]

The conflict over the location of the radio sources 'raged in England', according to the American astronomer Jesse Greenstein.[93] It was Greenstein's colleague Walter Baade, Hoyle's host on his wartime trip to Mount Wilson, who settled the argument: Ryle's colleague Francis Graham-Smith had asked Baade for help in identifying one of the sources they had found, an object in the constellation Cygnus that they called Cygnus A, and had sent him its position. Baade found that Gold had been right: the source had an extremely high redshift, which implied that it was indeed very very far away. Baade took the photographic evidence for this to the International Astronomical Union meeting in Rome in 1952, just a year after the Massey Conference, and showed it to Ryle in the lobby of the conference hotel. Gold witnessed the scene:

> Walter [Baade] explained what was on those pictures and Martin [Ryle], without saying another word, threw himself on the couch that was in that antechamber, buried his head in his hands and sobbed. Just imagine that situation. He was lying there for quite a while, absolutely completely uncontrollably sobbing. . . . It was just impossible to have him with this passionate pride overcome that situation – that he had so openly fought us at a big conference in London and then a year later he was clearly defeated, and that he never forgave.[94]

Hoyle would clash with Ryle many times in future years, and he dated the origin of the rivalry between them to that first exchange at the Massey Conference:

> Ryle began an attack that was to persist for almost two decades. . . . There is all the difference in the world between a critic saying 'I don't agree with you' or 'I get a different answer' and Ryle's habit of flat denunciation. On this occasion, he began, 'What the theoreticians have failed to understand . . .' (with the word 'theoreticians' implying some inferior and detestable species). Actually, as it turned out, the theoreticians hadn't failed to understand at all.[95]

4
New world

*Out of the doldrums; an American road trip; getting noticed
at Caltech; meeting Willy Fowler; finding a new state of
carbon; the cosmological controversy; religion and politics;*
A Decade of Decision

Hoyle's academic output for the early 1950s was relatively slight. He published only 13 new papers in the years 1950 to 1953, half the number for the years 1946 to 1949.[1] Nor were colleagues citing his papers in their own work as frequently as before. He was actively producing papers, but Hoyle felt he was suffering from 'a sort of revenge' by scientific colleagues because of his public profile and the success of the radio programmes. He had already felt there were battles to be fought for the acceptance of his work before he became a public figure; now there was this added censure from colleagues:

> 1951–52 . . . were bad years for me because I ran into a lot of trouble with the professionals – one way of taking a sort of revenge on me was to make it very difficult for me to publish my papers. They never succeeded if I was determined to print a paper – they never to my memory stopped me doing anything that I really wanted to print, but it became such a psychological effort that from about 1950, from the time I did those broadcasts, to 1952 I hardly published anything.[2]

Hoyle's publication record bears out his claim: his productivity reached its lowest – one paper in a year – in 1952. Perhaps reflecting publication time lags, the low point in citations by other scientists was reached in 1954.

Other high-profile scientists of Hoyle's generation have recorded similar experiences. Popularizing science was a risky activity, especially for a young scientist in a controversial field. According to its critics, popularization demeaned science; and it unduly magnified the work of the popularizer in comparison with that of his colleagues, and gave minority viewpoints undeserved prominence.[3] Hoyle knew that many scientists felt that way about popular science, but took the risk anyway:

> [Popularisation] was frowned on very heavily by the scientific establishment of the day, so it was a tug of war between what one might earn with a young family, and incurring the unpopularity. If you were incomprehensible to the public that was OK, but if there was any appreciation by the public that was regarded as very bad.[4]

But it may have been *The Nature of the Universe* that provided Hoyle with a lifeline. In November 1951, he received a letter from Jesse Greenstein at the California Institute of Technology, sounding him out informally about the possibility of Hoyle taking up a Visiting Professorship in the spring term (the Americans' 'winter' term) in 1953.[5] Greenstein had gained his PhD at Harvard University in 1937 with a thesis on interstellar matter, and then had concentrated on the spectroscopic analysis of starlight at the University of Chicago's Yerkes Observatory. He had moved to Caltech in 1948 to establish a graduate department of astronomy, to make good use of the new 200-inch telescope at Palomar. He was looking for some stimulating visitors, and had torn out and kept some of the articles from the *Harper's* magazine serialization of *The Nature of the Universe*. These were the only supporting documents he filed with his invitation to Hoyle.[6]

Caltech sits at the foot of the San Gabriel Mountains, in Pasadena, just north of Los Angeles. Instituted in 1920 under physicist Robert Millikan and privately funded, it was developed as a centre of excellence in scientific research, and boasted the world-beating telescopes at Mount Palomar and Mount Wilson that Hoyle had visited on his illicit detour during his wartime trip to the USA. The campus of buildings designed by leading contemporary architects, surrounded by sub-tropical trees and flowers in the California sunshine, was a far cry from Cambridge in November. Edwin Hubble was there, and Walter Baade, so prospects for astronomical discussions were good.

The post required of Hoyle three hours of teaching per week, and offered in return not just accommodation in the splendid and highly congenial Caltech faculty club, the Athenaeum, but also access to the observatories. Greenstein's invitation, which emphasized his own interests in interstellar matter, his group's work 'on the borderline between theory and observation', and their own lack of a theoretician, was one Hoyle found very attractive. Christmas was an inconvenient time to be negotiating a leave of absence at Cambridge, but by mid-January Hoyle was able to reply that he thought it likely he would be able to accept the invitation.[7] The trip was to be the first of many.

Hoyle later remarked that Greenstein's letter had saved his career:

> Then I got an invitation from the Americans . . . and the attitude was completely different, so I woke up again then . . . But if I'd had to stay in this country . . . I don't know what would have happened, whether I would have come to life again or not. . . .[8]

Hoyle headed off for the USA just before Christmas 1952, giving himself almost a month to reach Caltech. He had planned an adventure.

His flight was in several legs: he left Bournemouth on the south coast of England and travelled via Scotland, Iceland and Newfoundland in huge troop-carriers that had been transferred to civilian service after the war. Hoyle's first appointments were in Princeton, where he visited Henry Norris Russell and his wife, renewing their war-time acquaintance (Hoyle recorded in his diary that he was unable to break into the rapid flow of their conversation). Hoyle also met Martin Schwarzschild, who, like Russell, worked on stellar evolution. Schwarzschild was pioneering methods for using computers to undertake the protracted and complex calculations of the many changing physical properties of evolving stars. Hoyle was very impressed with Schwarzschild and captivated by his computers and their potential. This meeting was the basis for a fruitful collaboration: Hoyle and Schwarzschild would produce some important work on stellar evolution in the years to come.

Before Hoyle left Princeton, Schwarzschild lent him $1,500 that was overdue from *Harper's* for *The Nature of the Universe*. This generous act and the future collaboration came despite Hoyle's fears after a conversational *faux pas* regarding the war: Schwarzschild was German and Jewish, and his father, the astrophysicist Karl Schwarzschild, had died while on military service during World War I. Hoyle noted in his diary: 'terrible brick concerning Germany. Must eschew politics until know people well.'[9] Then, armed with a US driving licence and a second-hand Chevrolet, Hoyle felt that after the privations of the war and the drab years since, 'driving off into the blue across the United States made you understand what freedom really meant.'[10]

He headed south through Maryland and Virginia, where he thought American women were prettier than on his last visit during the war – maybe they were taking more care about their appearance, he noted, now that the men were home. As he drove, he prepared a lecture on relativity, the age of the universe, continuous creation and the weaknesses of Gamow's work on the evolutionary cosmology.

Then on across Alabama and Arkansas, where the racial mix of the southern states intrigued him. He noticed the intonation of the white people: 'speech Negroid, presumably children reared by Negro servants', and in 'Indian country' he noted that 'everyone takes Indians for granted as normal people.' Generally though, he found Americans 'very rude people. A high standard of life plus very low intellectual standard.' He bought presents for Barbara and the children, and kept detailed notes of the quality and price of his meals as he travelled, including an account of a 'breakfast disaster' where he had taken a swig of a dark brown liquid he took to be coffee, only to find it was maple syrup.

Hoyle reached the Grand Canyon on Christmas Eve, and spent Christmas Day walking to the Phantom Ranch at the bottom of the Canyon. There, for $15 – four times the price of his usual motels – Hoyle enjoyed a fine dinner and gambled on a game of cards with three men he described as 'cowpokes'. Despite not having much sense of the rules of the game, Hoyle came out enough in profit to replace the shoes he had wrecked on his walk to the Ranch and back, though the fact that he had won the money to pay for them was the only thing he really liked about the strange pointed-toed American shoes. On Boxing Day he reached Los Angeles, where he noted the 'low intellectual standard' of the newspaper and 'lunch, $1.10, poorly cooked.'[11]

Hoyle's arrival in Pasadena just after Christmas 1952 coincided with the meeting there of the American Physical Society, for which Hoyle had prepared his lecture. The Caltech news bureau – not an office to be found on British campuses then – led its press release about the meeting with the announcement that Hoyle would give a talk about 'The Expanding Universe.' It emphasized Hoyle's popular success at the BBC, and his international celebrity from *The Nature of the Universe*, which 'combines modern theories of cosmology with certain concepts and ideas of his own.'[12] The lecture was held at Pasadena Junior College, because the expected audience demanded a much larger theatre than was available on the Caltech campus.

After the lecture, which was well received by the packed auditorium, Hoyle was taken to a party at the home of William A. Fowler, a nuclear physicist in Caltech's Kellogg Radiation Laboratory. Willy Fowler had come to Caltech as a PhD student in 1933, and was a jovial man with a lively sense of humour who had retained his childhood passions for baseball and steam trains. Fowler and Hoyle got along well, and Hoyle subsequently went to Fowler's office for advice on a problem in nuclear physics. It was to do with the origin of the elements.

The idea that the chemical elements might be formed in the stars was not new. However, until the advent of quantum mechanics it had been difficult to see how particles could possibly have come together to form even the simplest elements. Gamow's theory still could only account for the elements hydrogen, helium, lithium and beryllium, and they would have been formed in the explosive first moment of the universe. The presence of the rest of the elements therefore remained a mystery. The solution to the mystery was, according to Hoyle, that elements evolved gradually and continuously in the interior of stars. This idea had received a significant boost from the work of Caltech astronomer Paul Merrill earlier in 1952: Merrill found the heavy element technetium in stars. Technetium is unstable, and ought to have long since broken down if it had all been made at the beginning of time. Its presence suggested that it had instead been formed more recently, and so perhaps was still being made in the stars. Hoyle teased Merrill, whom he considered rather conservative, for having made such a surprising discovery.[13]

If technetium was still being made, what was it being made from? There was still no mechanism for anything heavier than beryllium, and technetium is about twelve times as heavy. Hoyle thought he could see a way out: in an extreme environment – the interior of a supernova, for example – a particular nuclear reaction could occur which would lead to the formation of heavier elements. Two nuclei of helium (which each have four nuclear particles) would form beryllium (which has eight), and a third helium nucleus could join in to form carbon-12, taking the elements beyond the dead end in Gamow's sequence. This nuclear reaction was the 'passport to success' that had dawned on Hoyle during the war, when supernovae and the imploding bomb at Chalk River had come together in his mind. But now he needed Fowler's help.

In 1952, Cornell University physicist Edwin Salpeter had published a paper in which he described a similar process whereby three helium nuclei – also known as alpha particles – come together to form carbon. Salpeter had been thinking about this 'triple-alpha' process during a summer visit to the Mount Wilson Observatory in 1951, when the behaviour of helium in the cores of red giants stars was interesting many people there. For a while, Hoyle thought he had been trumped: Salpeter had published the idea and would get the credit. Hoyle 'was angry with myself for having let a good thing slip through my fingers, and I was still in a mad state when I arrived at Caltech in December 1952.'[14]

However, Salpeter had not solved the problem: his 'triple-alpha' process was very inefficient. Hoyle checked to see how Salpeter's process would work in the physical conditions in stars, and found that the helium there was not burning fast enough to produce the energy required to sustain the reaction. It simply did not produce enough carbon to account for the abundance of it in the universe. Hoyle had a second chance: 'Bad luck for poor old Ed, I thought to myself, through a happy hour or two.'[15]

For the reaction to happen efficiently – for the helium and beryllium nuclei to come together and then stay together as carbon – the carbon nucleus thus formed would have to be at a particular energy level. The unit of energy that scientists use in these circumstances is the MeV, and, for the carbon nucleus, Hoyle calculated that the required energy level would be 7.65 MeV above the lowest possible level, which is known as the ground state. Nuclei exist only at certain levels of energy, which are called states, and they jump from one state to another under certain conditions, but they do this in a single leap, not gradually – there are no intermediate states for the nuclei to occupy. Hoyle faced the problem that the energy state he needed at 7.65 MeV was not listed among the known states of carbon – it fell in one of the gaps in between. But he felt that it had to exist: Hoyle thought that since human beings exist to ask such questions about the universe – and they exist in their particular biological form because carbon exists in plenty – then the universe must be one in which carbon is readily made. Since there are heavy elements in the universe, and since, in Hoyle's scheme, matter must have been in his state of carbon in order for some of it to then be the building material for the heavier elements, then this state of carbon had to exist. The only way this could happen would be if, at the point where beryllium-8 joins with the third helium-4, carbon-12 could exist and be stable in a state at 7.65 MeV.

Fowler had already done some work on beryllium-8, which is unstable and breaks up into two helium nuclei. Hoyle asked Fowler to look for the state of carbon-12, and Fowler, Hoyle recalled, 'didn't exactly laugh'.[16] Fowler did not know that Hoyle had some training in nuclear physics; and he was sceptical about steady-state theory – it was not popular among nuclear physicists – and 'in particular I was sceptical that this steady-state cosmologist, this theorist, should ask questions about the carbon-12 nucleus.'[17] But Fowler took Hoyle to meet his colleague Tommy Lauritsen, who checked the known states of carbon and found that there had been one suspected at around 7 MeV, but that scientists at the Massachusetts Institute of Technology had looked for it and failed to find it. Therefore, it did not exist. According to Fowler, Hoyle was not persuaded:

Well, Hoyle just insisted – as I remember, we didn't know him all that well, and here was this funny little man, [who] thought that we should stop all this important work that we were doing otherwise and look for this state, and we kind of gave him the brush-off. Get away from us, young fellow, you bother us.[18]

But Hoyle was not to be brushed off. Eventually Fowler gave in, and, according to his colleague the Kellogg physicist Charles Barnes, who already knew of Hoyle from his student days at Cambridge:

Willy Fowler called a meeting of several of us in his office, which was cleaner than my office but smaller, and first of all Willy said I don't believe any of this but Fred wants to talk to you. So Fred made his presentation and I recognised right away that at that time we had three van der Graaf accelerators and the one that I was associated with was not really set up to [test] this immediately, but that the smallest of the three was actually perfectly set up to do this, if the people on it would do it – a professor by the name of Ward Whaling was in charge of that machine.[19]

Whaling, who was new to Caltech, had been impressed by Hoyle's public lecture and was keen to try out his equipment. Hoyle's idea was ideal material. Whaling bombarded nitrogen with helium nuclei, producing carbon-12 and alpha particles. He found that the alpha particles leaving the carbon-12 had less energy than the alpha particles that leave carbon-12 in its usual state. This implied that the carbon-12 he had produced was in a higher state than usual, and he found that this higher state was exactly as Hoyle had predicted. According to Barnes, 'it was literally just a few weeks and we had done the experiments and sure enough: a state in carbon-12 exactly where Fred said it would be.'[20] According to Fowler, 'we then took Hoyle very seriously.' Further experiments showed that any nucleus that broke down into helium nuclei could also be made from them, opening the way to the formation of many more elements.

The idea that from one's own existence one could infer such esoteric properties of the universe – an idea now known as the anthropic principle and considered with caution by many scientists – received in Hoyle's prediction of the state at 7.65 MeV its first, and some would say only, confirmation. But henceforth, 'Fred was always joyously received in Kellogg', according to Barnes,[21] and that first visit was the start of a long relationship.

However, even in those early days, Hoyle's reception across campus at Caltech was not uniformly positive. There was much negative reaction to steady-state theory from astronomers as well as nuclear physicists, and they would snipe at Hoyle during lectures, prompting him to retreat, check his thoughts, and return the next time with a response. Hoyle was however able to return to an idea that had bugged him for years, and with its champion, Edwin Hubble: they had many opportunities to discuss the age of the universe. Hubble and his wife Grace were keen on all things English – Hubble wore tweeds and smoked a briar pipe – and they invited Hoyle to visit them soon after he arrived in Pasadena. On that first trip, they would often spend Sunday mornings walking and talking. At the time, new observations were occasioning revisions of Hubble's measures of the expanding universe. As Hubble described the controversies in his own career, Hoyle decided that Hubble had made a number of compromises in his work to make it acceptable to sceptical colleagues, and some of those compromises had resulted in an age for the universe that was too small. Hoyle was reminded of his radar colleagues and their 'measures of hope': each measurement still useful enough on its own, but which together amounted to useless data. Hoyle set about revising the age of the universe, upwards.[22]

Hubble died later that year, unexpectedly, at the age of 64. He collapsed outside his house, having just driven home. Hoyle remained friends with Grace for the rest of her life: she was to outlive her husband by 35 years.

Hoyle drove back east from Caltech, across the mountains and through the desert, stopping at Chicago University to see the Indian theoretical physicist Subrahmanyan Chandrasekhar who was working on stellar evolution, and then on to Princeton to work with Schwarzschild. In Princeton he met Immanuel Velikovsky, a scholar of Jewish literature whose work on scriptures from around the world had led him to propose that cataclysmic events recorded the – floods, fires in the sky and so on – had been caused by astronomical events during a turbulent time for the solar system during the two centuries before Christ. Velikovsky's book *Worlds in Collision* had caused such offence among some astronomers that they had, in a concerted action, tried to limit his influence. One tactic was to enrol the universities into a boycott of his publisher, Macmillan – who had survived by the simple expedient of selling the rights to the book to a publisher that sold mostly outside of academia. Velikovsky was an old friend of Einstein, which gave him his entrée to Princeton's astronomy seminars. Hoyle tried to explain to Velikovsky that one had to follow the rules: in studies of the scriptures, perhaps the text was all important; but in astronomy the rules of mathematics were what really mattered. 'This made Velikovsky look sad,' recalled Hoyle, 'which is how we parted.'[23]

Steady-state theory was still nagging at astronomers. Its challenge to the evolutionary cosmology would not be resolved by compromise: the universe either had a beginning in time or did not, and no intermediate solutions were possible. Each theory claimed possible corroboration from observation: steady-state theory predicted values for a number of potentially measurable but as yet undetermined parameters, and evolutionary theory offered the traces of radiation from the explosion at the start of time. Astronomers were not looking for this radiation; Gamow thought it would be undetectable.[24] So the observations that could help decide between the two theories on this point had not been made.

There was one development in the 1950s: it concerned the redshift of distant galaxies, which had been attributed to the expansion of the universe. This expansion was supported by much observational work and was explained by both the evolutionary and the steady-state cosmologies, and it was taken to account for the shift towards the red in the spectra of light from distant galaxies. From the redshift could be calculated the speed of motion of the galaxies as the universe expanded. According to steady-state theory the galaxies should be speeding up, but in the mid-1950s, Hoyle heard that Alan Sandage of the Mount Wilson Observatory had found that the galaxies were slowing down.[25]

But otherwise, in the absence of much by way of evidence and with shortcomings in both theories, there was for a long time little that counted as debate on technical issues. For example, *Nature*'s account of the jubilee conference on relativity theory in Berne in 1955 reports that:

> There was no violent conflict of views arising from theories depending on unique or upon continuing creation of matter. Apparently the general attitude was that a verdict must await the outcome of further evidence. . . . But, indispensable as present theories are for giving direction to the work, certain of us expect that fresh concepts will sooner or later be needed.

The report also remarks: 'special reference was made to the ways in which radio-astronomy may soon provide crucial information.'[26]

However, a lack of debate over evidence should not be taken to imply a lack of debate: steady-state theory attracted much criticism during the 1950s. Views were mixed as to its influence; according to Gold, 'when we proposed the steady-state theory of cosmology, Bondi, Hoyle and I found all the official astronomers extremely hostile.'[27] Radio-astronomer Antony Hewish, who was impressed by the theory but did not believe it, recalled that:

It was the accepted theory in those days, oh yes: Bondi, Hoyle and Gold were a powerful influence. And Fred was the voice of astronomy in this country, and a lot of what he said was wrong but whenever he spoke on the radio people said 'oh yes, that's it.'

But the community of professionals interested in such matters was rather small – according to Hewish:

There weren't that many cosmologists around at that time. And there weren't very many astronomers and astrophysicists around at that time . . . if there were ten supporters of steady-state theory, that's quite a big number, actually . . . I mean, who was supporting the big bang? You couldn't say.[28]

Nevertheless, steady-state theory provoked debate in the broader community of scientists and laypeople for many years. Polls, debates and bouts of temper kept the theory in both the professional literature and the public eye, and the controversy was magnified by the publicity, raising hackles and fuelling debate. Ordinary people took sides: in rural Canada, a young lad was pinned up against a tree by his cousin and punched until he declared his allegiance one way or another.[29] At a night school in South Wales, a miner told his tutor after class that he had not understood the tutor's account of the evolutionary cosmology, but not to worry as he was 'a steady-state man' himself.[30]

Hoyle had many platforms from which to speak about the steady-state cosmology, both professional and public. Bondi and Gold both popularized it too – Bondi produced the influential book *Cosmology* in 1952 – but Hoyle seemed to garner most of the limelight. Indeed, he was so prominent as a popularizer of steady-state theory that some people thought that was all he was: even the Astronomer Royal, Sir Harold Spencer Jones, indicated as much to the *New York Times* when it reported his lecture tour of the USA in the spring of 1952. Continuous creation was a central theme of Sir Harold's tour, but he said that Bondi and Gold were the authors of the theory, and Hoyle its champion.[31] In the 1950s, the Astronomer Royal was hugely influential in British astronomy, and, if one wanted to achieve any-thing in the field, it paid to stay on his good side. But Hoyle could not resist correcting him: he wrote to Sir Harold, explaining that Gold had had the idea of the never-ending universe – Hoyle told Sir Harold the story of their trip to the cinema – and that Bondi and Gold had written up the steady-state aspects of the theory, while Hoyle himself took credit for the idea of the continuous creation of matter. Sir Harold said he had been misled by the two separate original papers, and by Bondi and Gold's paper having appeared before Hoyle's, but he undertook to remedy his error before his lecture was published.[32]

The continuing stream of popularizations about steady-state theory rankled with its critics.[33] George McVittie, the cosmologist who had supported Martin Ryle's criticism of Gold at the Massey Conference, even cited the 'hulabaloo about the new revelation', which he felt was squeezing out other work in cosmology such as his own, as one reason why he emigrated from the UK to the USA in 1952.[34] Herbert Dingle was still agitated about steady-state theory in 1953, when he became President of the Royal Astronomical Society. In his presidential address, fired up by Bondi's recent *Cosmology*, he singled out 'one writer' – Dingle does not name him, but the paper he cites is Hoyle's – for particular criticism.[35] Dingle is patently agitated about what he sees as a lack of scientific integrity in Hoyle's work:

> ... though some of the prominent contributors to modern cosmology have unconsciously realised that the nature of science has some special relevance to their researches, . . . they betray not only a profound ignorance, but also a lack of any serious reflection . . . they are not really concerned to know whether they are being scientific or not, but wish to dignify their own opinions, and discredit opposing ones, by invoking a name that commands general respect. . . .
>
> 'It is against the spirit of scientific enquiry,' says one writer, 'to regard observable effects as arising from "causes unknown to science," and this in principle is what creation-in-the-past [as opposed to continuous creation] implies.' . . . what conceivable misfortune could have led the writer to misinterpret both the words and the performances of almost all the great contributors to science. . . . After making this unfortunate remark he proceeds to develop his own idea, which turns out to be precisely what he has just opposed to the spirit of scientific enquiry. . . .[36]

Dingle argued that the problem for cosmologists was that since most of the cosmos is unobservable, cosmological theory is very difficult to test. To protect themselves from 'idle fancy', cosmologists therefore needed a 'strict adherence to the essential principles of science.'[37] Dingle concluded that continuous creation:

> ... appears ridiculous because it is ridiculous. . . . One naturally inclines to think that the idea of the continual creation of matter has somehow emerged from mathematical discussion based on scientific observation, and that, right or wrong, it is a legitimate inference from what we know. It is nothing of the kind . . . it has no other basis than the fancy of a few mathematicians who think how nice it would be if the world were made that way.[38]

Hoyle, Bondi and Gold had been aware from the outset that steady-state theory had shortcomings. Friendlier voices had also expressed concern: in 1950, nuclear physicist Wolfgang Pauli had reminded Hoyle that the physics of the creation of matter needed to be accounted for. Hoyle acknowledged that 'In 1948, I had no physical ideas about the details of creation. . . . So, in place of physics, I made a mathematical hypothesis.'[39] The mathematics, nevertheless, needed some support from physical theory. The key challenge was to find a mechanism for the creation of matter.

But aside from these technical points, the scientists in this debate were often responding to gut feelings, and emotional and aesthetic concerns played an important role in their scientific judgements.[40] This was not unusual: for example, in the 1930s, Eddington and Lemaître had corresponded in *Nature* on whether or not the idea of the cosmos having a beginning was 'repugnant' (Eddington thought so; Lemâitre did not).[41] According to Gold,

> Most of the responses [of our critics] were not based on anything we knew about the astronomical universe; they were mostly based on 'I find this much more acceptable than that.'[42]

In 1955, Hoyle's Caltech colleague Jesse Greenstein told him that he 'would love to have a gentle drizzle of matter being made . . . and leaking out into space, so that the universe will evolve with proper decorum towards the heavy elements, instead of with big explosions. It appeals to my aesthetic sense more.'[43] And McCrea, who had been instrumental in publishing Hoyle's paper on continuous creation, recalled that:

> Many scientists have confessed . . . that they found an aesthetic appeal in the steady-state concept as contrasted with a distaste for the notion of the big bang. . . . There appears to be nothing rational about a preference for a little bang for every elementary particle in the universe as against one big bang for the entire universe. . . .[44]

Newspapers, public lectures and popular books provided an outlet for these feelings in a way that scientific journals could not; and often, when the debate was quiet in scientific journals, it could be found in full flow in other, more public places. The cosmological theories became entangled with contemporary issues, and new perspectives were added as different features became important, especially in the popular press. The late 1950s and early 1960s were a boom time for science coverage, and for coverage of the physical sciences in particular;[45] and the theories received extensive press attention. Popular media have a broader scope than scientific journals; and in the newspapers or in a public lecture, scientists might take advantage of the relative informality to risk a more personal or impassioned criticism of their opponents. The popular press thus became a repository of attacks and reactions, arguments and allusions in the personal battles that constituted this debate.

One particularly fraught issue from the very earliest days was the philosophical basis of the steady-state theory. Although cosmology has a history as long as recorded time, its career as a science had barely begun. As recently as the 1930s the key questions in cosmology were not only 'what description does science provide of the origins of everything?', but also 'is science an appropriate tool for describing those origins?' According to one textbook published in 1931:

> The origin of the universe, it would appear, is not a problem that can be brought within the scheme to which science confines itself. Whether this indicates that the universe had a transcendental origin, or whether it indicates that science, as a scheme, is not yet completely coherent, is at present a matter of opinion.[46]

Hewish felt that there was still something of this attitude around when steady-state theory was published in 1948:

> The whole thing was so nebulous . . . cosmological theory was sort of there but people didn't bother about it because they didn't think you could really ever know the answer to these things.[47]

For this reason, cosmology provided much exercise for philosophers of science, some of whom, such as Dingle, had a general antipathy towards scientific explanations of origins – few cosmologists had gone unrebuked over the decades since the birth of scientific cosmology. Dingle was also particularly incensed by steady-state theory's cosmological principles. Bondi and Gold's cosmology was derived from these principles, and Dingle thought this methodology preposterous. In his presidential address to the Royal Astronomical Society in 1953, he claimed that:

> . . . in cosmology two impostors have usurped the throne of science, worn her crown and taken her name . . . one impostor is personal taste, . . . the other [is] pure reason. . . . How has it come about that . . . these pseudo-sciences have again come to life and threaten to deceive the very elect?[48]

Dingle was not alone in his criticism – cosmology briefly became a topic for discussion in philosophy journals, for example[49] – but his was the loudest voice to raise fundamental philosophical objections to steady-state theory.

The religious connotations of cosmology have been controversial for as long as cosmology has been part of science – Jeans' experience with *The Mysterious Universe* was typical – and this latest debate provided some more fertile ground for explorations of the issue.[50] The evolutionary cosmology, with its moment of creation, implied a Creator; whereas the steady-state universe, endless and regenerating, godlessly sustained itself. As physicist Stephen Weinberg later wrote, 'the steady state model . . . nicely avoids the problem of Genesis.'[51] But this apparent godlessness did not necessarily deter religious people, as Gold commented:

> . . . we had intensely religious people very much for us – others were much against us. The biblical religious people wanted a moment of creation, and obviously the big bang was their stuff.[52]

The religious implications of a creation moment had occasioned some resistance to the evolutionary cosmology among scientists when it was first proposed.[53] However, to others, this aspect was attractive. Lemaître was not just a mathematician: he was also a priest, and was easily recognizable in his long black robes and dog-collar,[54] but he claimed that his cosmology 'remains entirely outside any metaphysical or religious questions.'[55] However, he did not wish to deter anyone whose religious feelings might attract them to the idea:

> It leaves the materialist free to deny any transcendental Being. . . . For the believer, it removes any attempt to familiarity with God. . . . It is consonant with the wording of Isaias speaking of the 'Hidden God' hidden even in the beginning of [creation].[56]

Gamow was, like Hoyle, an atheist, but he was familiar with organized religion: his grandfather was the Metropolitan, the senior bishop, of Odessa Cathedral.[57] His satirical writings often featured God, and he used biblical language to make jokes about fellow cosmologists. He even published the endorsement of Pope Pius XII for the evolutionary cosmology at the start of a paper he published in *Physical Review* in 1952.[58] It would seem, then, that these men, irrespective of their personal beliefs, understood that religion formed one important context in which many people would judge their work.

Another important context was politics: communism became an issue. Gamow, as a Russian living in the USA, was assumed to be anti-communist, even though he had emigrated in search of a good place to do science, not to escape from politics.[59] Since the evolutionary cosmology's adopted homeland was the USA, it was therefore given an anti-communist tag. Steady-state theory was attributed, via similarly tenuous links, with a pro-communist position: for example, Hoyle visited the Soviet Union in 1958 and wrote, not uncritically, about his trip for the Sunday newspaper the *Observer*[60] – and he had spent his student days in Cambridge at a time when intellectual communism was fashionable.[61] There was a wave of scandals involving scientists and Cambridge men spying for the Russians: the nuclear physicist Klaus Fuchs, who had been part of the group at Birmingham University that Hoyle knew well, was arrested in 1950, and Guy Burgess and Donald MacLean, who had met at Trinity College Cambridge in the 1930s, defected in 1951. Bondi and Gold were perceived as left-wing because they were assumed to be in England as refugees from fascism, whereas they, like Gamow, were émigrés, and had left Austria before the Anschluss.[62] According to Felix Pirani, who studied under Bondi from 1951 to 1954, communism was so entangled with atheism that the godlessness of steady-state theory might have lent it its communist tinge, but few of the supporters of the theory were actually on the political left, and Hoyle, while an atheist, 'didn't have time to be a communist.'[63] Nevertheless, the association between the steady-state theory and communism was current for a while, and steady-state supporters were known as 'fellow travellers', a label that usually identified a communist sympathizer.[64] Hoyle's one-time supervisor Paul Dirac, who was a fellow traveller on both counts, was denied a US visa in the early 1950s – a common experience then for scientists suspected of left-wing political views.[65] Hoyle's frequent trips to the USA during the 1950s were dogged by visa problems, and once the US–Soviet space race was under way, bringing government patronage and close military links into astronomy in the West,[66] attitudes to communism became particularly pertinent for astronomers.[67]

Only the British humour magazine *Punch* reversed the political associations of the rival cosmologies: when it likened the debate to British party politics, the Conservatives championed steady-state theory, and Labour the expanding universe.[68] The fact that this configuration makes more sense in terms of the content of the theories themselves only serves to make the opposite case more striking.

Steady-state theory became known as a British theory: Gamow's friend and colleague the atomic scientist Edward Teller thought that steady-state theory appealed to the British 'because it has ever been the policy of Great Britain to maintain the *status quo*.'[69] Certainly most of its professional supporters were British or resident in Britain, and it received little attention from scientists elsewhere,[70] though it was familiar to the many readers of Hoyle's popular books around the world. The evolutionary cosmology, with its European heritage, crossed the Atlantic and became a US citizen, though Gamow remained proud of his theory's Continental ancestry: in 1956, an article he wrote for *Scientific American* magazine was accompanied by portraits of his intellectual ancestors, including the very recognizable Einstein, and Lemaître in his priestly robes.[71] Hoyle, however, was not one to claim any lineage, and he avoided mentors vigorously;[72] his theory was thus denied a supporting cast of intellectual forbears.

Evolutionary theory was a physical theory, and while it took some time to reach astronomers and be seen as a cosmology, it nevertheless came to enjoy the support, resources and prestige of the very powerful physics community in post-war America.[73] American science – albeit done by people of many nations – had used physics to win the war in a most spectacular way. The proponents of the evolutionary cosmology were all physicists, and largely nuclear physicists who carried with them the added prestige conferred by their association with atomic science. It was a time of optimism about atomic energy – atoms had brought world peace and would bring limitless power; what is more, the start of the new era had been marked by a huge explosion.[74] In 1950, the *Observer* commented that: 'The present popular interest in cosmology seems to have a connection with the atomic bomb.'[75] In Cold War America, where schoolchildren performed daily drills designed to protect them in case of an atomic attack, big bangs were an ordinary, ever-present part of everyday life.[76] The atomic age and its icons also had an enormous impact on the culture of science: they were highly visible in scientific magazines, which advertised nuclear fuel and prefabricated nuclear power stations; there were jobs to be had in new nuclear power stations and research centres; and there were calls for forward-looking scientists to design atomic-powered ships and aeroplanes. A recurring slogan was US President Eisenhower's 'atoms for peace.' The evolutionary cosmology, then, with its big bang, was a theory of the atomic age. Out of the turbulence of a big explosion had come a new universe expanding into a calmer future.

Steady-state theory could claim little association with the atomic age. Cambridge's early start in nuclear physics had been lost, and its reputation for physics was waning.[77] In any case, opponents of steady-state theory considered its proponents as mere mathematicians rather than physicists: Dingle had referred to the steady-state supporters as 'a few mathematicians', and Ryle, a physicist, had called them 'theoreticians.'[78] Hoyle had acknowledged that, in the absence of any physics of creation for his 1948 paper, he had resorted to mathematics.[79] He called physicists 'the tough guys of science ... [they] tear objects ... to pieces [to] find out how they work' – a tactic denied to mathematicians.[80] In the USA, mathematicians who were doing theoretical physics (as were Hoyle and his colleagues) would be housed within a physics department, along with experimentalists, whereas in England, they would be in a mathematics department, concerned only with theory.[81] If groups with opposing views are forced to confront each other each day, they soon have to find some common ground, and in the USA the theoreticians and the experimentalists, had they ever differed, came together behind the evolutionary cosmology. In the UK, the two scientific disciplines worked in separate institutions, and each could pursue its own line without constantly treading on the toes of the other, allowing two opposing theories to flourish in parallel.[82]

There was a strong biological flavour to steady-state theory, perhaps betraying Hoyle's long-standing interest in biology.[83] The steady-state universe is dynamic: matter is continually created and destroyed. It is cyclical, replacing lost energy with newly created matter. Unlike the dissipative, degenerate, violent and essentially physical big-bang universe, the steady-state universe is organizing, generative and constructive. According to Gold, some steady-state supporters liked the idea that:

> . . . the universe that we knew was constantly being replenished and sort of . . . biological – I don't mean literally biological, but the population-like situation that makes the new all the time and lets the old go away in some way.[84]

The steady-state theory essentially describes a kind of organic universe: indeed, in 1956 Hoyle described it as 'a kind of cosmic ecology.'[85] And in 1962, he claimed that only a steady-state universe could support life.[86] This was to become a recurring theme in his arguments for the theory.

While the principles of steady-state theory were simple and the popularizations of it engaging, the technical works describing the mathematics were very difficult for most people to understand. A *Daily Express* reporter enthusing about Hoyle confessed that 'I understand only about one word in ten of what [Hoyle] is saying',[87] and the *Sunday Times* noted that a talk for an audience of scientists 'left most people floundering.'[88] Hoyle himself acknowledged that continuous creation 'may seem a very strange idea and I agree that it is.'[89] There were few such comments about evolutionary cosmology.[90] It had a longer history, and had settled reasonably comfortably into a culture that had experience of coming to terms with evolutionary ideas. While both cosmologies presented distance and time on scales that were difficult to conceptualize, evolutionary theory at least allowed the consolation of limits – a finite volume and a start in time. The steady-state cosmos, with its endless circularity of energy and matter and its infinities in space and time, offered little that could easily be contained within the scope of the imagination, or defined by analogy with the mundane.

Poor steady-state theory: it was on shaky philosophical ground; it was new, irreligious, communist, faintly biological and British; it had not enough 'breeding' and too much mathematics, and it was too difficult to understand. Evolutionary cosmology had become a theory of the New World and of the nuclear age: a simple, explosive, progressive, capitalist theory with a fine pedigree and powerful allies in physics and in God. All these issues provided grist to the mills of those who would push for their own cosmic story; and a range of welcoming media gave them space in which to do it, on show to everyone.

Deep in the politics of cosmology, Hoyle decided to air his thoughts on the politics of Britain. In 1953 he published a book called *A Decade of Decision*, which offered an analysis of the problems of the post-war situation followed by solutions which, Hoyle believed, would be more effective than those that politicians were trying to foist on an ill-informed electorate. Hoyle's publisher, Heinemann, had sought advice from expert readers before going ahead with the book, but their reports were far from positive: the readers thought the book could do a great deal of damage to Hoyle's reputation. One reader, an Oxford statistician, noted that in reviews of *The Nature of the Universe*, which was about Hoyle's own specialist area, he had been charged with over-reaching himself; now, in this book, he was in an area in which he was not trained and 'his remarks are rarely profound and sometimes plain silly.' Hoyle would be very vulnerable, and he would be caught out. But Heinemann had arranged a deal with a science book club, and sales were assured: they published the book, and in 1954 it sold 6,000 copies.[91]

The book was dedicated: 'To those who are determined that Britain shall not decline into apathetic dependency.'[92] Hoyle claimed that his motive for writing it was duty to his country; in the preface he wrote:

> There comes a time in life when a man feels under an obligation to the country that reared him, and to the people that he grew up amongst. It may seem strange that this should have led to the writing of a book that seeks to emphasise our economic and physical insecurity. But in recent years I have come to feel more and more urgently that public commentators are not bringing this insecurity home to us with sufficient vividness; of our being tied to a wasting economy; of what the fate of this tightly packed island in an atomic war would very likely

be; of the menace of the world's rapidly rising population; and of the threat of an ultimately disastrous decline in world-wide resources.

He recognized that society does not usually turn to scientists for solutions to social problems:

> [When] scientists . . . keep silent on social and political matters, they are accused of being indifferent to the abuse of their discoveries. When they speak out, they are usually told that their views are too 'calculated', or that they are too 'naïve', or that they are guilty of oversimplifying difficult problems. . . . This has not dissuaded me . . . because the successes . . . achieved by the people who claim to understand the subtlety of human problems have not been impressive. This being so, I do not think that the non-expert need be unduly bashful about putting forward a new outlook on human affairs.[93]

A Decade of Decision offered drastic solutions to what Hoyle saw as the most important problems of the time. For example, to solve the over-population of Britain, he suggested that every year one million people should be deported to the Dominions. They would be happy to go, he believed, if the outcome of a nuclear attack on Britain were explained to them.

Hoyle's public profile ensured that the book was widely reviewed. The *Daily Telegraph* wrote: 'The distinguished Cambridge astronomer's excursion back to earth is not very helpful from a practical standpoint, but the book is eminently worth reading for its lucid statement of a serious problem.'[94] The *Economist* was less complimentary, and fulfilled the publisher's readers' predictions:

> When a man of outstanding intellectual attainments uses them to cultivate a field in which he has no special expertise the results may be illuminating or plain silly. Mr Hoyle's *Decade of Decision* is both. . . . One may wonder at the naivety which allows him to ignore . . . everything that is organic in the relations of interacting economic units; all considerations of scale, specialisation and indivisibility; all that makes the difference between the tabula rasa of his imaginary empty Britain and the infinitely complex palimpsest of a modern industrial society.[95]

The *Sunday Times'* reviewer was the science-trained novelist and civil servant Charles Snow, who was shortly to become famous for his 'two cultures' thesis, in which he argued that many of the world's problems were due to a breakdown in communication between the sciences and the humanities. According to Snow's review, 'Mr Hoyle is the kind of figure that this generation badly needs.'[96] In a novel published the following year, Snow wrote that:

. . . the physicists, whose whole intellectual life was spent in seeking new truths, found it uncongenial to stop seeking when they had a look at society. They were rebellious, questioning, protestant, curious for the future and unable to resist reshaping it.[97]

Indeed, Hoyle was not the only scientist to publish a programme for social and political reform in the 1950s, but it may have served his own particular ends at the time. Perhaps he felt that as an anti-establishment scientist, and as the loudest proponent of a cosmology tainted by allusions to communism, he should protect his foothold in the USA, increasingly gripped as the country was by an anti-communist fervour that was denying visas to left-wing scientists, by exhibiting his conservative politics. Or he may just have thought, like the physicists in Snow's novel, that he could tackle anything the world could throw at him.

5
Under fire

Fowler, the Burbidges and the origin of the elements; a
challenge from Ryle; Frontiers of Astronomy;
cosmological controversy

At Caltech, Willy Fowler had been impressed by Hoyle, and, inspired by the carbon-12 work to think further about the evolution of stars, he had arranged to spend the academic year 1954–55 in Cambridge on a Fulbright scholarship. It was his first trip to Cambridge, and in order to explore more widely he bought a car – a convertible Hillman 10, like Tommy Gold's but without all the mechanical problems. He enjoyed touring in it so much that he had shipped it home at the end of the trip. In Cambridge itself he found 'a city of very old culture', and 'a really terrific experience ... really one of the great years of my life.'[1] Fowler gave a seminar in the Cavendish Laboratory on the nuclear physics experiments that his colleagues at the Kellogg Radiation Laboratory had done to find Hoyle's new state of carbon-12, and in the audience was a young astrophysicist called Geoffrey Burbidge – the man who, as a student, had handed round the question slips at the Massey Conference in 1951. According to Fowler:

> I discussed the experiments ... in which we have shown that Hoyle's state really could do the trick because [carbon] could break up into three alphas and thus it could be formed from three alpha particles. So I gave a colloquium on that and discussed the nuclear physics, and some of the astrophysical interpretations, which I had learned from Hoyle. Well, the next day, Geoff Burbidge came into my office ... Geoff has, as everyone knows, very striking physical characteristics, very heavy jowls, at that time he looked like a very young Charles Laughton and in many ways acted like Charles Laughton the great actor, anyway, Geoff Burbidge came into my office, introduced himself, said he had been at my colloquium, and that he found it very interesting, but he really couldn't care less about this complicated process by which helium is converted into carbon.[2]

Geoffrey Burbidge was deeply interested in the nuclear processes that underlay the production of the heavy elements in stars. He thought that the proportions of different elements that we see in the universe were a guide to the processes that made them – common elements should be more readily made than rare ones. Burbidge was working on this problem with his wife Margaret: they had met as graduate students at University College London just after the war. Margaret had graduated in astronomy there in 1939 – her studies had included lectures on atomic physics from Herbert Dingle at Imperial College – and during the war she had been charged with keeping the University of London Observatory running. With the astronomers dispersed on war projects, she had plenty of time on the telescope there, and in the good observing conditions while London was blacked out during the Blitz she wrote her PhD thesis on the beryllium content of stars.[3] She was therefore already a very experienced observer when she met the more theoretical Geoffrey when he arrived at UCL as a graduate student, and they formed a powerful partnership that was to destined to last for more than half a century.

By the time they met Fowler in Cambridge, the Burbidges had already worked in France and in the USA. In 1953, during post-doctoral appointments at Yerkes Observatory at the University of Chicago, they attended a conference organized by the nuclear scientists Maria Mayer and Harold Urey on the origin of the elements. The conference considered the growing body of evidence about the composition of stars, and pondered how such different stars – some rich in heavy elements, some with hardly any – could have come about. George Gamow was there, making jokes about the problems he had with producing heavy elements, and Margaret Burbidge was reminded of Hoyle, and a talk he had given at the Royal Astronomical Society in 1946 about his work on the structure of galactic nebulae, when he discussed the physical conditions – and his implosion mechanism – for nuclear reactions in stars. Margaret brought the data from her observations in the USA with her when she accompanied Geoffrey to his next job, in Cambridge.

Geoffrey Burbidge's new job was at the Cavendish Laboratory. Despite his experience of Martin Ryle at the Massey Conference, Burbidge had taken a job in Ryle's radio-astronomy group. He was interested in the mechanism by which radio sources produced their radiation, which he suspected would turn out to be a nuclear process. The radio-astronomers, however, favoured a theory based on the behaviour of ionized gases, and Burbidge's concentration on the nuclear option isolated him within the group.[4] And as Fowler noticed, 'Ryle and Geoff just didn't get along':

Whether this is because Geoff was too independent or whether as Ryle might say, he just thought Geoff was talking nonsense and wasn't too good a theorist, which I am sure he believed – I don't know. But very clearly, Geoff did not work in to Ryle's group. The blunt fact is Ryle and his group like to do their own theory. Geoff has told me that Ryle just wouldn't make available to Geoff the observational results on which Geoff could do some theoretical calculations. But one way or another, of course he started to work with me on this element abundance problem, and so he spent full-time on that.[5]

Fowler also noticed the tension between Hoyle and Ryle in 1954, during his year at Cambridge:

It was apparent to me even then in Cambridge that Fred Hoyle and Martin Ryle did not see eye to eye. Ryle was convinced even then that his observations in radio astronomy were incompatible with steady-state theory – that he was seeing even then evolutionary effects that could only be explained in terms of an [evolving] Universe. Hoyle took the attitude that in so far as he could get his hands on Ryle's data – which of course eventually Ryle had to publish – Hoyle always had some way of making a result compatible with the steady-state. . . . it was very apparent that there was no love lost between Hoyle and the Ryle group.[6]

But there were other things to do. The Burbidges were studying Margaret's Yerkes data to see what it suggested about the abundances of heavy elements in stars, those beyond the stumbling block in Gamow's scheme. Fowler joined them in trying to understand how these abundances could have occurred. Their thinking centred on the idea of nuclei capturing neutrons, and thus building up to form heavier elements. The more likely processes, with the more stable products, would lead to greater abundances. This work followed on from some ideas that Hoyle had published in 1954.[7] Describing the neutron capture processes meant integrating a protracted sequence of differential equations which Margaret Burbidge calculated on a noisy little mechanical machine operated by pulling levers (Fowler called it 'the Babbage machine' after the nineteenth-century computing pioneer). The three wrote two papers together during Fowler's year in Cambridge, and Hoyle often joined them, 'adding together', recalled Margaret Burbidge, 'one piece after another of the puzzle' of the abundances of the elements.[8]

One year was not enough time to solve this great problem, and as Geoffrey Burbidge's appointment at the Cavendish was coming to an end and he and Margaret were looking for work, Fowler made enquiries at Caltech about the possibility of them joining him there. The problem was that the Caltech Observatories, unlike those at Chicago, did not appoint women, which made it difficult to find a job for Margaret despite her excellent reputation. The reasons given for excluding women were to do with the provision of toilet facilities and that the male technicians did not expect to take instructions from women. But Fowler was able to make a joint arrangement for the couple whereby Margaret could accompany Geoffrey to a job that gave access to the telescope at Mount Wilson, and Geoffrey could accompany Margaret to her job at Kellogg – in effect, the opposite of what their qualifications and interests would have suggested. However, this gave both scientists access to both environments, which suited them well. Because women were not allowed in 'the Monastery', the Observatory's residence, and to avoid treading on any sensitive toes, the Burbidges rented a small cottage on Mount Wilson, and at first took their own food rather than eating with the other staff, but they were both made very welcome at the Observatory and they soon joined the nightly 'lunch' with the other astronomers. Between observing runs, they headed down the mountain to Pasadena, and cranked out their calculations on the origin of the elements.

In Cambridge, Hoyle was busy keeping steady-state theory in the public eye. He found an opportunity to write about the origin of the universe in February 1955, when he was asked to review Patrick Moore's *Guide to the Planets*. Moore, a self-taught astronomer, had yet to become well-known as the presenter of the BBC's TV programme *The Sky at Night*, which began in 1957. Hoyle devoted five paragraphs of his review to Moore's book, and spent the other nine discussing his own views on cosmology. He excused this dismissal of Moore's straightforward account of the planets by implying that it was old-fashioned: 'Astronomical research on the planets is nowadays very largely concerned with their origin.' Hoyle described the controversies of the time, and was apparently aware of the criticism he had attracted for appearing so sure of his views in public. In his own defence, he attributed the same fault to his opponents:

> A plethora of theories has been put forward in recent years. An enormous diversity of opinion still persists because those of us who are actively concerned all possess the faculty of being able to see the other fellow's shortcomings very clearly, but we are not too much disturbed by our own.[9]

What Hoyle saw as another fellow's shortcomings were shortly on display. In May 1955, just before Fowler and the Burbidges left for California,

Martin Ryle announced that his research in radio-astronomy had disproved steady-state theory.

Radio-astronomy involves collecting and analysing radio-frequency signals from space. Some researchers used parabolic dishes to collect the radiation; others, such as the Cambridge group, also used devices called interferometers: aerials strung out like washing lines between posts hammered into the ground, which located sources by detecting interference among the signals. The radio-astronomers were at first not taken seriously by other academics in Britain, either in astronomy or in Cambridge, where their hand-made aerials looked rather unprofessional. Radio-astronomy was viewed by some as a rather lowly occupation – a branch of engineering, or something technicians could do – and optical astronomers thought that radio telescopes were decidedly blunt instruments.[10] The radio-astronomers at the Cavendish Laboratory were all physicists, and did not think of themselves as astronomers.[11] According to Hewish, the former radar colleague who had joined Ryle's group in Cambridge in 1948:

> We didn't care what it was called. I thought it was physics, I still do. I think astronomy is physics really, it's just physics on a large scale. . . . We didn't regard ourselves as astronomers. We didn't know any astronomy when we began . . . we didn't know where things were in the sky . . . it was definitely physics.[12]

Astronomer William McCrea recalled that 'in the early 1950s . . . no-one seemed to think that radio-astronomy might help to clear the mist'; after all, some of the results of radio-astronomy were so surprising that 'astronomers had to practise hard before breakfast at believing it.'[13] The technique was regarded with great suspicion by the Royal Astronomical Society, though after the presentation of some interesting papers the Society came to be an enthusiastic supporter of the new recruits.[14] But when Hubble came to England in 1953 on what turned out to be his last visit, his view was not unusual – radio-astronomer Bernard Lovell asked him:

> Would he like to come and see us here [at Jodrell Bank] or would he like to go to Cambridge, and he said well it's very kind of you but he really didn't think that our work had anything to do with his discipline, astronomy. Which must be one of the greatest errors of judgement ever made in science.[15]

Now that the radio stars were known to be very distant, finding them meant a major increase in the size of the observable universe, and since very distant astronomical objects are in effect sending messages from the very distant past, they offer information about the history of the universe.

Radio-astronomers therefore thought that they could test the steady-state cosmology. As Hewish saw it:

> Bondi, Hoyle and Gold came up with this very neat idea. It gets round awkward questions: if there was a beginning, what was before the beginning and all that – the philosophy of it. Somehow it is neater if the universe is continuous. I never thought that way myself because it seems to me it just spreads creation over, dilutes it. If you suddenly produce a particle where there was nothing before you've still done something a bit strange, rather than having it all done at once. But it is a very nice neat theory and it was very beautiful in the sense that it was testable. Is the universe in a steady state, or isn't it? If you can answer that question definitely you've done something useful.[16]

According to Francis Graham-Smith, who had joined Ryle's group in 1947, the radio-astronomers' assault on steady-state theory came about as a result of some surprising observations:

> To start off with, there was as far as we knew one source in the sky, Cygnus A, ... We then put up an interferometer to have a look at this and discovered another one, ... Well that's interesting, now let's build a larger array and you see more, and then you say: well, gosh! you build a larger array and we see more and after a while you've got a catalogue of sources, and some of them are identified and some of them are not. And you say: well, are these things in the galaxy or are they at large in the universe beyond the galaxy? And then you start thinking a bit about cosmology.[17]

According to Hewish:

> The breakthrough came when the American astronomers identified the first radio galaxy [Cygnus A] as being at the limit of detection with optical telescopes. It was quite clear that radiogalaxies were much further off than anything else, and therefore gave an instant test, for example, of the steady state-theory of cosmology, because you are looking much further back than anyone else had done and you can really begin to answer the questions is the universe changing with time or is it in a steady state. That was such an obvious possibility and Martin Ryle latched on to that instantly.[18]

It appeared therefore that a catalogue of radio stars might offer some insight into cosmological theories – theories until then largely beyond the reach of observational testing. So Ryle began cataloguing radio stars in what was to become a series of surveys, numbered in sequence and labelled 'C' for Cambridge. The steady-state theory was enjoying such prominence, and the astronomers who thought that it was wrong had no evidence for their belief, and so Ryle thought he would provide it. If he could, Ryle would demonstrate the great value of radio-astronomy, and more particularly of the group at the Cavendish, to the astronomical community. Ryle could also exercise some of his personal feelings towards the steady-state authors. Steady-state theory therefore offered perfect opportunities: for Ryle to establish himself both as a scientist and in Cambridge; and for radio-astronomy, and, in particular, radio-astronomy in Cambridge, to establish for itself a powerful and pre-eminent role in the world astronomical community.[19]

Compiling the catalogue had meant making a survey of the sky to find objects emitting radio waves. Ryle's group spent two years measuring the energy output of these sources at a particular frequency to provide a scale comparable to the magnitude measurements used for visible stars, which indicated their brightness. This output, the flux density, was given the symbol S, and was used as a measure of distance – the brighter the source, the nearer it was likely to be. Ryle's latest survey, 2C, had found 1,936 sources. Some 30 of these were found to have dimensions which suggested that they were within the galaxy – too close to be interesting from a cosmological point of view – but that left 1,906 sources that were distributed across a range of distances, but all a very long way away. The analysis compared the number of sources, N, to their brightness, the flux density S, and the resulting graph of $\log N/\log S$ gave a clear and visible message: it looked like the right-hand half of the silhouette of a bell, and its slope was Ryle's weapon in his attack on steady-state theory.[20]

Counting astronomical objects of any particular type in this way is a standard technique. Dividing the universe into concentric shells around the observer, and counting the number of objects in each shell, provides a means of comparing numbers nearby with numbers far away – which provides a comparison of the recent and the distant past. The geometry of spheres dictates that if there are equal numbers of objects in each concentric shell, meaning that the universe has stayed the same over time, the graph that relates numbers to distance will have a slope of -1.5. So in a steady-state universe, -1.5 would be the slope of Ryle's graph. A flatter graph would not help to distinguish between the rival cosmologies, but a slope any steeper would mean that the universe had changed over time, and so was not in a steady state. Ryle announced, in a prestigious lecture at Oxford University, that his graph had the precipitous slope of -3.0: the universe was therefore not in a steady state.

His group's theoretician, Geoffrey Burbidge, had had no idea that the work was being taken in that direction:

I was working in the Cavendish, and they kept all that from me. They were secretive. Ryle did not allow any of his young men to tell me what they were doing – they were counting the sources. Finally I was told after 18 months or a year that Ryle had this amazingly important result in cosmology which I didn't know anything about and didn't care about. But this is what they had been doing and I should go next week to Oxford – they were all driving over to Oxford to hear Ryle give the Halley lecture, at which he would present this result. This was the first result in which he said steady-state is dead. I was so angry at being kept in the dark that I wouldn't go. I wasn't part of this, but I was there to do theory, and I was very angry at being treated this way because he was deciding what I could see and what I could not see. I know Margaret went over to hear it, but I didn't go, I wouldn't go. . . . The fight went on from there.[21]

Ryle's lecture was considered significant by other astronomers and by the press. According to *The Times*:

Discoveries on the grand scale about the numbers and distribution of 'radio stars' and their possible significance in deciding between different theories of the universe have been made by the radio-astronomy section of the Cavendish Laboratory, Cambridge, under Mr M. Ryle, FRS. . . . On 'continuous creation' theories . . . there is thought to be no way in which the distribution of 'radio stars' can be explained.[22]

According to Graham-Smith, Ryle's announcement:

. . . became recognised as being a very important statement – the first observational statement in cosmology for about 25 years. So at that point people took an interest in it, saying could these observations possibly be right. And then all hell got let loose, because when the observations became interesting, people started challenging them.[23]

So was the theory wrong, or the observations? The amenability of steady-state theory to observational testing had been flagged as an asset by its authors, and was also welcomed by its critics. Bondi in particular held in very high regard the falsification criterion of the philosopher Karl Popper, who asserted that science progresses not by proving correct theories but by disproving wrong ones. A good theory is one that is potentially disprovable, and it can be considered to be correct so long as it is not disproved.[24] According to Bondi, 'it is the natural fate of testable theories to be disproved, that is what science is all about, but it is much better to be testable than untestable.'[25] However, the failure of Ryle's data to match up to the predictions of steady-state theory indicated another asset, according to Gold: if the theory was assumed to be correct, it could be used to distinguish good data from bad data such as Ryle's. An exchange at a conference around this time clearly shows Bondi's strict Popperianism, compared to Gold's more pragmatic approach. They were in a discussion, which was recorded as follows, with F.J. Ernst of the University of Wisconsin, who worked on general relativity:[26]

> Ernst questioned if a steady-state theorist would drop the generalized Cosmological Principle if observations did not fit the steady-state theory.
>
> Bondi replied, 'like a hot brick.'
>
> Ernst continued with the statement . . . and asked where one could go from there.
>
> Bondi said he would drop the subject.
>
> Gold remarked that the steady-state theory is certainly successful in the sifting of observational data.[27]

So while data could be used to test a theory, a theory could also be used to test data. Indeed, in 1955 when Ryle made his claim, the supporters of steady-state theory were convinced that the data had to be wrong. Hoyle paid it serious attention: a fortnight after Ryle's announcement, he wrote to Jesse Greenstein at Caltech:

> The cosmological situation is indeed very exciting just at the moment, with Ryle's new results on the counts of radio stars. How these are to be interpreted does not seem yet clear to me, but the preliminary assessment does *not* look favourable to the 'steady-state' theory. It is still early to be sure, however.[28]

Hoyle believed that there were problems with Ryle's counting of the radio sources, and he wrote to the radio-astronomer Bernard Mills of the Australian Radiophysics group, another product of the war-time radar programme. While Ryle's aerials were laid out in a rectangle, Mills had been experimenting with aerials in the shape of a cross. Mills felt that the signals he was able to extract from his cross allowed a much more reliable measure than was possible with Ryle's rectangle. Mills' recent data from the cross suggested that the slope of $\log N/\log S$ should be a little steeper than -1.5, but not nearly as steep as Ryle's -3.0. Mills confirmed Hoyle's suspicions and contacted Ryle, suggesting that the Cambridge instrumentation had introduced some confusion that Ryle's analysis had failed to untangle. Ryle ignored Mills' advice, and subsequently published his data in *Scientific American*. A correspondence in that journal followed between Mills and Ryle, and after that any constructive relations between the Cambridge and Sydney groups ceased.[29]

Despite the criticisms, the Cambridge radio-astronomers were sure that their survey was valid. Hewish thought it:

... the first real evidence that the universe was changing in time. And that was unavoidable as far as I could see. In any new science with a controversial aspect, with different theories in the field and different groups doing things in different ways – you had the Australians on one side doing it with different instruments from our own and getting slightly different results – there is bound to be a tension, and that comes out in conferences too. But there is no doubt whatever that while Martin [Ryle] may have sometimes overstated the case, he was getting real evidence from the very beginning that the steady-state theory had to be wrong, because the universe was changing in time. It just was not uniform in distance, the results didn't fit, and that's what he kept saying. And that was actually right as far as I could see.[30]

When Mills published his own catalogue in 1957, it was very different from Ryle's. According to Mills:

> ... there is a striking disagreement between the two catalogues. ... the Cambridge survey is very seriously affected by instrumental effects which have a trivial influence on the Sydney results. We therefore conclude that discrepancies, in the main, reflect errors in the Cambridge catalogue and, accordingly, deductions of cosmological interest derived from its analysis are without foundation.[31]

Mills published a value for $\log N/\log S$ of -1.8 ± 0.1; still not quite -1.5, but nevertheless rather better for steady-state theory than -3.0. He later noted that of Ryle's sources in the 2C survey, at least 1,600 out of the 1,936 sources were 'fictitious.'[32]

It was difficult to resolve the disagreement between Sydney and Cambridge, as they were using different instruments and looking at the different skies at the opposite ends of the Earth. But a problem did come to light in Cambridge: some 'fictitious' sources had arisen because the signals from individual sources were hard to tell one from the other. The signal from an individual source ought ideally to look like a sudden pulse of peaks and troughs on an otherwise flat line. But some of the pulses were very broad, and looked like they might contain more than one overlapping signal, indicating more than one source. In Cambridge, they were trying to pick the individual pulses out of a signal that was almost continuous peaks and troughs, with no flat line between them to distinguish one from the next, as Hewish explains:

> ... if your sensitivity is enough and you're picking up [sources] all the time and what you're getting is a record that looks like [a continuous series of peaks and troughs], then what you do is get a research student to come along and say that looks like one [source], and that looks like one and there's another one; and you can do that over-enthusiastically if the sources are overlapping. So it was the

positions and intensities of the sources that were going wrong. . . . really, if you have a record like that you should only go for the really big ones: . . . if you get a really big one then that's clear, and you say 'here's one which we can say definitely is in about this place and about this size, but over here it gets a little bit difficult to say what's going on.' I think there was a bit of over-enthusiastic interpretation in this. And so these will have the wrong positions and intensities but not totally the wrong numbers because clearly there have to be [sources] there to produce these interfering records. So to say that the others are rubbish is wrong: what you are really saying is that you can't quite separate one from the other, so that the placing won't be quite right and the intensities won't be quite right.[33]

But Hewish was sure that even if their methods needed more work, their conclusion was valid:

[2C] was the first major sky survey and it was over-interpreted. . . . But it wasn't just the source counts that were the evidence; it was also [the] work on how to interpret these overlapping records that also gave very clear evidence that the universe couldn't go on absolutely uniformly as it was predicted to do by the steady-state theory. And that was unassailable evidence too, and Martin Ryle was very conscious of all these things, and the evidence added up, and it was impossible to reconcile with steady state. That's always stood the test of time I think.[34]

The challenge from Sydney did nothing to help relations between Ryle and Hoyle, which had been poor for some time. Ryle's temper was severely strained. Graham-Smith was now Ryle's brother-in-law, as Ryle had married his wife's sister, and he had got to know him well:

Martin Ryle began to feel that people were criticising him personally for what the group was doing, and [he] reacted. He was an individualist . . . so he didn't take this too easily. And so there were the cosmological theorists, Hoyle, Bondi and so on, seeing this going on, and saying 'those radio-astronomers don't know what they are doing.' And the radio-astronomers reacted very badly to that attitude and said of course we know what we are doing, there is something wrong with what the Australians are doing. . . . Cambridge was very touchy and jealous about the Australian radio-astronomers. That was a very real embarrassment, and it was fought on both sides I am quite sure. It was insularity in Cambridge, not believing that anybody could do anything half as well, same old story.[35]

Ryle's temper boiled over. Peter Scheuer, who was responsible for the mathematical analysis in Ryle's group, emerged from a meeting with Ryle with a throbbing weal across his forehead: Scheuer later told a colleague that 'Ryle threw a heavy glass ashtray across the room and hit him in the forehead and knocked him out. Peter had the impact of that ashtray on his forehead for some time.'[36] Mills' 1957 catalogue revived the bad feeling, unfortunately for Brian Robinson, a student visiting from Australia at the time:

> Late one afternoon in 1957 I was walking up the main stairs at the main entrance to the Cavendish Laboratory. Martin Ryle was coming down the stairs. I noticed under his arm a preprint, bound in a way I knew well as a [Sydney] Radiophysics Preprint. Martin stopped to speak to me about something. And soon he had grasped me by the shoulders, and was banging my head against the wood panelling.[37]

Robinson recalled noticing, out of the corner of his eye as his head thudded repeatedly into the wall, that a large oil painting of Lord Kelvin was wobbling precariously above him, and he worried that it might become dislodged and fall and kill him. He thought that Ryle was venting his frustrations with Mills on him simply because he was the nearest available Australian.[38] Robinson survived to enjoy a long career in astronomy back in Australia. Other students decided not to endure Ryle's temper, and his rivals recruited many good Cambridge students as a result.

Ryle did not attend many conferences, nor did he join committees. His group was tightly knit, and less communicative than some others at the Cavendish; and visitors, as Gold had found some years previously, were not always welcome. Ryle went to great lengths to protect the privacy he had established for his laboratory, perhaps to conceal his personal behaviour as well as to guard his data. Yet for many people who came into contact with Ryle during the 1950s and 1960s, his erratic behaviour was par for the course – it was worth it, just to be with such an inspiring scientist. David Edge, who was a doctoral student working on the 2C survey and who later became a leading figure in the sociology of science in the UK, knew that if his data were in inconvenient places on the graph of log N/log S, he could expect to bear the brunt of Ryle's temper until they were where Ryle needed them to be. But Edge believed that Ryle's genius excused his unusual behaviour: 'We put up with it', he said, 'that was Martin.'[39] Fifty years on, Hewish was sceptical about the reports of Ryle's outbursts:

> If you spoke to any of his research students, they'd say what a wonderful chap he was to work with. He felt very strongly about things, and if he was angry with somebody, he would show it. But he wasn't the least bit vindictive. . . . If you made a fool of yourself, he was inclined to be cross. Beating people up would have been quite out of keeping. He was a very gentle man but with an extremely

strong social conscience. If he thought somebody was doing something wrong which would have social consequences, he would have righteous indignation, let's put it that way. Did I ever see him lose his temper? I don't honestly think I did. The fact is he developed a coherent group in the Cavendish who were devoted to him and still are. The group feeling persisted long after he departed – he generated an extremely good friendly group that worked together, no question.[40]

The controversy over $\log N/\log S$ drew attention to the cosmological potential of radio-astronomy, though in 1955 Ryle was generally not widely credited with having achieved anything decisive. The rivalry between Cambridge and Sydney ensured that both groups paid intense attention to the detail of their observations and analysis. New radio-astronomy groups were starting up: by now, Caltech had hired John Bolton, a Yorkshireman and Cambridge graduate who had worked alongside Ryle and Lovell during the war and had then gone to Australia to work under radar pioneer Edward Bowen.[41] But steady-state theory was safe for now: it had been tested, and it was the test, not the theory, that had been found wanting.

Hoyle never stopped writing. His academic publications had recovered from the lull of early 1950s, and in June 1955, just a few weeks after Ryle's announcement, Hoyle published a book called *Frontiers of Astronomy*. Like many of his books, it was dedicated to his wife, Barbara. With the children now away at boarding schools, Barbara acted as editor for much of his writing. In the book, Hoyle assessed the current state of astronomy, including ideas about the origin of the elements, and the evidence for and against the various theories of the origin and evolution of the universe, starting from 'Oddities about the Earth' and ending with 'The Continuous Origin of Matter.' Hoyle brought in the latest research, and alerted readers to developments that had occurred since *The Nature of the Universe* five years before. The book became highly influential for many in the next generation of astronomers with its startling new synthesis of mathematics, nuclear physics and astronomy – many senior astronomers of 50 years later remember experiencing a kind of Damascene conversion when they read it[42] – but at the time, reviews were mixed. The Astronomer Royal, Sir Harold Spencer Jones, concluded in the *Sunday Times* that 'Hoyle has done a job that was well worth doing, and has done it well.'[43] McCrea, writing in the *Spectator*, went so far as to say that 'Mr Hoyle has written one of the most remarkable books in the story of modern science. For its significance in the progress of modern thought, it may come to be ranked with, say, Charles Darwin's *Origin of Species*.'[44] Like Darwin's book, *Frontiers of Astronomy* was, however, one long argument: for steady-state theory. So some reviews were very critical: a particularly irate 'D.W.M.' wrote in the *Financial Times*:

> Many a scientist and astronomer reading this book will rise in wrath against the new and extended gospel which its author preaches within its covers. Part of this reaction will undoubtedly be due to jealousy, for Mr Hoyle can write with a flair for catching the human imagination that is vouchsafed to few scientists, but much of the criticism, dull though it will seem to the author's avid readers, will be justifiable scepticism.
>
> 'It is true', the author states in his concluding paragraph, 'that we must not accept a theory on the basis of an emotional preference,' yet, he continues, 'it is not an emotional preference to attempt to establish a theory that would place us in a position to obtain a complete understanding of the Universe. The stakes are high, and win or lose, are worth playing for.' This is scarcely the approach of either philosophers or scientists. It is a clear case of putting the cart before the horse, a proceeding that makes it very difficult to win stakes of any kind, let alone stakes where man's opponent is the mystery of the Universe and the harnessing of its forces and hazards.[45]

Among the 500 pages in *Frontiers of Astronomy* were three on the origin of life.[46] Hoyle began by calling this 'a problem for the biologist and the biochemist rather than for the astronomer,' but suggested that important aspects might be overlooked 'if the astronomer should keep himself entirely out of the problem.' Hoyle then described a process in which 'simple chemicals,' under the action of ultraviolet light from the Sun, formed 'molecules of moderate complexity.' These then aggregated, 'as a film on the surface of a solid particle,' to form 'molecules of great complexity.' These molecules then either broke down into their constituent atoms, or they became capable of replicating themselves. In this way, life arose – in interplanetary space:

> It has always been supposed that life originated on the Earth. The physical and chemical requirements must, however, have been more favourable for the building of complex molecules *before* the Earth was aggregated. The Earth intercepts only a tiny fraction of the ultra-violet emitted by the Sun, whereas the gases out of which the planets condensed intercepted a large fraction of the ultra-violet. The energy source was therefore much greater before the planets were aggregated. . . .
>
> An interplanetary origin of life would have seemed impossible in the days when it was believed that the Earth was formed in an entirely molten state, for the associated high temperature would have destroyed all complex organic

molecules. Now we realise that the Earth must have accumulated from a multitude of cold bodies it is no longer possible to be so sure of this.[47]

The following summer, *Scientific American* showcased the cosmological debate. Eight years after the publication of the steady-state theory, players on both sides were clearly still willing to argue their case before a broad international readership.[48] The issue, which was dedicated to: 'astronomy as it is related to cosmology,' featured articles by Hoyle and Gamow, Caltech staff Alan Sandage, Walter Baade and Willy Fowler, and Martin Ryle and Herbert Dingle, among others. Gamow's article preceded Hoyle's, and claimed to present a theory that 'most cosmologists believe', and made much of its intellectual pedigree.[49] Gamow described the debate as 'the major present issue in cosmology.' He dismissed steady-state theory, but acknowledged the complexity of the issues involved: he suggested that the matter would only be resolved 'by extended study of the universe as far as we can observe it.'

Hoyle's article began: 'Some cosmologists dissent from the evolutionary view.'[50] Hoyle discussed and compared the steady-state and evolutionary theories, and while he claimed that 'radioastronomy offers the exciting possibility of something close to a direct test,' he dismissed the radio-astronomical data produced so far that questioned steady-state theory, and set more store by the redshift work of Allan Sandage at Caltech. Sandage's article, however, which followed Hoyle's, presented data that 'do suggest that the steady-state model does not fit the real world, and that we live in a closed, evolving universe.'[51] Ryle's article, the one in which he chose to ignore Mills' criticisms, concluded that his data 'should help make possible a decision between the evolutionary and the steady state theories of the universe . . . our present conclusions support the evolutionary view.'[52]

Dingle reiterated his philosophical objections to steady-state theory, and in particular to Bondi and Gold's cosmological principles. These principles made heavy demands on the cosmos: they required it to be the same in whichever direction one looks, and for this sameness to persist over time. According to Dingle, the steady-state argument was that the continuous creation of matter must occur because if it did not the perfect cosmological principle would be violated – the sameness would not persist over time; 'and the only reason for supposing that this cannot happen is that a few mathematicians would not like it. It seems insufficient reason.'[53]

Despite being the chief proponents of the rival cosmological theories, this encounter in *Scientific American* was one of the few in which Hoyle and Gamow were in direct conflict. Scientific players in the cosmological debate seem to have taken little notice of the scientific arguments for and against the rival theories, and did not spend much time scrutinizing each other's arguments. In Hoyle's 1950 broadcasts – and with his first use of the term 'big bang' – he had said that he found the idea unsatisfactory before he had studied it in detail;[54] and later he wrote: 'I have always disliked the big bang.'[55] Gamow rejected steady-state theory without studying it: he felt it was 'artificial and unreal.'[56] In 1964 he wrote a book chapter comparing the two cosmologies, and explained that most of the chapter would be about evolutionary theory because he knew more about it than he did about steady-state theory.[57] According to Bondi's student Felix Pirani:

> I heard Gamow talk – at Imperial College. . . . we used to go to each other's lectures and say something at the end like 'that was interesting.' . . . it's polite on the surface but what it says is 'this guy's an ass.'[58]

Critics of steady-state theory often confused Hoyle's ideas about the continuous creation of matter with Bondi and Gold's steady-state universe – a confusion that irritated Bondi: apart from the insult to his own standing as a cosmologist, it implied that the critics were not familiar enough with the two aspects of the theory to be able to tell them apart.[59] As Hoyle later acknowledged, 'after fifty years in the game, I have come to doubt whether scientists ever read each other's papers, especially if they disagree about something judged to be important.'[60]

Reading each other's papers was one thing; discussion was quite another: Hoyle and Gamow had spent some time together, talking over questions in astronomy and cosmology, in the summer of 1956, just before the publication of the *Scientific American* special issue. Hoyle had gone to Caltech that spring to fulfil his duties as Addison White Greenway Professor of Astronomy, and Gamow was visiting there at the same time. They got on well.[61] Along with Fowler and the Burbidges, they borrowed one of the old wooden buildings at the Scripps Institute in La Jolla, overlooking the Pacific just outside San Diego, and talked about the origin of the elements. Gamow argued for the origin of matter and the production of the elements at the beginning of the universe, and the other three pressed the case for an ongoing production of elements within stars. Hoyle recalled the meeting as 'a good-humoured confrontation,'[62] but there was to be no agreement. Burbidge, Burbidge, Fowler and Hoyle would continue on their mission to make the elements in the stars.

6
New Genesis

The origin of the elements and B²FH; The Black Cloud;
Clarkson Close; the Plumian Chair; the age of the universe;
Ossian's Ride; Alchemy of Love

B y 1956, Hoyle's academic career was flourishing: ten academic
papers would be published that year, and the work was to attract
considerable attention in the academic press. In the spring, he
headed for Caltech.

After their collaboration with Fowler in Cambridge, Geoffrey and
Margaret Burbidge had joined him in California, where they continued to
work on the origin of the elements. With Hoyle there too, they pondered
the nuclear reactions that might produce elements in stars. They were
encouraged by a recent review article, 'The Abundance of the Elements' by
the nuclear physicists Hans Seuss and Harold Urey, whom the Burbidges
knew from their time in Chicago, which much clarified the abundance data,
and indicated that an important process in the formation of new elements
must be that nuclei catch and hold on to neutrons.[1] Margaret Burbidge was
pregnant, and so had abandoned her observational work which had meant
clambering up and down ladders at the telescope; now she could concentrate on analysing spectra for the elements work. Geoffrey Burbidge was
browsing through data from the ongoing H-Bomb tests at Bikini Atoll,
which had recently been released as part of an international effort by the
nuclear powers to see how nuclear energy might be used for peaceful purposes. Burbidge noticed that the unstable element californium-254, which
is made in nuclear explosions, had a half-life the same as that of the decay
of a supernova, which had recently been determined by Walter Baade.
Although californium later turned out to be a false lead, these parallels were
enough to convince the four colleagues that they were on the right track: the
extreme temperatures and pressures of nuclear explosions, with their bursts
of high energy neutrons, that were making new elements on Earth could

also be making elements in the stars. Work, according to Margaret Burbidge, 'was proceeding marvellously.'[2]

Hoyle's new state of carbon meant that elements with nuclei made of units of four particles could be accounted for easily enough through the coming together of alpha particles, each made of two protons and two neutrons. Carbon at atomic number 12 plus an alpha particle in the right conditions – temperatures of thousands of millions of degrees in stars – gives oxygen at 16, and so on, across the table of the elements. But this process leaves out the elements whose atomic number is not a unit of four, and it does not produce the very heaviest elements either: alpha particles carry a positive electrical charge, and if too many come together they tend to repel each other, flying apart rather than sticking together. So another process was needed to bring in more particles to build up the bigger atoms, and it had to be one that did not also bring with it yet more positive electrical charge.

Atomic explosions generate new heavy elements and new isotopes because they produce, in conditions of great heat and pressure, torrents of very energetic neutrons. These neutrons have no electrical charge. When they collide with the nuclei of the lighter atoms, they can be absorbed by the nuclei to make heavier elements. The conditions within some stars meant that this same neutron process could be happening there, adding more to the range of elements. But again, not every element can be made in this way. Once more, another route was needed, and the four friends decided on a process happening very slowly through collisions with low-energy neutrons. But yet again, not every missing element could be formed this way: still another process was needed. Some elements needed more protons, so a process that generated fast protons was proposed. This involved hydrogen, which has a nucleus consisting of one proton, at temperatures of billions of degrees, such as are found in supernova explosions.[3]

With ideas like these in mind – there were eight processes in all – the detailed work of plotting the nuclear processes could now begin. For each element, what ingredients were needed? Where might those ingredients be found? What might bring them together, and in the right proportions? What conditions – temperatures, pressures, electrical charge, energy states of the nuclei – would be required for the new nuclei to form, and then be stable and persist? How likely were these processes to happen in each case, and did that likelihood square with the abundance of each element throughout the cosmos? The task was long and detailed, challenging Burbidge, Burbidge, Fowler and Hoyle with questions in astronomy, nuclear physics and mathematics. Graduate student John Faulkner watched the four astronomers at work together:

> One of the things that impressed me ... was how much the Burbidges really enjoyed their science. They got one heck of a kick out of doing it. I had the pleasure of seeing the four of them working together – of Willy Fowler writing things on the board; of some calculation being needed and Willy whipping out his slide-rule, and Fred doing it in his head because he knew the logarithm tables intimately. They had this friendly competition in which Fred would often beat Willy – who was an absolute whiz with a slide-rule – to the answer. And the Burbidges adding in everything from their expertise. It was a true four-person collaboration – a wonderful thing to see in operation. I have never seen a collaboration between scientists as exciting as that one. They always enjoyed their science so much, and it was a wonderful thrill to see how much they got out of it.[4]

Margaret Burbidge gave birth to a daughter, Sarah, just as the work was drawing to a close. Hoyle had gone to England for July and August and had returned, on a special leave of absence from Cambridge, with Barbara and their daughter Elizabeth; and with Fowler's family they all enjoyed feeling that something exciting and very important had been achieved. Fowler and Hoyle celebrated the last results by indulging their great enthusiasm for sport at Pasadena's big football game on 1 January 1957. According to Hoyle:

> We pretty well had a royal flush, and it became only a question of how the results should be presented. Ambitions grew, and as our standards rose, more time was needed. At first we thought to finish the work by June 1956, but it would not be until the last month of the year that the end would really be within sight. I stayed in Pasadena with my family until after the Rose Parade, and until after Willy (wearing what an Englishman would call a pork-pie hat) had some-how managed to urge his old Hillman through heavy traffic to the Rose Bowl game. I recall him having a fierce duel of words concerning the potency of the Hillman with a crowd of girls in another open car.[5]

Iowa beat Oregon State 35–19, and the theory of stellar nucleosynthesis was complete. Margaret Burbidge had contributed observational data, analysis and endless calculations; Geoffrey Burbidge the theoretical astrophysics, Fowler the nuclear physics and Hoyle, according to Fowler, 'worked on everything.'[6] The four friends had first published a qualitative account of their work in *Science* in 1956, and now they produced a quantitative paper that ran to over a hundred pages in the prestigious journal *Reviews in Modern Physics*.[7] Although this journal publishes reviews, that is, overviews and assessments of the state of knowledge of a particular topic, the paper, which became known as B²FH after its authors, not only reviewed the contributions already made by others to the theory of stellar nucleosynthesis, but also contained the substantial body of original work that Burbidge, Burbidge, Fowler and Hoyle had undertaken in Cambridge and Caltech over the last three years. It is considered by cosmologists to be a major achievement in astrophysics, and the high point of Hoyle's scientific career.[8] Gamow, whose own theory on the origin of the elements had produced only the first four, later wrote a satirical piece in the style of the Old Testament called 'New Genesis', in which God also failed to create the heavier elements. To make up for his mistake 'in a most impossible way . . . God said: "Let there be Hoyle." And there was Hoyle. And God looked at Hoyle . . . and told him to make heavy elements in any way he pleased.'[9]

The work on the origin of the elements brought Hoyle an acknowledgement from the Royal Society: in March 1957, he was elected a Fellow – the mark of distinction to which British scientists aspire. The *Guardian* reported that Hoyle was being recognized for 'his work on stellar constitution, on nuclear reactions in stars, and on cosmological theory.'[10] Hoyle had been waiting for some time for his FRS:

> I confess to experiencing some irritation when, year after year in the 1950s, I was not elected to the Royal Society of London. My wife knew I was under some tension . . . [and] burst into tears when she read . . . that I had at last been so elected. Unfortunately the same tension must darken many a household where science has come down to roost.[11]

B²FH was not perfect: in the opposite of Gamow's achievement, it did not produce the very light elements. Burbidge, Burbidge, Fowler and Hoyle proposed a mysterious 'x-force' to account for some of them. Helium remained a mystery however: the process suggested for making it might well work, but could not possibly have made as much helium as there is in the universe. Hoyle would pursue the light elements at Caltech and later back in Cambridge: he and Fowler soon started a project they called 'Deuteronomy': an appropriate title for the next step after their new genesis, and a reference to deuterium, an isotope of hydrogen. So while B²FH showed that the universe did not burst forth fully formed at the start of time, it still left big-bang theory to explain the light elements. Thus it was an idea from which both sides of the cosmological debate could take some comfort, and it was not controversial. Perhaps because of the immediate recognition for B²FH, Hoyle did not campaign for it in the way that he did for steady-state theory. Later in life he credited Caltech with the power to see that the theory was taken seriously: 'on that occasion,' he said, 'I had Caltech on my side, and that made a huge difference when it came to convincing everyone.'[12]

Walter Baade organized a conference on the evolution of stars in Rome for May of 1957, and Fowler and Hoyle presented B²FH there. Martin Schwarzschild and Georges Lemaître were among the colleagues attending the meeting, and they discussed the cosmological implications of the new work. Hoyle also argued that computers would be vital to their research in the years to come.[13] The meeting was so convivial that it came as a great relief when Pope Pius XII offered to pay the wine bill. Afterwards Fowler and Hoyle and their wives spent a week relaxing at the Bay of Naples, feeling very pleased and knowing that an extraordinary episode was over.[14] Fowler responded to Gamow's 'New Genesis' with a 'New Exodus' of his own:

> On the seventh day, God rested. Hoyle wandered in the Garden and came upon nuclear physicists pruning the trees of knowledge. And Hoyle tasted the fruit of knowing through experiment. God awakened but, forewarned of the anger of God, Hoyle consumed the fruit, even the core thereof but for one seed, and exodused.
>
> And Hoyle came into the land of the sun and planted therein the seed. From the seed grew a new tree to the delight of those who, even unto this day, try to figure out exactly how the genesis was done.[15]

Just as B²FH was coming to fruition, so was another project: Hoyle had become a novelist. He had been preparing for this for some time. During the war, his colleague Cyril Domb had been impressed by Hoyle's mental energy: his relaxation seemed to come not from resting but from switching to a new challenge. So Domb had been surprised to see Hoyle reading science fiction in the evenings, and had asked him why he was wasting time on 'low-grade literature.' Hoyle explained that he was reading not for relaxation but for profit: 'These people don't know any real science,' he said, 'and they make money from writing this stuff. I do know some science, I should be able to do very much better.'[16]

His first novel, *The Black Cloud*, was published in the autumn of 1957.[17] Hoyle told a reporter that he had written it because he had nothing to do while it rained during a recent holiday (the family had spent five weeks in Wales the previous summer).[18] He wrote by hand, on big pads of lined paper, and Barbara typed the manuscript. The *Daily Telegraph* announced 'Science fiction writers receive a highbrow recruit today. . . . Dr Hoyle calls the story a frolic, but not an inconceivable one.'[19] Although, in a short preface, Hoyle disclaimed any association with his characters' views, his friends recognized how much of Hoyle's personal experience was mirrored in his novel: according to Bondi, 'everything you need to know about Fred Hoyle is in *The Black Cloud*'.[20] Like much of the science fiction of the period, this is a story of invasion: in this case, the invader is the Black Cloud of the title, which makes its first appearance in the novel when a young astronomer in California notices a new dark circle in the constellation Orion. He tells his senior colleagues who track and measure the black object, which they decide is a cloud of interstellar dust heading for Earth. Unbeknown to them, in Britain amateur astronomers have noticed perturbations in the orbits of the planets, and they report the matter to a rather formal and hierarchical meeting of the Royal Astronomical Society. A Cambridge astronomer, the tall, dashing maverick Chris Kingsley, calculates that the perturbations are due to the gravitational effects of a massive object in Orion. Once the American and British data are brought together, a relaxed but efficient international team led by Geoff Marlowe at the Mount Palomar Observatory at Caltech tracks the Cloud's progress.

The Cloud is expected to blot out the sunlight on Earth, so emergency measures have to be taken, which means that governments have to be informed. The British government insists on secrecy, and in anticipation of this Kingsley has sent the news to top astronomers all over the world. The Secret Service traces the letters and 'invites' all these astronomers to come to Britain, where they are effectively imprisoned in a 'research establishment' – a country house in the Cotswolds. From there they monitor the Cloud, and determine that it consists of granules of ice on which are deposited organic molecules. The conglomerations of molecules can communicate with each other, so they are in effect components of a single organism. Kingsley realizes that while its components might be inanimate, the Cloud itself is alive – a conclusion that brings about radical changes in the scientists' handling of the situation. Kingsley explains how he came to his shocking realization that life had arisen elsewhere in the universe:

> Well, it's pretty obvious really. The trouble is that we're all inhibited against such thinking. The idea that the Earth is the only possible abode of life runs pretty deep in spite of all the science fiction and kids' comics. If we had been able to look at the business with an impartial eye we should have spotted it long ago. Right from the start, things have gone wrong and they've gone wrong according to a systematic sort of pattern. Once I overcame the psychological block, I saw all the difficulties could be removed by one simple and entirely plausible step.[21]

The scientists establish contact with the Cloud, which has stopped by the Sun in order to absorb energy. This disturbs the weather dramatically, plunging the world into darkness, cold and chaos. The governments of the world pressure the scientists to send the Cloud away; but they refuse, as they are trying to learn from it. The Cloud functions electromagnetically, and is allergic to nuclear radiation, so when the governments on Earth fire nuclear missiles at it, it deflects the missiles and sends them back to Earth. The astronomers cut their communications with the politicians and run the project themselves in an attempt to win the trust of the Cloud and avert a global catastrophe. Not only that, but they have learnt from the Cloud that it is extremely old, and that there was no beginning to the history of its species ('Kingsley and Marlowe exchanged a glance as if to say ". . . That's one in the eye for the exploding-universe boys." '[22]). Thus it has answers to many of the questions about the universe that have been vexing the astronomers, and they are keen to learn.

The scientists try to work out a way of communicating with the Cloud more efficiently, since they are running out of time – it has decided to leave the solar system, leaving the world to recover from the catastrophe. The scientists build an apparatus that sends the Cloud's knowledge directly into a human brain. The first person to be wired in dies, and the scientists assume that this person lacked adequate brain capacity, so the top scientist – Kingsley himself – volunteers. He dies too, but in his last moments of clarity he reveals that he has had a glimpse of extraordinary new knowledge. Kingsley's colleagues suspect that he died because his brain was washed clean – the idea of 'brain-washing' was a common one in the ideological conflicts of the real world of the 1950s – but the resident doctor suspects otherwise:

> 'No. . . . There was no washing. The old methods of operation were not washed out. They were left unimpaired. The new was established alongside the old, so that both were capable of working simultaneously.'
>
> 'You mean as if my knowledge of science were suddenly added to the brain of an ancient Greek.'
>
> 'Yes, but perhaps in a more extreme form. Can you imagine the fierce contradictions that would arise in the brain of your poor Greek, accustomed to such notions as the Earth being the centre of the Universe and a hundred and one other such anachronisms, suddenly becoming exposed to the blast of your superior knowledge?'
>
> 'I suppose it would be pretty bad. After all we get quite seriously upset if just one of our cherished scientific ideas turns out wrong.'[23]

So it was the shock of hearing ideas that contradicted his own world view that was too much for Kingsley. His colleagues concluded that they should have wired up someone with nothing in his brain, such as Joe Stoddard the gardener, whose otherwise baffling dramatic role in the novel thus finally becomes clear. This story introduced a theme that was to recur in Hoyle's criticism of the scientific establishment: scientists rarely change their minds, because the effort needed to replace knowledge acquired over years and held with conviction is too great.[24] For Hoyle, the suspension of disbelief was essential for the acquisition of new scientific knowledge. The layman, as exemplified by the gardener Joe Stoddard in *The Black Cloud*, on the other hand, whose ideas about the world are, according to the scientists in Hoyle's novel, more tenuous, may well be better able to abandon them and embrace new ideas.

The Black Cloud is not a great work of literature: the writing is clumsy and clichéd, and its characters are strident and shallow. Typically of Hoyle, meals are described in detail. But the plot stampedes its herd of astronomers at breakneck pace through a series of engaging ideas and situations that seemed only too real and only too possible. The novel was a great popular success, and was soon reprinted; and translation rights were sold around the world.[25] It was also adapted for radio, and broadcast for 'Saturday Night Theatre' on the Home Service. The *Daily Telegraph*'s announcement of the radio play referred to Hoyle's radio success of 1950: 'Fred Hoyle, the Cambridge astronomer whose astronomy broadcasts made such a profound impression some years ago, has unexpectedly blossomed as a writer of science fiction.'[26] Film producers vied for the rights: one offered Hoyle a film with the action set in the USA, which would have entertainment as its primary aim, but which at the same time 'should add up to a powerful argument for recognition by the peoples of the western world that the rate of conquest of this vast new area of discovery depends upon the amount of money and drive that the individual countries put into it.'[27]

Reviewers of the novel commented that the label 'science fiction' was not entirely appropriate: according to the *Spectator*, *The Black Cloud*:

> . . . bears about the same relation to the rest of science-fiction as antibiotics to homeopathic remedies: one supposes it to be the real, as well as the latest thing. Approximately it is not really fiction. . . . I believe [Hoyle] absolutely when he says it could all happen. . . .[28]

The *Evening Standard* reported that on the question of life in space, 'modern astronomy is in step with science fiction here,' and quoted Hoyle as saying:

'I think it would be extraordinarily surprising if there were no other planets with life on them.'[29]

The influence of Hoyle's academic work is apparent in the novel: steady-state theory aside, the idea of a cloud of interstellar dust coming between the Sun and the Earth and upsetting the weather was the subject of a paper written by Hoyle and Lyttleton as long ago as 1939.[30] But how did other scientists react to *The Black Cloud*? Some scientists certainly read it, for they cited it in academic publications.[31] According to Hoyle:

> I don't know [whether my colleagues read it]. That never worried me because I just wrote it for amusement, so I wasn't really much concerned with what people thought about it. . . . one scientist certainly read it, . . . that's Wolfgang Pauli [the distinguished pioneer of quantum mechanics]. He apparently thought he saw very deep psychological aspects in that book which I never intended, and I ran into him at an international conference in 1957, and he had a dreadful cackle and he was famed for his devastating remarks. And soon as he saw me he said 'Ha! Hoyle! I just read your novel, and I thought much more about it than I'd ever thought about your scientific work.'[32]

According to Felix Pirani, many of Hoyle's colleagues 'did read it, and talked about it.'[33] They would have recognized the rivalry in the novel between radio-astronomers in Cambridge and Sydney; they might also have recognized Kingsley's rant to a fellow astronomer about politicians:

> . . . in spite of all the changes wrought by science – by our control over inanimate energy, that is to say – we still preserve the same old social order of precedence . . . politicians at the top, then the military, and the real brains at the bottom. There's no difference between this set-up and that of ancient Rome. . . . We're living in a society that is a monstrous contradiction, modern in its technology but archaic in its social organisation. For years the politicians have been squawking about the need for more trained scientists, engineers and so forth. What they don't seem to realise is that there is only a limited number of fools. . . . People like you and me, we're the fools. We do the thinking for an archaic crowd of nitwits and allow ourselves to be pushed around by 'em into the bargain.[34]

Kingsley blamed a gulf between 'the literary mind and the mathematical mind . . . that's where the real clash lies.'

Willy Fowler could see himself in one of the characters, a Caltech nuclear physicist 'who was overshadowed by a brilliant young theoretical physicist [Kingsley]. I suspect I was the older man who lost all the arguments with the younger man about the nature of the Black Cloud. You have two guesses about who the younger man was.'[35] Other scientists since have noted the impact the novel made on them: for example, evolutionary biologist Richard Dawkins has reservations about it, but:

> . . . one can forgive much for the sheer scientific intelligence of the exploration of the nature of the Cloud. This book certainly made me think more clearly, as a biologist, about individuality and the blurring of parts and wholes in living structures. I believe it was reading *The Black Cloud* that first brought home to me the importance of information theory and the essential inter-changeability of coded information through different physical media. Finally, the Cloud's own humility in the face of the fundamental laws of the universe has left me with an undying reverence for what is called 'The Deep Problems.' Splendidly gripping, *The Black Cloud* is one of science fiction's great classics. . . .[36]

Hoyle and his family had lived for ten years in Quendon, a small village outside Cambridge, chosen because they could not, just after the war, afford a house in Cambridge itself. But late in 1957 they were able to move into one of a small cluster of houses that was being built on St John's land just across the playing fields from the college. They paid for it with the money Hoyle had been saving from his income from *The Nature of the Universe*. The house was in what then seemed a rather futuristic style, with its sharp geometry and expanses of glass (it now looks rather typical of the 1960s), and journalists remarked that it seemed appropriate for an author of science fiction. The house had wide windows overlooking the garden, and Hoyle put an armchair by the living-room window and thus created his office. Because of a back problem Hoyle much preferred to work in an armchair with a notepad on his knee, than at a desk, and the habit became a kind of trademark. Hoyle also rarely went to the university, so the great majority of his correspondence was received at, and sent from, home. The address '1 Clarkson Close, Cambridge' was to become as familiar to the astronomical community as it was to readers of the letters column of *The Times*.

From this new base, Hoyle enjoyed 1958: it was, in his opinion, 'a vintage year.'[37] It was a time of relentless activity, eclectic interests and a high public profile. The year began well with an academic honour: in January, the doorbell rang at 1 Clarkson Close, and a university messenger handed over a short note from the Vice Chancellor offering Hoyle the Plumian Chair in Astronomy and Experimental Philosophy at Cambridge – the most senior academic post in British astronomy. Although he knew that the post would bring with it some responsibilities such as committee work and trips to

London that Hoyle did not relish, he was glad to accept.[38] Since his reputation as a cosmologist was controversial, and there were other candidates for the chair, colleagues believed that the 'very solid piece of work' in B[2]FH had earned him the post.[39] The appointment was widely reported: the *Guardian*'s announcement identified Hoyle by reminding readers that he 'became popular as a radio broadcaster in 1950. . . .'[40] The *Daily Telegraph* pointed out that Hoyle differed on a number of questions from his predecessor Sir Harold Jeffreys, and mentioned Hoyle's claim that Earth will boil in 10,000 million years' time – which will allow him, the *Telegraph* suggested, plenty of time to work on his defence theories. For some years now, Hoyle had been calling for more money for his colleagues the rocket scientists – Harrie Massey was one – and the international tensions of the Suez Crisis and since had provided an opportunity for Hoyle to argue that the lack of funds for rocket science left a serious gap in the British deterrent.[41] Hoyle now told the *Telegraph* that current defence policy was foolhardy, and that counter-attack by long-range rockets was the only way to defend Britain.[42]

The *Evening Standard* also featured the new professor. Reporter John Thompson met Hoyle in Senate House in Cambridge, and noticed that an older don looked disapprovingly in their direction when Hoyle announced that the Earth was 'going to boil':

> I sympathised with the elderly don's uneasiness. A prediction like this takes on an extra dimension of alarm when Fred Hoyle utters it. He is the sort of short, comfortable fellow who would never, you feel, say anything just for effect. You remember that his father worked in the Bradford cloth trade, where the atmosphere discourages flamboyant talk. His own brilliant career as an astronomer has never disturbed his level, steady Yorkshire accent.[43]

Thompson gave Hoyle a further opportunity to express his concern over scientific funding:

> When Hoyle talks about outer space he is as unemotional as a bowling club secretary reading his annual report. But one subject makes his blunt and cheerful features grow solemn, and his eyes gleam indignantly behind his spectacles. That is when he talks about the shortage of money for scientific research. Through such research lies the way to a vast increase in industrial prosperity: that is his theme – that, and the 'sheer idiocy' of cheeseparing. 'Industry is not a finished thing. We have seen only the beginning of the first part of the first scene of the first act.'

Thompson questioned whether the 'man-in-the-street' would be willing to pay up, but added: 'Hoyle thinks that the facts are so plain and urgent that they only have to be understood to get action. But then he is a scientist, not a politician.' Thompson concluded that Hoyle was '... the modern astronomer in essence: a working class boy who, powered by his own brains, has soared to the scientific stratosphere.'

Hoyle's work at Caltech produced 17 papers in 1958, and the first results from the Russian Sputnik, the first artificial satellite that had been launched in 1957, were keeping astronomers busy all over the world. Hoyle was travelling again: in March, he went to the IBM research laboratory at Poughkeepsie, north of New York City, to use a new fast computer there for calculating the evolution of stars.[44] As his work with Schwarzschild had proved, with a computer, the prospects for understanding stellar evolution, and many other complex astrophysical processes, were brightening considerably.

Computers that could solve real, practical problems had first been produced during the war, and used for code-breaking, calculating missile trajectories and other military problems. These were one-off machines, each built for a specific purpose and running according to its own unique system. In the late 1940s, researchers in universities and in industry had looked to stretch the potential of these machines, and to find ways to develop them commercially. Manchester University had the first working computer in a British university by 1948 (four years after the USA's first, at Harvard University), and it took delivery of a commercially produced computer, the Ferranti Mark 1, in 1951. From then on, Manchester had provided computing services for researchers and companies across the UK. Cambridge University, however, had its own machine, EDSAC, built in 1949 by a team led by Maurice Wilkes at the Mathematics Laboratory.

Similar developments had been taking place in the USA, where some universities and government research laboratories, including the US Weather Bureau and the Jet Propulsion Laboratory near Caltech, were running state-of-the-art computers. These early computers were pro-grammed using punched cards or paper tape, and each relied on thousands of vacuum tubes – valves – which tended to explode or burn out, limiting the capacity and the running time of the machine. The computers were also huge, and could weigh around 100 tons. Occupying vast halls, they could perform thousands of additions, and hundreds of multiplications every second. By 1953, there were 100 working computers around the world, performing calculations for research and administration. For Hoyle, the Jet Propulsion Laboratory's computer was a major attraction of Caltech.

The invention of the transistor in 1948 suggested new possibilities for computers. Through the 1950s, spurred on by the military and scientific demands of the Cold War, the USA invested heavily in computer research. In 1958, the first computer to use transistors was built there, by IBM in Poughkeepsie. The new machine, the IBM 7090, was 100 times faster than any previous machine, and Hoyle had been determined to get his hands on it as soon as possible. He had persuaded the Rockefeller Foundation that it was essential for his work, and thus had mustered the funds to go to Poughkeepsie to use it.[45]

Back from the USA, Hoyle headed for Paris and a conference on radio-astronomy, where he was in charge of one session. Caltech astronomer Rudolph Minkowski clearly felt that there was some potential for friction here, as he reminded Hoyle before the meeting that the idea of the session was not to discuss results so far 'which are really very shaky', but to 'make it clear to the radio-astronomers which observations are of vital importance for cosmological questions.'[46] Then on to Brussels, where Hoyle and Barbara went to the World's Fair, a great expanse of exhibits and amusements: Barbara, who was recovering from an operation to remove her appendix, enjoyed herself enormously, while Hoyle hated the noise and the crowds.

They were in Brussels because Hoyle had been invited to participate in the Eleventh Solvay Conference there. The Solvay conferences had been instigated in 1911 by the Belgian chemist and industrialist Ernest Solvay, who had intended them as forums in which the best minds in physics might discuss Solvay's own theories of gravitation and matter – theories that he felt were being unjustly ignored by the physics community.[47] Solvay died in 1922, but his legacy allowed the conferences to continue, and at each meeting small groups of invited speakers discussed important issues in physics. The topic for 1958 was 'The structure and evolution of the universe', and among the guests were Hoyle, Bondi and Gold, and Lemaître, Alan Sandage from Caltech, Pauli, Baade, Bernard Lovell, and Harvard astronomer Harlow Shapley. Gamow claimed that he had been invited to the conference by Pauli, who was a member of the organizing committee, but that the invitation had been withdrawn by Cambridge physicist Lawrence Bragg, the committee's chairman, on the grounds that all the places had been filled. Gamow believed that he had not been invited 'since I was an opponent of the steady-state theory'; but there were plenty of other critics of the theory who were invited. Gamow thought the theory a very English theory, so perhaps he suspected Bragg of chauvinism.[48] Bragg was a great supporter of Ryle in Cambridge, and so was not likely to have been protecting Hoyle, Bondi and Gold from Gamow.[49]

The meeting opened with Lemaître's paper on 'The primeval atom hypothesis', the cornerstone of the evolutionary cosmology; then came Swedish physicist Oskar Klein (of the Klein–Gordon equation) on the development of systems of galaxies, Hoyle on steady-state theory, Gold on the nature of time and American physicist John Wheeler on general relativity. Then Englebert Schücking and Otto Heckmann from the University of Hamburg launched a methodological broadside against Bondi, Gold and Hoyle for abandoning a principle that had, they claimed, guided scientists since Newton's time:

> A theory constructed on a sound foundation of empirical data ought not to be discarded unless new facts turn up that cannot be fitted into the framework of this theory. . . . We believe . . . it is sound policy to refrain from theorizing along the lines of Bondi, Gold, and Hoyle until there is strong empirical evidence for continuous creation of energy and momentum.[50]

The conference then turned to experimental matters, and to the evolution of galaxies and stars. Hoyle gave a second paper, on the origin of the elements, and from the published *Proceedings* it would seem that it was this work, rather than steady-state theory, that was prominent in the minds of the conference participants. Schücking and Heckmann aside, there seems to have been little by way of a fight. However, such proceedings tend to be polite records, and perhaps Shapley's vote of thanks to Bragg for his chairmanship points to more heated exchanges: 'You have maintained a neutral – I might say neutron – pose during the turbulence, during negative and positive charges, the explosions and implosions of gas and argument.'[51] But if the cosmological debate was being played out in Brussels, it is not readily apparent from the record. Hoyle's account of events consists entirely of sharing jokes with Pauli.[52] For steady-state theory, the late 1950s would turn out to be the calm before the storm.

After Brussels, Moscow: Hoyle recalled that Gold persistently searched their hotel for surveillance bugs, and the food was dreadful.[53] He did not enjoy his insight into the Soviet system (though he kept the little metal name badge made for each conference delegate[54]) – he thought it displayed a very poor philosophy of 'good intentions excuse bad results,' in which context he made reference to the National Health Service in Britain which had kept Barbara waiting for her appendix operation so long that they thought it would explode.[55] Hoyle also noted that 'I have no regard at all for Lenin, whom I regard as one of the world's most appalling little squirts, but for Marx I have a fair respect.' The trip caught the attention of the press, and Hoyle took the opportunity once again to talk about science funding in

Britain.[56] He told the *Observer*, for example, that funding for science was not simply an internal problem for the scientific community, but something for which society needed to take collective responsibility:

> All should be openly discussed. Newspapers ought actively to consider what proportion of our national income we should be spending on research and development. The state of our laboratory equipment should be made widely known, not only as compared with America and Russia, but also compared to technologically efficient small nations like Holland. If we find the situation to be bad, there should be no attempt at face-saving. Our one resolve must be to make a rapid, decisive assertion of our former scientific skill and greatness, whatever this may cost.[57]

Science was already high on the public agenda. Sputnik and International Geophysical Year, a worldwide programme of scientific research and public events about the Earth, pushed space and astronomy to the forefront. Even reports of discussions at learned societies became commonplace in the newspapers. For example, after a Royal Society meeting to discuss the origin of cosmic rays, Hoyle was reported as saying that a telescope on a satellite 'could be extremely useful' in solving the mystery[58] (the first such device, Ariel 1, was launched by the UK in 1962, and did provide data on cosmic rays); and a talk Hoyle gave at the Royal Institution on the possibility of people travelling to the Moon was reported in *The Times*. The idea was, according to Hoyle, 'lunatic fringe stuff'; he suggested that television cameras should be sent instead.[59] Hoyle could invariably think of a better way to spend money than space travel. He also spoke publicly about nuclear physics: he joined in the correspondence in *The Times* about the effect of atomic weapons tests on the weather,[60] and he contributed an article to the *Sunday Times* explaining stellar nucleosynthesis. Hoyle described how the elements are all made from hydrogen, but concluded by saying that the question of where the hydrogen comes from 'is now becoming one of the most urgent problems of science.'[61]

Hoyle had not forsaken the culture of letters: in Cambridge in the summer of 1959, he performed as Bottom in an amateur production of Shakespeare's *A Midsummer Night's Dream*, which was organized by Barbara. A French film crew came to interview him in his dressing room.[62] Colleague Leon Mestel thought Hoyle gave a 'magnificent performance', and noted that Bottom 'is not content with the assigned role of Pyramus, but wants to be considered for all the other parts in the play within the play. It reminded [me] of how in addition to his major areas of research . . . Fred had a finger in most other pies on the astrophysical menu.'[63] And Hoyle had written his second novel: *Ossian's Ride* was published in 1959, and it told the story of Thomas Sherwood, a naïve Cambridge mathematics graduate who is recruited by the Secret Service to investigate some mysterious industrial development in Ireland. At the heart of the mystery is a company called I.C.E. – a dig, perhaps, at ICI, Imperial Chemical Industries, a company riding high at the time. I.C.E., the Industrial Corporation of Eire, has been moving huge structures around at speeds that suggest that a new power source has been harnessed. Hoyle had enjoyed holidays in Ireland, most recently in 1957, visiting Kerry and the Dingle Peninsula, and in *Ossian's Ride*, he describes Sherwood's adventures there, battling stormy seas, crossing challenging terrain, and dodging a cast of villains. Sherwood eventually discovers I.C.E.'s secret: it is run by aliens, who use nuclear fusion to power their machines. But it was not the aliens themselves who had invaded Ireland: instead, they had sent their knowledge, for which selected humans provide a repository and then turn into action. The idea of the transforming, disruptive potential of knowledge was one that was to recur in Hoyle's writing, both in fiction and in non-fiction. Astronomer Jesse Greenstein at Caltech told Hoyle that the characterizations and style were a great improvement on *The Black Cloud*:

> I thought *Ossian's Ride* is extraordinarily good. I read it through as quickly as I could, found it very interesting, found in it many of your original ways of looking at the world, and also found the English style much improved over your first piece of fiction. The characters now have some depth and individuality, and I am amazed how you find the time to learn and do so much.[64]

Hoyle sometimes tried to do too much: around this time, he realized that he had underestimated the amount of work needed to complete the computer models for the structures of evolving stars – the site of his nuclear synthesis.[65] The work was never completed: the job of Plumian Professor had brought with it administrative responsibilities within and beyond Cambridge, and Hoyle was irked by the demands these were making on his schedule.[66] Over the next couple of years, though, he did find time to estimate the age of the universe – a matter that had nagged at him since his wartime discussions on cosmology with Bondi and Gold. B[2]FH had concluded with an observation on the age of the elements based on the relative proportions of uranium isotopes:

> [W]e call attention to the minimum age of the uranium isotopes. . . . We feel that this minimum age is significantly greater than the currently accepted value for the geochemical age of the universe.[67]

Hoyle and Fowler addressed this problem by calculating the abundances of uranium and a similar element, thorium. In March of 1960, the *Guardian*'s scientific correspondent reported a Royal Society conference at which Hoyle presented his revised age for the universe: he had calculated the age of the Universe to be 15 billion years, much greater than the 1.8 billion that he had found so unpersuasive during the war. The *Guardian* said:

> The implications of this new work are further to strengthen the belief that it is wrong to think of the universe as having a beginning in time – an instant before which it did not exist. Rather, belief will be strengthened in the newer doctrine – also partly due to Professor Hoyle – that the universe has no beginning and no end, and that the matter of which it is made is constantly being created throughout its whole volume in a continuous fashion.[68]

By 1960, then, with steady-state theory twelve years old, there were still some commentators willing to give it its chance.

Hoyle also found time to promote a new venture: he had written the libretto for an opera, in collaboration with the American composer Leo Smit, whom he had met in California. Hoyle wanted to see the opera, *Alchemy of Love*, in production, and aimed high: he sent it to major opera houses. The replies were swift and decisive: the director of London's Sadler's Wells Theatre had chosen not to look at the work. John Christie, who owned Glyndebourne, the country house in Sussex, and had founded the celebrated opera festival in a theatre he had built there, had been a science teacher at Eton College before he inherited the house. But he too returned Hoyle's work, with an explanation written by his secretary:

> Glyndebourne is not in a position to undertake productions of operas by unknown composers, however meritorious. Our very rare excursions into the field of contemporary music are made only after exhaustive discussion, and then with composers who have already proved themselves in the operatic field. In these circumstances, Mr Christie feels that we are not able to help you and Mr Smit.[69]

Perhaps the run of good fortune was coming to an end.

7
Eclipsed

Funding for science; the politics of science; an idea for an institute; the Department of Applied Mathematics and Theoretical Physics; another challenge from Ryle

In 1959, Gallup fielded an opinion poll among astronomers about cosmology, and the results showed considerable ambivalence about the big-bang and steady-state theories:[1]

	Yes	*No*	*No comment (%)*
Did the universe start with a big bang?	33.3	36.4	30.3
Is matter continuously created in space?	24.2	54.6	21.2
Will [the debate] be resolved by 2000 AD?	69.7	9.1	21.2
Will the answer be given by radio-astronomy?	60.5	9.1	30.4
. . . or by a telescope on a satellite?	33.3	21.2	45.5

Unanimity was reached on only one point:

Is [this poll] helpful to scientific progress?	0.0	100.0	0.0

Had Martin Ryle been polled, he would no doubt have been among the 60.5 per cent who had high hopes of radio-astronomy. Although a Cambridge colleague of the steady-state authors, he had already clashed with them, publicly, many times. But the long history of animosity would be trumped by events in 1961.

Radio-astronomy had a new high profile in the early 1960s: it had been established in the public mind amid great drama in 1957 when it was used to track Sputnik, the Russian satellite with its twin auras of scientific progress and Cold War espionage. In Cambridge, Ryle had attracted public attention as the first person in Britain to pick up the signal from Sputnik, and it was reported that his immediate reaction had been romantic: he had telephoned his wife.[2] The space race had begun, and press activity had been intense, Ryle's colleague Antony Hewish recalls:

> The moment that the position [of Sputnik] had been announced, that we'd actually got it and knew where it was, . . . the place filled up with the media . . . and serious science became just about impossible – the place was full of reporters. You can't keep them out and they are tripping over things. Every time that Sputnik went over there was a sort of mini-press conference, . . . just for an artificial satellite.[3]

The Jodrell Bank radio-astronomers too had made their mark: director Bernard Lovell had completed the construction of his radio-telescope in 1957, and the great dish became a symbol of the age. In 1958 Lovell had given the BBC's prestigious annual Reith Lecture. This long-running lecture series aims to advance public understanding and debate about significant issues of contemporary interest, and Lovell spoke on 'The Individual and the Universe' – his was one of six science-related Reith lectures of the 10 broadcast during the 1950s. He was to be knighted for services to astronomy in 1961.

In this atmosphere, Ryle was planning another assault on steady-state theory. His first challenge, in 1955, had not been successful: astronomers' scepticism about the technique, and the differences between the results obtained in Cambridge and those coming from Sydney, were enough to leave steady-state theory to fight another day. Once it had become apparent then that the Cambridge group had indeed made systematic errors in their analysis, they had become all the more determined to justify their existence, and to beat their new rivals in Australia, by demonstrating that steady-state theory was wrong. Several years of fierce competition had ensued between the two groups, and other new groups in Europe and the USA had joined in the work of cataloguing radio sources. By 1960, however, radio-astronomy was running out of money: in a *Guardian* cartoon, a ragged Lovell scuttled after well-equipped representatives of the armed forces, holding out his telescope dish as a begging bowl.[4] The radio-astronomers needed a way of showing that they deserved more public money.

The funding and organization of science generally in Britain was a matter of increasing concern to the government after the heavy investment of the 1950s, and the newspapers were keen to carry stories about the problem. The Labour Party was growing in influence in opposition to the

Conservative government under Prime Minister Harold Macmillan, and science was high on its agenda. In 1961, Labour published a report called 'Science and the Future of Britain', in which it declared that if science was to be used to best advantage for the nation, 'the present near-anarchy in the administration of science must end.'[5]

This chaotic administration was a legacy of war-time, when scientific experts had become influential in political and military circles and had been seen as key players in the war effort. After the war, an independent committee of distinguished scientists was established by the Department of Scientific and Industrial Research to continue this contribution from science to national life. From 1947, this committee, the Advisory Council on Scientific Policy, considered issues such as the relations between science and central government, and science and the military. It was also concerned with expanding the universities and increasing the number of scientists in Britain: it suggested raising the school-leaving age so that more students would qualify for university courses, and it wanted broader curricula in science, so that students could move between disciplines as national needs and scientific trends changed. However, according to the science journalist and policy critic J.G. Crowther, 'the effect of the Council's advice was not equal to the quality of many of its ideas':[6] it tended to be a commentator, rather than a decision-maker or opinion-leader. Decisions about science were still being made through a variety of channels, and scientists were fighting with a variety of weapons for the same pot of money. In 1956 the Department of Scientific and Industrial Research was given powers and funds to support research in the universities, and to set up specialist institutes at universities for research in new fields or with commercial potential – a move some universities saw as unwelcome political intervention that would compromise their academic freedom and traditional administrative autonomy. In 1960 a Minister was given responsibility for science: Quintin Hogg, the Lord Hailsham, was the first in the post. This development was welcomed by the Advisory Council on Scientific Policy, which declared itself happy to advise the Minister and to be concerned with the 'strategy of scientific affairs.'

The crucial issue was that the pot of money was not as large as it had been during the 1950s, and tough decisions would have to be made about which projects could be funded and which could not. Scientists engaging in new areas of research would be taking money away from existing areas, and new projects might not be possible if on-going work was still consuming funds. On top of the needs of the existing professional community, the growing numbers of science students would soon be expecting jobs in research. Space and astronomy were expanding areas, and it was difficult to see how the growth could be sustained, especially in an area with little industrial potential. So how should the money be shared out? The Advisory Council on Scientific Policy said that 'the problem of priorities in science and technology lies at the heart of national science policy, and therefore of our national destiny', by which it meant that science was so important to the nation that the Government should decide. But the politicians and civil servants were being bombarded with requests for money, all of them from experts in their own fields – Professors, Heads of Departments, or chairs of committees at learned societies – and all making claims about the great value of their projects. Some requests were coming from new fields that no-one knew how to categorize: should 'biophysics' be paid for from funds for biology, or physics, or medicine? While the Advisory Council on Scientific Policy pondered 'strategy', the Department of Scientific and Industrial Research had to handle the day-to-day decisions, which it tended to do by referring projects to the Treasury. At the Treasury, a civil servant surrounded by piles of research proposals would have to decide which could be afforded and which could not, usually without any knowledge of the science or the people involved. Nobody was enjoying this situation: the scientists felt that unqualified people were deciding their careers for them, and the civil servants, with no policy to guide them, were conscious of their own inadequacy for the task. Decisions were delayed on ever longer timescales. And in this 'near-chaos' that the Labour Party had identified, individual scientists made their own contacts, cultivated friendships in high places, campaigned openly and fought private battles over public money.

Hoyle, as Plumian Professor, was a member of the British National Committee for Astronomy at the Royal Society, which made recommendations for the funding of research in astronomy, and of the Research Grants Committee at the Department of Scientific and Industrial Research, which distributed money across the sciences. He had his own views about how science contributed to national well-being. In January 1961, in an article for the *Observer* called 'Science: a vocation in the sixties', Hoyle wrote:

> There is an increasing understanding that changes in the world spring more from scientific discovery than from any other cause. On a less global scale, here in Britain we have come to realise that our economic future is bound up with the development of a thriving technology. Whether we like it or not, we all have a big stake in our technological success, or failure.[7]

Hoyle then asked whether technological security would require more scientists, and looked at the differences between the education of arts and science students – a topic that had bothered Kingsley in *The Black Cloud*, and had been brought to public attention recently by C.P. Snow with his 'two cultures' thesis, which found wide popular resonance. Arts students, Hoyle wrote, do not specialize until the last possible moment. But:

> The situation is completely opposite in science. For the able boy, already certain of his approximate vocation, specialisation cannot come too early . . . try to instil an appreciation of the arts into a young scientist, to 'round' him, or broaden him, or otherwise alter his cultural shape, and your efforts will very properly be wasted. An impassioned lesson on the beauty of the English language will fall on deaf ears – the boy will be pondering some geometrical theorem or perhaps the behaviour of a new brand of bug.

Hoyle argued that young scientists initially did not concern themselves with practical questions, but that over a period of about 10 years they learned to write and to handle administration – talents arts graduates seemed to have as soon as they graduated. Educationalists complained about the apparent backwardness of science graduates, but according to Hoyle this was a necessary consequence of their subject, and was overcome, except not soon enough. The only problem scientists then had is that they could not express themselves: this was a direct consequence of the fact that scientific conversation was not socially acceptable in polite circles, so the scientists did not get many opportunities to talk about their work outside of their professional world. Hoyle thought that this was fair enough: scientists would have to learn not to use jargon if they wanted to contribute to everyday conversation.

Hoyle claimed in his article that the many arts graduates who administer science often learn much about science, and are very sympathetic to it. Unlike C.P. Snow, he saw no problem in the lack of science graduates in the administrative hierarchies, but he would have liked to have seen a more even partnership between the arts and sciences. However, the absence of scientists from government was a cause for concern:

> It is a curious anomaly that in a world changing apace as a result of scientific discoveries not a single first-flight scientist can be found anywhere in the innermost councils of any Government. It seems to me difficult to dispute the absurdity of this situation. The problem is to find some means for its correction.

Hoyle's solution was to inject the use of scientific methods into politics:

> There has in recent years been a regrettable tendency to build a mystique around the methods of science. Actually science has only one very simple method. It can be stated in two phrases: Policies (theories) are to be judged by their results, and *by their results alone*.

If we were to judge policies on their results, argued Hoyle, we might have to reject some policies to which we were emotionally attached. Scientists, however, were not swayed by their emotions, and so would be better at changing bad policies than politicians.

The bad policies Hoyle had in mind related particularly to the funding of science. He was frequently in the press, calling for more money.[8] He drew attention to the better facilities for scientists in the USA which made necessary his long absences from Cambridge, and lent his voice to predictions of a damaging 'brain drain' as scientists deserted Britain for a better deal elsewhere. A programme of temporary scientific postings in the USA had been developed after the war to allow British scientists to catch up with the latest ideas and technology, the idea being that they would then bring their new knowledge and skills back to Britain. But by 1956 Britain was losing around ten per cent of its science graduates to the USA, and of the postgraduate students who took up short-term posts there, forty per cent never returned.[9] After Sputnik in 1957, having been beaten into space by the Russians, the USA had created many new jobs in the sciences, with much better salaries than in Britain, thus creating plenty more tempting opportunities for British scientists.

The Americans did not escape Hoyle's criticism, however. He was particularly dismissive of the space race, which he thought was diverting minds and money from more important pursuits, and showed how politicians were getting the priorities wrong. This was a recurring concern: in 1962 Hoyle would tell *The Times* that the results of the radio-astronomy being done in the UK and Australia were worth more than the whole US space programme: 'If I were an American I would be alarmed at the amount of money being spent on this kind of research, but it is the Americans' prerogative how they spend their money.'[10]

The high profile of radio-astronomy, and the space research that was also attracting public attention, had brought the state of all branches of UK astronomy into the policy spotlight. The Department of Scientific and Industrial Research had given large sums of money to astronomy in the late 1950s: a total of almost a quarter of a million pounds, with one third going to optical astronomy and two thirds to radio-astronomy. Cambridge had received eight times as much money for radio-astronomy as it had for optical astronomy. Funding for theoretical astronomy was harder to quantify, both nationally and in Cambridge, because it was undertaken by physicists and mathematicians as well as by astronomers. In November 1959, the Advisory Council on Scientific Policy noted that there were astronomy departments in one third of UK universities, but that astronomy was not being taught in physics departments as it was in other countries. Optical astronomy, with its strong historical links to navigation and the calculation of tides, as well as more recent surveillance and detection work, was still concentrated around the rather unadventurous Royal Observatories, and was still largely under the control of the Admiralty. The Astronomer Royal, the South African Richard Woolley, who had been appointed in 1955 after successes in Australia, was well known for the military style in which he ran Royal Greenwich Observatory at its new home, Herstmonceux Castle, in the Sussex countryside near the south coast of England. Astronomers in Britain found it particularly irksome that while the UK was leading the world in radio-astronomy, they had had to pass their discoveries of radio sources to the Americans for optical analysis. Some of these analyses had major cosmological implications, and the Americans were getting the credit for work the British wanted to be able to do themselves. The Advisory Council on Scientific Policy noted that 'There has been some difference of opinion among astronomers themselves about the relative roles of radio and optical astronomy in recent times,' but it concluded that the UK had lost its lead in optical astronomy.

The Council met to discuss the problem a few months later, in March 1960. It concluded that despite the Royal Greenwich Observatory's new Isaac Newton Telescope (which, although it had been authorized in 1946, was still being built), the British were not keeping up with the excellent new telescopes in California, for example, and optical astronomy was not attracting good students. Many UK astronomers had close and productive links with American observers, but the transatlantic wartime alliance was losing its political significance: Britain was looking to bolster other connections. A possible new telescope for the southern hemisphere sky was under discussion: it could be a European project located in South Africa, or a Commonwealth project in Australia. The UK already owned and ran the successful Radcliffe Observatory outside Pretoria in South Africa, and the competition between South Africa and Australia for resources for observing was fierce. In Australia, where the astronomers were very enthusiastic about the project, a Commonwealth telescope would mean that the UK would be the richest partner and would have to contribute the lion's share of funds; while in South Africa, a European collaboration would see many rich partners of which the UK would be only one. However, the Council noted 'discouraging political developments in South Africa': this mild phrase referred to events just a few days before the meeting, when 56 people had been killed and 162 injured by armed police during a demonstration against the pass laws. Demonstrators had deliberately come out without their passes

– the identity papers black South Africans were required to carry whenever they left home – and were in the process of turning themselves in at Sharpeville police station when armed police opened fire on the crowd. The Sharpeville Massacre was a defining event in the anti-apartheid struggle. In the UK, in a memo submitted to the Council dated 1 January 1960, Woolley had already noted the 'extreme reluctance' of British astronomers to go and work in South Africa, 'because of the political and racial conditions which obtain there.'[11] After Sharpeville, Australia looked set to win the new southern hemisphere telescope.

The needs of theoretical astronomy were also considered. Woolley's memo noted that 'some astronomers' – Hoyle had written to him about this late in 1959, from an address in Pasadena – wanted an institute of theoretical astronomy, and that one of their arguments for concentrating resources in this way was 'the great cost of modern calculating machines.' Hoyle had circulated a document he had written outlining a plan for investing in theoretical astronomy.[12] He explained:

> The present proposal arises from the conviction that astronomical studies will increase markedly in scientific and popular importance during the next decade. Space programmes in America and Russia, involving extremely large financial expenditures, must, it seems, guarantee the correctness of this supposition. The question arises of what contributions Britain can appropriately make to this development. Probably the most effective; the quickest, and the least expensive would be a full deployment of our resources in theoretical astronomy.

According to Hoyle, the attractive features of such a policy were that 'the moulding of observational data into viable scientific theories commands high international scientific prestige. A strong school of theoretical research would command far greater respect in world opinion than would a comparatively minor program of satellite launching.' Theoretical astronomy was also much cheaper than launching satellites. 'Past achievements of British science, from Newton to Eddington, provide a strong historical tradition that would favour the success of such a project,' and the rigorous mathematical training of students in the UK fits them effectively for work in theoretical astronomy (and much better, 'it is widely admitted in the United States', than the 'more qualitative curricula of the American colleges').

Hoyle laid down two conditions for the development of theoretical astronomy: firstly, that 20 or so active workers should be brought together in one place, and secondly that they should have 'a suitable electronic computer.' There were enough theoretical astronomers with international reputations in the UK, and some good younger workers, but they were scattered. Hoyle explained how the USA shows us that 'a group of scientists – provided it is not too large – will always achieve very much more when working together than it will as separate individuals. And experience also shows that habitual cooperation is not possible unless scientists are in almost daily communication with each other.' Gathering the scientists together in this way 'would also make reasonable and feasible the provision of an electronic computer':

> The computer is now just as necessary to the theoretical astronomer as the telescope is to the observational astronomer. Without adequate access to such an instrument, theoretical astronomy, so far from flourishing in the next decade, would almost certainly decline sharply away.
>
> To continue working effectively over the years, provision would certainly be needed for the continual intake and promotion of young men. This, I believe, requires that the institute be attached to a university. The institute should, moreover, take an active part in university teaching, particularly at post-graduate level. In view of the very high quality of students trained in the Mathematical Tripos, my own preference would be for an association with Cambridge University.[13]

Hoyle then suggested a budget: about a million pounds to buy the computer and the building; and £50,000 a year for running costs (to be subsidised by the sale of computing time). There would be 20 staff and 20 graduate students. He concluded:

> Britain as a nation must make some effective response to the incidence of the 'space age'. One response that might well prove highly effective would be to give full emphasis to our potentialities in theoretical astronomy. Whereas America and Russia are giving their main attention to gadgets, let us give ours to people.[14]

Although it was not expressed in personal terms, the plan was for an ideal working environment for Hoyle himself. According to Geoffrey Burbidge, Hoyle had seen that senior scientists in the USA were rather better off than he felt he was in Cambridge. In the USA, senior scientists often had groups of people around them, and they were well supported by grants from the federal government. 'Fred thought: he was well-known, he was good at it, and he was entitled to some support.'[15]

The Advisory Council on Scientific Policy concluded that adequate computer facilities should be provided for theoretical astronomy.[16] It passed Hoyle's idea for an institute to the British National Committee for Astronomy, of which Hoyle was a member. The Committee resolved in September 1960 that such an institute should not aim at establishing a monopoly on theoretical astronomy and would not need exclusive use of a computer. But the general idea of bringing theoretical astronomers together was good, as was the idea that there should be space for overseas visitors and for British scientists on sabbatical. A sub-committee was established to write a detailed proposal: it consisted of Hoyle as convenor, his former colleague Hermann Bondi, now at King's College London; his old friend Ray Lyttleton of Hoyle's own department, Applied Mathematics and Theoretical Physics at Cambridge; steady-state supporter William McCrea of Royal Holloway College of the University of London, and D.H. Sadler, a positional astronomer who worked at the Nautical Almanac Office and was General Secretary of the International Astronomical Union. The five men met at Burlington House on Piccadilly, the then home of the Royal Society, and Bondi turned their discussion into a draft proposal to be submitted by the British National Committee for Astronomy to the Advisory Council on Scientific Policy.[17] The proposal differed from Hoyle's plan on a few points: perhaps reflecting the ongoing cosmological controversy, it suggested that a range of viewpoints should be represented at the new institute, to stimulate discussion and to benefit PhD students who might have studied previously under only one theoretical astronomer; it proposed a small senior staff to allow room for a changing population of young researchers and visitors; and it stressed that research and teaching should not be separated. The proposal also expanded Hoyle's list of possible locations from one to two: it suggested that the institute be in either Cambridge or the new University of Sussex. It expressed the hope that such an institute could help to establish a new university like Sussex nationally and internationally; also, it would be an attraction for 'some of the best intending undergraduates, thus, instead of scraping from the bottom of the growing barrel of intending university students, the college could gain a good selection from the top.'[18]

The British National Committee for Astronomy met within days of Bondi's submission. The chair of the Committee was Sir William Hodge, Lowndean Professor of Astronomy and Geometry at Cambridge (though he was, according to Hoyle, 'all geometry'). Hodge was also Master of Pembroke College Cambridge, and Physical Secretary of the Royal Society. He was disdainful: he said that the argument about computing power was ill-founded, and that the proposal threatened the University of Cambridge as it would bring government interference in what ought to be university business. If theoretical astronomy was so important, ought there not to be a case made for a Cambridge department of theoretical astronomy? Hodge said he would not put the proposal forward to the Advisory Council on Scientific Policy.

Civil servant Roger Quirk, who had attended the meeting on behalf of the Department of Scientific and Industrial Research, reported back to colleague Frank Turnbull: 'I then went out with Professor Hoyle, who was in a state of great dejection and concern. He said he thought he might as well take his "single ticket to America".' Quirk reported that Hoyle had told him, in strict confidence, that Hodge had always been opposed to the plan, and Hoyle had been concerned that Hodge might 'sabotage it'. Turnbull was inclined to agree with Hodge that the case for the computer was over-stated: he thought that while such an institute could easily use a dedicated, high-specification computer all for themselves, most of what they would be doing with it could be done on a machine that cost a quarter of the price. Hoyle, Quirk noted, felt that the issue about computing resources need not be crucial, but there would still need to be 'some sort of organization for theoretical astronomy which is different from an ordinary University department,' because 'for this specialized and extremely difficult subject, there is no room for a large number of university departments with undergraduate teaching.'

One of Hoyle's reasons for wanting an institute was so that he could function without recourse to a university department. His own, the Department of Applied Mathematics and Theoretical Physics, was a place Hoyle did not care to visit. This Department was a relatively new institution. It had been formed within the Faculty of Mathematics, at the instigation of George Batchelor, one of its lecturers, shortly after Hoyle took up the Plumian Chair. The founding principle was that while pure mathematicians work alone, applied mathematicians work in groups, and so the applied mathematicians within the Faculty could usefully be brought together. Having a departmental label would also help them make links into other subject areas, such as engineering and physics, where their mathematics might be applied.[19] Although at the time the most senior member of staff was Hoyle as Plumian Professor, Batchelor was appointed Head of the new Department. The post was, however, to be a rotating one, and the rotation would occur every five years.

Hoyle and Batchelor did not get along. According to Geoffrey Burbidge, Batchelor was Hoyle's 'sworn enemy'.[20] Hoyle objected to what he saw as Batchelor's poor handling of staff and students, and to the changes Batchelor was making in the mathematics syllabus. Hoyle thought that Batchelor was making the Tripos too difficult – and too many good students were being demoralized and deprived of credit because of it. Batchelor was not, as Hoyle noted, 'a Cambridge product' – he had come to Cambridge from the University of Melbourne – and so he had not experienced the Tripos as a student.[21] Hoyle also felt that Batchelor's own mathematics, which was in fluid dynamics, was outdated, and of little use to researchers doing original work.

The poor relationship with Batchelor was one reason why Hoyle spent very little time in the Department. He usually had 15 or so PhD students during the early 1960s, and, according to one of them, there were times when Hoyle would appear for 'no more than 60 seconds, and he would ask how things were going. He didn't want to be seen or caught around the Department much.'[22] Instead Hoyle worked at home, in his armchair by the big window in his living room at 1 Clarkson Close, where Barbara Hoyle acted as gatekeeper, the student recalled:

> She was always very very protective of Fred's time and his solitude – she regarded her task in life as to put up a kind of invisible wall around Fred so that he would be protected as much as possible from the troubles and the interruptions of the ordinary world. . . . if you as a student wanted to see him, she ran interference – you virtually had to prove to her every time you wanted to try to see him . . . that it was something that he would want you to get in touch with him about, or that you were under instructions to get back to him about a task that he had set you.[23]

So if Hoyle was to create his ideal working environment, it could not be in a Cambridge University department where one might be subject to a meddlesome Head or students with trivial problems. Quirk urged Hoyle to press his case for an institute. He should think about one that could accommodate visitors and 'put up such a scheme through the Cambridge University machine. Hoyle said that it would "almost certainly be sabotaged by large vested interests." '[24]

Despite the negative mood, Quirk asked other astronomers for their opinion on the idea of an institute. He spoke to Harrie Massey of University College London, the host of the conference in 1951 where 'the theoreticians' had come under attack. Massey suggested that the Department of Scientific and Industrial Research could fund an institute at Cambridge under Hoyle on the model of the specialist research units being established in various universities, including Cambridge, by the Medical Research Council, the body that distributed government funding for medical science. These research units established mutually beneficial relationships with the universities with which they were associated, but were free of undergraduate teaching and ran on public money. Many had been set up around the country since the war, including, in 1947, the Medical Research Council Unit for Research on the Molecular Structure of Biological Systems, at Cambridge. But Massey also argued that the UK did not need to concentrate all its theoretical astronomy in one place: there were already thriving groups around the country, such as his own at University College London.[25]

A few weeks later, Quirk heard from Woolley, the Astronomer Royal, who had introduced Bondi to the Principal of the new University of Sussex at Brighton. Suddenly there seemed to be momentum in this option, while the Cambridge plan had stalled. Woolley told Quirk that any development in Sussex would need the approval of the British National Committee for Astronomy as it would be important to be sure that astronomers would be willing to move to Brighton. He also suggested that the scheme might bene-fit from the support of Lord Hailsham, the Minister for Science, who had strong links with the area: the town of Hailsham is not far from Brighton.[26] Quirk assured Woolley that he would mention it to Hailsham, and that 'meanwhile I wonder what Fred Hoyle would think of this Institute being in Sussex rather than Cambridge?'

Hoyle was aghast. Of the two possible locations that had been discussed, surely Cambridge was far better than Sussex? He was still reeling from this blow when Ryle launched his latest salvo in the war on the steady-state cosmology.

Ryle had submitted a paper to the Royal Astronomical Society in January 1961, presenting his latest survey of radio sources and his assessment of its implications for the steady-state cosmology. He was confident that this time, his data would stand the test, and steady-state theory would be defeated.[27] Hoyle knew that something was afoot when, at tea in the Cavendish Laboratory, Ryle gave him a glimpse of a new graph of $\log N/\log S$. Hoyle started to prepare his response, along with his new student Jayant Narlikar. Hoyle then received a telephone call from Mullard Ltd, the electronics company, which was funding Ryle's research, inviting him and Barbara to attend a press conference that Mullard was hosting in London, at which Ryle would announce 'new, hitherto undisclosed results that [Hoyle] might find of interest.'[28] On the same evening, Friday 10 February 1961, Ryle would make a formal presentation at the Royal Astronomical Society under the chairmanship of the President, another Cambridge Professor, Roderick Redman of the Cambridge Observatory. When the Hoyles arrived at the Mullard building in Shaftesbury Avenue – a street better known for theatre than for science – Barbara was escorted to a seat among the journalists, and Hoyle to a chair on the stage. He sat there while Ryle delivered a lecture in

which he announced that: 'These observations . . . appear to provide con-
clusive evidence against the steady-state model.'[29] Ryle had been counting
radio sources, and looking further and further out into space in order to find
them. The sources were plotted on a new graph of $\log N/\log S$, and the
slope had been measured, and it was -1.8 – steeper than steady-state theory's -1.5. Ryle explained to the journalists that the further out he and his
team had looked, the more sources they found; and since observations of
the far universe are essentially pictures of the distant past, Ryle concluded
the universe had once been more densely populated by radio sources than it
is now – and therefore is not in a steady state.[30]

There was a press embargo on this conference, and on the press release
that had preceded it: all reporting should have waited until after Ryle's
formal presentation at the Royal Astronomical Society that evening; but
the embargo was broken by the London evening papers that appeared
that afternoon. The *Evening News* announced 'The Bible was right,'[31] and
the *Evening Standard* claimed ' "How it all began" fits in with Bible story'.[32]
Ryle was furious when he encountered the headlines on the short trip
from the Mullard press conference across Piccadilly Circus to the Royal
Astronomical Society, and it was reported that he 'barked out a rude
expletive' at the *Evening Standard* journalist who had broken the embargo.[33]
Hoyle was furious too. He felt he had been 'set up' at the press conference:
the new evidence had been sprung upon him, in public and without
advance warning.[34] He did not attend the Royal Astronomical Society
meeting that evening, as he was giving a lecture at the University of
London, but he had deputized Narlikar to respond on his behalf. Narlikar
had come to Cambridge from India on a scholarship, and, like Hoyle, had
achieved distinction in the Mathematical Tripos. He had wanted to work on
steady-state theory for his PhD, but Hoyle had advised him against it,
because 'it's too controversial a theory and a research student shouldn't be
involved in controversies.'[35] Narlikar had been working closely with Hoyle
since his recent glimpse of Ryle's new graph, and now, after only six months
as a postgraduate student, he was suddenly in the thick of it.

Ryle presented his paper at the Royal Astronomical Society, where,
according to Narlikar, the atmosphere was 'very charged up'.[36] Despite
being sat on the stage at the Mullard press conference, Hoyle was not cited
in paper: the only reference to the work of the steady-state proponents was
to Bondi's popular 1952 book *Cosmology*, a new edition of which had
recently been published. So it was Bondi, who was at the time Secretary of
the Royal Astronomical Society, who was called on first to respond to Ryle.
As Narlikar recalls, Bondi was 'quite cynical in a typical way':

He said that in 1955 Ryle had made the same announcement and he had got a very steep slope, 2.5 or 2.3. Three years later he brought it down to 2.2. Now he is talking 1.8, which he claims to be different from 1.5. So, Bondi said, 'I am willing to wait another three years for him to come to 1.7.' Ryle got very annoyed.[37]

Then it was Narlikar's turn to respond, which he did by arguing that there were other ways to account for Ryle's distribution of sources which were compatible with steady-state theory. Narlikar had been 'very scared' about giving the response, but Hoyle had reassured him:

> [He] said that 'if you just stick to what we have done and if you feel you have done a correct job there's no need to be afraid of anything', and that gave me confidence. He also told me to time it for 8 minutes because two minutes are maybe gone in interruptions and it shouldn't happen that you spend all the 10 minutes on background without coming to the main thing. So I went into the thing well practised and it worked out all right. I gave the talk and showed some slides. The advantage was that subsequently I didn't feel afraid of any controversy.[38]

Ryle was explicit that his work was about testing steady-state theory, and had little to say about big-bang cosmology: the title of his paper was 'An examination of the steady-state model in the light of some recent observations of radio sources,' and the abstract read 'Special consideration has been given to the predictions of the steady-state model.'[39] It had been the defeated Hoyle, not a triumphant big-bang proponent, who was spotlighted at the press conference. According to Ryle's colleague Francis Graham-Smith, Ryle was unconcerned about Hoyle's feelings: all that mattered was that Hoyle was wrong.[40] Hoyle knew that Ryle was gunning for him:

> From about 1955 onwards Ryle had the idea that by counting radio sources as a function of their fluxes he could disprove the steady-state theory. His programme, which he pursued relentlessly over the years, does not seem to have been directed towards any other end. There was no question of establishing the correct cosmology, but only of disproving the views of a colleague in the same university. . . .[41]

Ryle's colleague Antony Hewish had a different explanation of Ryle's interest in steady-state theory:

> . . . when you have conclusions which matter so much, people naturally get excited. . . . Perhaps it's a little quieter these days and slower, but in those days you wanted to be first in the game, and when it comes to 'is the steady-state

theory right or wrong?', that's a very big question. It mattered a great deal to Fred Hoyle. To a theoretician, it matters a great deal if the theory is right or wrong. Some of what he said [on the radio] was his opinions, they weren't facts, but he rather gave the impression that this is what the universe is like. Martin Ryle, because he's done lots of hard work and sweated to get apparatus built and the apparatus then gets different results, is very keen to put that side of the question. These bloody theoreticians, as he would say, who haven't done a stroke of work in their lives except sit at a desk with a sheet of paper in front of them. . . .[42]

In 1958, it had been Hoyle who was billed as 'the modern astronomer.'[43] So when Ryle presented this latest challenge, some journalists anticipated public sympathy for Hoyle: the morning after the press conference *The Times* found space to reassure readers that 'Professor Hoyle is not depressed.'[44] But he had been famous for a decade, and new sciences and new times were producing new popular heroes. In 1961, the media homed in on Martin Ryle.[45]

Hoyle was later to record in his autobiography that Ryle was tall and charismatic;[46] he, on the other hand, tended to attract adjectives such as 'comfortable', 'cheerful' and 'stocky.' Ryle could be extremely charming, and even *Paris Match* magazine noted his film-star looks: he was described as 'style sportif', as opposed to Hoyle's 'style Pickwick', and he 'seemed to have stepped straight out of a Hollywood film about atomic science.' Hoyle, on the other hand, was described as having spectacles that gleamed like a pair of flying saucers, exuding a mixture of intelligence and myopia.[47] In another magazine article about the latest developments in science, the main picture was of Ryle, daringly casual in an open necked-shirt with the sleeves rolled up.[48] In one magazine, Ryle was nominated by a panel of actresses as one of the seven most romantic men in the world; the report included the same rather glamorous and obviously posed photograph of him that had appeared in many newspapers.[49] The newspapers also published action shots of Ryle, lecturing in front of a blackboard (or in one case, pointing at a globe) or windswept in a field with his telescope in the background. So although Ryle disliked any sort of public occasion and rarely presented his work himself except at professional meetings,[50] he did become a popular media subject.

That there was a press conference at all for Ryle's announcement, though common practice for scientific announcements now, was very unusual at the time; and that Hoyle should have been invited along to be publicly humiliated caused much bad feeling. Hoyle was, according to the *Daily Express*, 'hopping mad' at Ryle's 'cheap success'; other scientists, including

Bernard Lovell and the American Olin Eggen, then Chief Assistant to the Astronomer Royal, expressed dismay at the circumstances of the announcement, which would, according to Eggen, have 'put Phineas Barnum to shame.'[51] The *Evening News and Star* reported that 'Hoyle . . . is joined by other scientists in complaining that the publication yesterday of Professor Ryle's claims was made before data worked on by the team was circulated to other astronomers. They say this move was "unacademic and irregular".'[52] Lovell, however, also told the *Daily Telegraph* that Ryle's work was 'splendid . . . and a great credit to British radio astronomy.'[53]

The impetus for the press conference had come from Mullard, which had provided Ryle with a press agent.[54] The agent did a thorough job: in the extensive coverage, some of which had gone to press before the press conference, the Mullard company is mentioned frequently.[55] The political magazine *Time and Tide* reported some friction between journalists occasioned by rumours that some of them had been contacted by Mullard sooner than others, giving them a better chance to prepare (Mullard denied this).[56] Ryle disapproved of the way Mullard handled the press, and seems to have been innocent of its plans: he only found out after the press conference that his paper had been sent to journalists in advance. According to one report, Ryle 'turned to the Mullard press officer with a querulous request to know what had happened.'[57] He also claimed, a few days later, to be in 'such a "state of demoralisation" over the rumpus that he almost wished he had never made the announcement';[58] and he regretted the way his work had been 'painted so black and white.'[59] Ryle did, however, continue to give interviews over the next few days, even to the *Evening Standard* journalist whose pre-emptive article had earned him a burst of Ryle's temper on the day of the press conference.[60]

Although the evening newspapers of Friday 10 February, briefed by Mullard for articles written before the Royal Astronomical Society meeting, did not mention Hoyle,[61] the next morning *The Times*' science correspondent reported that Hoyle had 'conceded' at the press conference that Ryle's results did indeed imply that the universe is not in a steady state. However, *The Times* also reported that Hoyle had 'within the past fortnight modified his version of the steady-state theory so that it agrees with Professor Ryle's evidence.'[62] Hoyle was interviewed for the Sunday papers: the *Sunday Dispatch* announced, in a front-page headline, ' "Creation" professors in bitter row: Hoyle slams Ryle on cheap success.' Hoyle talked angrily in the article about how Ryle had ignored their earlier discussions and had refused people access to his data.[63] Another newspaper likened the row to a boxing match:

Many will think Ryle's upper-cut to Hoyle's jaw at the Royal Astronomical Society on Friday a knock-out blow. But though Hoyle has been stopped dead in his tracks and is at the moment desperately covering up to gain a breathing space, I doubt whether we have seen the last of this stocky astronomical fighter from Yorkshire.[64]

8
Fighting for space

Responding to Ryle; trouble for the institute; problems with computers; A for Andromeda *on television;* Rockets in Ursa Major *on stage; the institute goes to Sussex*

Ryle's press conference and the defeat of steady-state theory filled the newspapers over the weekend. By Monday 13 February, Hoyle and Ryle had reportedly spoken on the telephone and apologized to each other: however, the *Daily Express* noted, 'the big bang row is still on.'[1] Now, the controversy was between the two men. Several newspapers published photographs of Hoyle and of Ryle – no one seems to have been able to capture them both in the same photograph – all of which show Ryle looking elegant and composed, while the *Daily Express* published a ten-year-old photograph of Hoyle at the height of his radio fame, and three different photographers managed to catch him scratching his head.[2]

In the late 1940s and early 1950s many commentators on steady-state theory had mentioned the fact of Hoyle's relative youth (he was 33 when the theory was published; and Bondi and Gold were still in their twenties), perhaps to imply that his work might not be reliable.[3] By the 1960s, when 'youth' had more positive connotations, the ages of Ryle's team of graduate students were mentioned – they were in their mid-twenties.[4] Ryle's team also included another modern feature: a woman scientist. This 'brunette', Patricia Leslie, aged 25, 'the wife of a research graduate,' reportedly commented: 'Exciting, isn't it?'[5] Press coverage also included striking images of the exploding universe, and photographs of radio-telescopes strung out across the Cambridgeshire countryside.[6] Nowhere in the coverage were there any illustrations to lend explanations or emphasis, accurate or otherwise, to accounts of Hoyle's arguments.

Soon Hoyle and Ryle were sharing the front pages with the launch of a Russian rocket to Venus.[7] The two stories fuelled each other: according to the *Daily Mail*, this 'hop and a skip to the Morning Star' will 'begin to answer many of the questions that are troubling astronomers such as Martin Ryle and Fred Hoyle. . . . As astronomers wrangle over their starry theories of creation the Russians leapfrog out 26,000,000 miles to see for themselves . . . and get the facts to clinch all arguments'[8] The *New Statesman* suggested that the 'Ryle–Hoyle dust-up' was given 'extra point' by the Russian rocket.[9] In another Cold War reference, a *Daily Mail* cartoon showed an H-bomb reacting with less than surprise as it read the newspaper reports that the 'world began with big bang'.[10]

Press coverage was dominated by talk of 'the creation.' While the broadsheet newspapers of February 1961 remained aloof on this point – the *Guardian* curtly dismissed the matter, saying there was no chance that Ryle's work would 'validate or invalidate the Christian account of this cosmological event'[11] – in the popular press, Ryle was upstaged only by an omnipresent God. Even the earliest reports, for which the source must have been Mullard, had carried headlines such as 'The Bible was right.'[12] The *Evening News* and the *Daily Express* both printed what appear to be the opening lines torn out of the Book of Genesis.[13] The *Sunday Express* claimed that Ryle's work was proof of the 'God-given nature of man':

> . . . man himself is born of God . . . man is God's own image. . . . consider the work of Ryle's small team. From this tiny clouded planet they have reached out to the edge of space and the beginning of time. . . . Could there be any surer proof of the God-given nature of man? . . . there is something about man's inspired, unceasing quest for truth that is indeed divine.[14]

The *Daily Sketch* juxtaposed comments from the Bishop of Manchester and Hermann Bondi under the headlines 'Man of God' and 'Man of Science.' The Man of God said: 'The scientists say the universe began with an explosion. The Christian says that the explosion was caused by God.' The Man of Science said: 'I can admire the flexibility of mind of anyone who can regard this as confirmation of a Biblical statement that the world began 5,000 years ago, and at the same time accept the theory of the scientists that it began 10,000,000,000 years ago.'[15]

This religious emphasis annoyed Ryle, who was a Humanist.[16] The day after the press conference he had told the *Daily Mirror* that he 'would not agree that his findings strengthen the religious theories of creation.'[17] He told the *Sunday Dispatch* that 'The Biblical story has no relevance in our work. . . . It's a bit hard to have words put in one's mouth . . . "The Bible is right" and all that rubbish.'[18] He even pointed out, to the *Daily Express* and the *Daily Mirror*, that there was a discrepancy between his work and the Biblical account: Genesis claims that God created the universe out of nothing, whereas he was suggesting it was produced by the explosion of compressed matter.[19] Despite these protestations, Ryle was still accused of having made the extrapolations into religious territory himself. Malcolm Muggeridge, who had accused Hoyle of a similar misdemeanour after his radio programmes in 1950, wrote:

> I am fascinated by theories such as Ryle's. . . . The idea of examining galaxies as they existed several thousand million years ago . . . enthrals me. At the same time I would respectfully suggest that those who map the heavens are as liable to draw fanciful conclusions from the data they collect as the people who first mapped the earth. Astronomy . . . constantly extends our knowledge of how the cosmic clock works, but it can shed no more light than does the book of Genesis on who wound it up and to what end.[20]

Hoyle, despite his atheism, seems to have felt obliged to respond to the apparent advantage conferred by Ryle's unwanted holy ally: he told the *Sunday Dispatch* that there was nothing in steady-state theory that need undermine belief in the biblical Creation.[21]

It took the popular press to take a philosophical view: in a front-page editorial, the *Daily Mail* sighed:

> It has its funny side. Here are two Cambridge professors arguing about the size and shape of the universe and what happened 10,000,000,000 years ago. As if either of them knew! It is as though a couple of erudite ants in the cranny of a castle wall were quarrelling about the age of the building and the name of the man who designed it. . . . We are told that [Ryle's research] solves the mystery of the universe and how it all began. Nothing of the kind. . . . we still know nothing of the Creation. . . . What sort of answer you get depends on the age you live in. Before the 16th century people thought the sun and stars revolved around the earth. . . . No-one has greater respect for the astronomers than we have – but they do change their minds so much. What the scientists tell us must be a matter of faith. We take it on trust – yet we repeat it as if it were proved, eternal truth. The human mind cannot comprehend the Infinite, which is why, striving after it, most of us find comfort and peace in religion. The Bible version of the making of heaven and earth is simpler than the scientific. And just as likely to be true.[22]

Hoyle's fight with Ryle was being reported in the USA, where columnist Art Buchwald was moved to epic poetry, which began 'Said Ryle to Hoyle / "Please do not boil / The World began with a bang" / Said Hoyle to Ryle / "Well boil my bile / Your theory doesn't hang." '[23] The story was followed by colleagues at Caltech, and when Willy Fowler next wrote to Hoyle, he concluded his letter with a pun: 'To hell with Rile.'[24]

While Hoyle's anger at this time was attributed by most journalists to Ryle's attack on his theory, the *New Statesman* took a different line:

> I am not surprised that Hoyle is hopping mad. He has had the space-gun jumped on him. I'm told he was himself planning a breakthrough into the publicity knot-point network – a region where he is much at home. . . . He was getting ready for a statement when Ryle hit the headlines.[25]

Hoyle did indeed have relevant work in progress: Narlikar's response to Ryle in the discussion at the Royal Astronomical Society meeting on 10 February had been a preliminary account of the paper he and Hoyle would present in April. When Hoyle replied to Fowler's letter, he explained:

> . . . truth to tell, I have been too mad during the last six weeks to risk writing to any of my friends. The Ryle cosmology business was sprung on me in a very personal way in a series of lectures and at a press conference in London and on TV. The effect was to force me to drop all other work in order to stage a counter-demonstration, which will be mounted at the [Royal Astronomical Society] on 14 April. The work of the last six weeks looks good and there is a good chance that the theory will come out on the credit side.[26]

Hoyle and Narlikar presented this new work at an open meeting of the Royal Astronomical Society, which Ryle attended.[27] (Maintaining the fighting metaphor, Fowler commented afterwards: 'I hope you laid about you properly.'[28]) Hoyle and Narlikar pointed out Ryle had said nothing about the physical properties of the radio sources he was counting, and that:

> Manifestly there is a paradox in this, for one cannot count objects without some close specification of what are the objects to be counted. The paradox is actually resolved by the circumstance that the mathematical analysis . . . used by the Cambridge radio-astronomers itself imposes physical properties on the radio sources. If it should turn out that the properties thus imposed are indeed satisfied by the radio sources, then the claim that observations contradict steady-state theory may well be correct.[29]

Hoyle and Narlikar then demonstrated that if Ryle were to aggregate his data with that from other observatories, the result might well confirm steady-state theory. As Hoyle explained to Jesse Greenstein shortly after his presentation of this work:

> . . . the situation is unusual from a theory point of view. All observers, every-where in the universe, disagree with the normally calculated steady-state number count, but in the mean their results agree with the theory.[30]

In a follow-up paper, published in 1962, Hoyle and Narlikar demonstrated that radio source counts were consistent with the cosmological principle (and were therefore consistent with steady-state theory); that is, they were uniform irrespective of the position of the observer.[31] But after all the publicity of the February meeting, there would be little limelight left for Hoyle and Narlikar. According to McCrea, Hoyle and Narlikar's audience was not persuaded by their arguments: 'most of those present were mystified,' he noted, and the event had 'served merely to confuse people into thinking that Ryle's work after all had not invalidated straightforward steady-state cosmology.'[32]

Through it all, the *Yorkshire Post* stuck by the local boy, and reported that Hoyle's amended steady-state theory adequately accounted for Ryle's data: so 'it is not possible to say which theory is in the stronger position.'[33]

While steady-state was under attack, Hoyle was battling to protect another brainchild: it looked likely that his idea for an institute for theoretical astronomy would be snatched away from him, and handed to the University of Sussex. In March 1961, at a meeting of the sub-committee that was assessing the idea for the British National Committee for Astronomy, Hoyle made clear his views on that development: still smarting from his very public humiliation at Ryle's hands a few weeks previously, Hoyle resigned as convener of the sub-committee in favour of McCrea, because of the 'unlikelihood of the expansion of theoretical astronomy in Cambridge.' At that same meeting it was resolved to recommend the establishment of the institute within a university, possibly one of the new universities, such as Sussex.[34] Quirk reported to his colleague Turnbull:

> Prof. Hoyle expressed the utmost gloom at this prospect. He says he thinks there is not the slightest chance of an Institute in Brighton being able to build up a really high status for many years. At the moment, his department at Cambridge attracts distinguished research students from many countries, and this is a great asset for UK astronomy. He thinks that shifting to Brighton will dissipate this asset.[35]

Despite the tensions around Hoyle in Cambridge, he was committed to the University. Geoffrey Burbidge, who was visiting Hoyle every summer in the early 1960s to work on radio sources, noted that for Hoyle, Cambridge was 'the place' It attracted the best students, and Hoyle had never studied or been a faculty member anywhere else.[36] But Hoyle was sure, as he told

Quirk, that the scheme could not be realized Cambridge, even if the petty infighting and politics could be overcome: there were too many people there who were terrified both of letting the government tread on their patch, and of civil service scientists, like the nuclear physicists at Aldermaston, who might steal their ideas. Quirk gave Hoyle some stern advice: politicians and civil servants found it 'tiresome . . . to hear behind-the-scenes mutterings from the Universities.' Hoyle should put his fears in writing, and he should stress the scientific disadvantages of Brighton. Hoyle was clearly still pre-occupied with Ryle's challenge, for before they parted he gave Quirk 'some interesting points about the arguments on Cosmology.'[37]

The Department of Scientific and Industrial Research was becoming increasingly bemused about the progress of this issue. It had expected the outcome of the recent committee meetings to be proposals for the development of radio-astronomy, not theoretical astronomy. But in April, Sir Alexander Todd, Professor of Organic Chemistry at Cambridge, a recent Nobel laureate and the long-standing Chairman of the Advisory Council on Scientific Policy, told the Department that a proposal for radio-astronomy could be generated elsewhere, and suggested that Sir Harry Melville, its Permanent Secretary, might undertake this.[38]

In May 1961 Hoyle wrote to Greenstein, explaining why he would be unable to come to Caltech in the autumn:

> . . . national science politics are in a rapid state of change in Britain. Last year we did very well in astronomy, and there is a good prospect of still better things, if certain critical discussions go the right way. Everything is rather in the melting pot just at the moment. Possibly things will straighten out before September, but this I doubt, since events proceed more slowly than one tends to expect, particularly with the summer a dead period so far as committees in London are concerned. The crisis will probably be reached about October, and I would not like to be away at that time. (A bad decision over computer policy was taken last Fall in my absence, and it has taken me six months of uphill work to get it reversed.)[39]

In June, Quirk attended a meeting of the British National Committee for Astronomy to discuss the new telescope for the southern hemisphere, and once again the idea of an institute for theoretical astronomy arose. The committee heard that Hodge might have changed his mind, and would now consider having an institute in Cambridge, which made the case for Sussex rather weaker in comparison. The Committee was now recommending 'that the Cambridge possibility should be investigated first', and 'if it appears likely to succeed it should be pursued vigorously.'[40] A memo attached to the minutes of this meeting reiterated Hoyle's original arguments, stressing Britain's tradition of, and genius for, turning observation into theory. A new point was raised that perhaps related to the tensions in the Department of Applied Mathematics and Theoretical Physics: that university chairs for theoretical astronomy tend to be chairs in applied mathematics, and could therefore be held by mathematicians with no interest in astronomy. It would not necessarily boost theoretical astronomy, therefore, if an applied mathematics department were created or expanded. On the problematic issue of computing power: 'Professor Hoyle has shown how important problems in his field demand a massive computer.' A further memo by Hoyle spells this out:

> One can adopt two quite different points of view in assessing the computational needs in any subject. The minimum necessary to keep the subject alive in the face of international competition can be estimated, or, quite differently, the extent of the computing facility inherently desirable can be used as the basis of the estimate. The second of these points of view has been adopted in arriving at the present estimated needs of astronomy.[41]

Astronomy is different from other sciences where matters can be scaled down or simplified to make computation simpler:

> The astronomer obviously cannot control the boundary conditions that enter his problem; they are controlled by the state of the world. The situations that actually arise in nature are nearly always more complicated than similar situations in the laboratory, where the experimenter can usually contrive simplifying arrangements. In mathematical terms, it is often possible to reduce the number of independent variables that enter a laboratory problem, as compared with a similar problem in astronomy – the astronomer has no convenient 'boxes' in which to confine his system It is clear that only a rudimentary state of development could prevent astronomy from having very large computational requirements.[42]

Hoyle then outlined two problems in theoretical astronomy, one in stellar evolution and the other in stellar dynamics, both of which would require years of computing time.

Britain was trailing behind the USA in computing, and Hoyle was still relying on the transistorized IBM 7090 machines in the USA. His work with Fowler on galaxies had been very demanding of computer power, and Fowler had spent some time marshalling computer resources for them. They were able to use the 7090 that had arrived at the Jet Propulsion Laboratory in the spring of 1960, and Fowler had applied for funds for a 7090 for Caltech.[43] Otherwise, they relied on older IBM machines at the other universities and research centres elsewhere in California. Hoyle had occasionally to be reminded that his invitations to Caltech depended on his contributing some teaching there, as he tended to fill his schedule with computing time at the Jet Propulsion Laboratory.[44]

There had been some progress in the UK though: in 1956, Manchester University had begun to develop a transistorized computer, MUSE, and when the Ferranti company became a partner in the project, the computer was renamed ATLAS. ATLAS would be up and running by 1962, at a thousand times the speed of the Manchester Mark 1. With newly developed software, ATLAS was to become, in 1963, the most powerful computer in the world. It was to be outpaced in 1964 by IBM's new compatible machines, the System/360, which allowed researchers to take a project from one machine and continue it on another, and by the CDC 6600, designed by Seymour Cray, which was the first parallel processor. These machines were to dominate for many years. In 1964, a state-of-the-art computer from IBM, or its British counterpart ICT, would cost around £250,000 – at a time when a junior researcher's salary would have been less than £2000 per year.

Hoyle was sure that the investment was necessary. Of the two problems that he had spelt out, the problem in stellar evolution would require 'three years of <u>continuous</u> operation of a computer of the speed of ATLAS.' For the problem in stellar dynamics, 'the computing time required increases as the square of the number of objects considered'. Star clusters in our galaxy, of 100,000 to a million stars, would require about three years on ATLAS. Whole galaxies would have to be treated as groups of star clusters because ATLAS could not cope with more than 100 million individual stars.[45]

Hoyle's frustration with the computers he could use at Cambridge is illuminated by a story told by his graduate student John Faulkner. When Hoyle was working on the first automatic computations of stellar evolution with Brian Haselgrove of the Mathematics Laboratory, they used the hand-made EDSAC which was a mass of thermionic valves. These valves frequently fell victim to the erratic voltage supplied by the Cambridge City Electricity Company. Faulkner recalls:

> You had to work with EDSAC as a hands-on machine. The most important thing ... was how to turn it off. There was a high voltage and an irregular voltage electricity supply, and ... if these went off in the wrong order, every valve would blow and it would take two days to replace [them]. I never did understand why there wasn't one master control that you could turn off: instead, when you were taught how to use EDSAC you were told that there were three switches and you had to turn them off in a certain order. Any rational person

would have put these on the control panel in the order one, two, three; but that was too easy for Cambridge. So they had to be turned off in some bizarre order – one, three, two. What would happen would be that the Cambridge City voltage would start to fluctuate; at some point some automatic procedure would predict that the voltage was about to dip too low; and everyone in the room was supposed to dive for the control and remember to turn it off in the right order, rather than the wrong order in which case you still had the high voltage on and everything blew.[46]

Hoyle had found a pungent metaphor for the less attractive qualities of EDSAC, according to Faulkner:

Fred used to tell us that EDSAC, having been designed by people who belonged to the huntin', shootin' and fishin' set, was built rather like an English hunting dog, namely that if the input occurred here, naturally the output would be diagonally at the opposite end of the room. [And] like a dog, and if you weren't too careful with your inputs, then the outputs bore a strong resemblance to one another. Even though you might want to take results and feed them in for the next stage in the calculation, it didn't matter – the output was over there. So it was necessary for [Hoyle] to collaborate with Brian Haselgrove, so that while one of them was feeding stuff in, the other was collecting the output at the opposite corner of the room . . . rapidly winding it up into a great big ball and running around to feed it in here. . . . Fred told us was that what limited the length of time that the evolutionary calculations could be completed for was how long they could do this ridiculous race around EDSAC for.[47]

While Hoyle was running out of patience with the Cambridge computer, Quirk was keeping an eye on the debate over which of Sussex and Cambridge would be the location for a new institute of theoretical astronomy. After attending a meeting of the British National Committee for Astronomy in June 1961, he told Turnbull:

The real question which obsesses them is whether the place should be Cambridge or Sussex. . . . What most of them, I think, feel is that it would be hopeless to try and start up a new Institute in Sussex, if there was a possibility that the vigorous Professor Hoyle, and the University of Cambridge, wanted to do something themselves. They therefore think that, on balance, Professor Hoyle ought to have a shot at trying to get Cambridge to build this place up and capitalize on the international reputation of Prof. Hoyle and Cambridge. If Cambridge, for some reason or another, do not want to go ahead (as to which Sir W. Hodge was quite enigmatic), most of the astronomers would go all out for Sussex. . . .[48]

The formal report of the meeting noted that 'The National Committee is divided in its views on which location is preferable.' The enigmatic Hodge came clean in a report from the Council of the Royal Society (to which the National Committee reported) to the Department of Scientific and Industrial Research that he wrote at the end of June 1961. Hodge said that the sub-committee had decided to ask Hoyle to further the possibility of founding an institute at Cambridge, and that if that failed, Sussex would be the chosen location. The British National Committee for Astronomy itself, however, had 'considered the arguments put forward in favour of Sussex to be impressive, and recommends that Sussex be approached.' Hodge concluded:

> Council believes that the development of theoretical astronomy should take place within the existing university framework, since it must be closely related to developments in related branches of science, and since it should also be fully integrated into the university teaching programme. Council therefore considers that the best way of proceeding would be for an existing or new university to make a special effort to develop a school of theoretical astronomy. Council has considered the subcommittee's proposal that the University of Sussex would be a suitable location. It is impressed by the arguments put forward in favour of this location, and recommends that the report be sent to the Vice-Chancellor of the University of Sussex, and that he be invited to consider developing a strong school of theoretical astronomy at Brighton.[49]

There were good reasons to take Sussex seriously. It was a new institution and growing rapidly, and a new institute could be added easily without disrupting the existing infrastructure. It was near the Royal Greenwich Observatory at Herstmonceux, and this could bring benefits to both the Observatory, with its new Isaac Newton Telescope, and the institute. It would make the institute more attractive to have a good telescope nearby. The two places could share a large computer, making better facilities available to each than they could merit on their own. And there were good transport links to London, itself a centre for theoretical astronomy, and to the meetings of the Royal Astronomical Society there. However, none of these reasons was good enough for Hoyle.

Despite the blow from Ryle, Hoyle was still a major public figure. The BBC, in the wake of the success of its 'Quatermass' science-fiction series in the mid-1950s, had wanted to make a television serial based on *The Black Cloud*, but Hoyle had sold the rights to a Hollywood studio. He assured the BBC, however, that he had plenty of ideas for a new series. The BBC introduced Hoyle to a scriptwriter, John Elliot, and the result was *A for Andromeda*, which was broadcast in the autumn of 1961.

In *A for Andromeda*, scientist John Fleming and his colleagues are setting up a radio-telescope in a military-run research station. The telescope is supposed to be used for tracking missiles, but when they switch it on, Fleming picks up a signal from the Andromeda galaxy which turns out to be the plans for building a very powerful computer. The scientists build the computer, which issues a recipe for a life-form. The scientists follow the recipe and produce a dividing cell, but Fleming suspects alien evil-doing and destroys it. He is then removed from the project, and the scientists make another cell which divides to form a one-eyed organism they call Cyclops. The computer they have built has a pair of terminals which attach neatly to Cyclops, who starts inputting data about Earth and its inhabitants.

The computer needs data faster than Cyclops can provide it, and so it entices research assistant Christine to the terminals, where she is electro-cuted as the computer uses her to write the blueprint for a human being. The scientists then make this being too, and called her Andromeda. She grows in a few weeks to adulthood, learning quickly – they read science books to her – and showing a remarkable talent for computing. But the computer then starts destroying the scientists. A worried Fleming, observing this from the sidelines, befriends Andromeda, and she explains that her job is to replace humans with a race of superior beings: she tells Fleming that humankind is about to become extinct, just like the dinosaurs. Thus the novel is a story of alien invasion, but as in *The Black Cloud* and *Ossian's Ride* it is an invasion of information rather than of spaceships and little green men. Fleming and Andromeda battle for the moral high ground, and eventually Andromeda decides to help Fleming destroy the computer and burn the plans the scientists used to build it. They then escape to the coast with the military on their tails.

Hoyle originally conceived of Andromeda as androgynous, but Elliot, more in tune with the demands of television audiences, insisted that she be a woman. But they could not agree on an actress to play the part. As a last resort, Hoyle and producer Michael Hayes went to the end-of-year performances at the Central School, a London drama college that has trained many of Britain's most talented performers. There they spotted a cool young woman who immediately struck Hoyle as the perfect Andromeda. Her name was Julie Christie, and she took the role. Her career took off so dramatically from that point that the BBC was unable to afford her when they came to make the second series, *Andromeda Breakthrough*, a year later. Then, the role was played by another young actress who would enjoy a long career in British theatre and television, Susan Hampshire.

Both *A for Andromeda* and *Andromeda Breakthrough* were also published as novels.[50] The *Sunday Mail* described the book of *A for Andromeda* as 'science fiction at its best,' and the *Weekly Science Diary* called it 'A first-class novel, original and disturbing. It is a brightly written, really exciting tale with the added inducement of scientific accuracy.'[51] Critical reaction to the television programme was mixed. *The Times* announced: 'Science fiction serial starts well.'

> Although it is encouraging to have the authority of Professor Fred Hoyle for the scientific credibility of . . . *A for Andromeda*, . . . evidently it is the skill of Mr Hoyle the novelist which will mainly be called upon to hold our attention. . . . In the first episode last night it was well in evidence.[52]

Despite the fact that its critic was a fan of Hoyle's work, the *Daily Telegraph* was less impressed: L. Marsland Gander wrote: 'As a devotee of Prof. Hoyle and a keen student of disembodied intelligence I felt impatient . . . I am too well acquainted with [his] work to be disappointed at this early stage, but the temptation is great.'[53]

After the first episode, Hoyle was happy enough with his programme. He wrote to Greenstein:

> Our British audiences are willing, still, to watch a seven-part thriller, which gives one enormously greater scope than a story which must complete itself in a single hour. It is early yet to know how it will go, for these series tend to pick up towards the end. First reactions were about as we expected – people, generally, willing to go on watching! We started fairly pianissimo, so as to get the longest possible crescendo.[54]

A few weeks later, Barbara Hoyle rang the BBC to ask for viewing figures, and was rewarded by the news that by the end of the sixth episode, 25 per cent of the UK population aged 5 and over were watching *A for Andromeda*.[55]

While the series was on the air, the *Daily Mail* made the connection between the plot of *A for Andromeda*, in which living creatures were made from scratch in the laboratory, and the announcements from Russia and Italy that a fertilized human egg could be grown for a short time outside the womb:

> These days are tough times for the science fiction writer. Jules Verne and H.G. Wells have long been left in the wake of the rocket, the H-bomb and the atom submarine. Now our own Fred Hoyle is desperately trying to keep a jump ahead of his fellow scientists in the laboratory.[56]

So is *A for Andromeda* science or fiction, asked the *Daily Mail*? Hoyle replied that the knowledge needed to make a human being from a genetic recipe was a great many years away. But he added: 'our science fiction is based on sound scientific fact. . . . I see no reason why we should not keep just that leap ahead.' Indeed, Hoyle's friend the biochemist John Kendrew and his colleagues Jim Watson and Francis Crick would shortly bring Nobel Prizes to Cambridge for their work on the biochemical basis of life.

A for Andromeda kept Hoyle in the public eye through the autumn of 1961, but perhaps more usefully for his political campaign, he was featured in an '*Observer* profile' in October.[57] *Observer* profiles are short biographies of well-known people from a variety of fields: other people featured that year were the recently elected US President John F. Kennedy, sculptor Henry Moore, philosopher Bertrand Russell, civil rights leader Martin Luther King and the UK's Minister for Health Enoch Powell. *Observer* editor David Astor required his profile writers, who are invariably anonymous, to be 'a modest, fearless, self-effacing recorder of how things really are.'[58] The article gave an account of Hoyle's career and personality, and began with an assessment of what the public expect of their scientists:

> Though we are approaching the three-hundredth anniversary of the fall of Sir Isaac Newton's apple, the British public still expect their scientists to be men apart, largely incomprehensible, their operations somewhere outside the scope of anything the average person expects to understand.

The author did not think that Hoyle could shatter this perception: he was, after all, still a scientist, and in many ways fitted the common stereotype:

> . . . he ruminates through mathematics problems while driving his sports car (this rather worries his wife). . . . [He is] deeply . . . engaged in his subject, . . . there has been no interruption, not excepting holidays and weekends, to his life's work of cosmic interpretation.

There is something unusual, almost mysterious, about the way in which he works:

> Hoyle . . . [is] the speculative antennae, the reconnoitring force which can see unusually far ahead. . . . His special gift . . . is a strong intuition about the way the laws of physics operate. . . . [A theory] seems to take his imagination so much that he is able to feel the very process . . .

But Hoyle was also an antidote to the alienation many people felt from science:

> In theory, this feeling of estrangement should be slightly eased by the news that a science fiction serial . . . which the BBC began televising last week, was written by the distinguished incumbent of one of our most venerable science Chairs. . . .

is this the ultimately revealing common touch? In practice, the story is likely to remind people only that the author, Fred Hoyle, the Plumian Professor of Astronomy and Experimental Philosophy at Cambridge, is an exceptional kind of scientist.

Yet despite his special attributes, Hoyle was also unworldly, even childlike:

> In some ways he still seems the school prodigy: certainly the caricaturist would note the eager sixth-former that moves the middle-aged man, the occasional naivety, the lack of any guarded sophistication, the gaps in his general culture which appear a natural complement to the fierce and uncompromising concentration on his subject.

The radio talks of 1950 were singled out as the launch of Hoyle's popularizing career, and the profile suggested that his popular work had masked his scientific achievements:

> For scholars, popular achievements have their risks; and with Hoyle the public displays have tended, like those strange white clouds obscuring Venus, to hide the merits of the solid body behind. . . . It is never easy to isolate the particular talents of any one scientist, . . . But it is fair to say that Hoyle's gifts remain unique, and that some of the most striking recent advances in our knowledge about space or the construction of matter wouldn't have been possible without him.

While *A for Andromeda* was showing on television, Hoyle had ventured into a new medium: he had written – 'on graph paper, of course', claimed the *Sunday Times* – a play for children. *Rockets in Ursa Major* was to be staged at the Mermaid Theatre, in the City of London, at Easter of 1962. The theatre was new: it had been built by subscription in 1959 – 60,000 people each paid for one brick – and was run by its founders, the actors Bernard and Josephine Miles. Just before the play opened, the *Sunday Times'* reporter went to meet Hoyle the playwright in Cambridge:

> He likes working in the middle of things, and nothing can disturb him when he's hard at it. As we talked, his wife put his collar straight, his mother looked for the teapot, a carpenter clumped up the wooden stairs then round an open gallery, and the poodle Sam chased Tatty the cat. . . . He says he gets more plaudits from his scientific colleagues for his novels ('my ragtime stuff') than his science.[59]

The mother was not Hoyle's: both his parents had died in the mid-1950s, his mother of complications after an operation and his father, rather more slowly, of cancer. It was Barbara's mother, Christine Clark, known as Gram, who lived with them at Clarkson Close, and like Barbara she became well known to Hoyle's colleagues. Correspondence handled by Barbara often included news of Gram.

On the opening night of *Rockets in Ursa Major* in London, the *Daily Express* reported that the 'zestful scholar Fred Hoyle yesterday forsook the chair of Plumian Professor ... at Cambridge, where he is regarded as one of the brightest stars in the mathematical firmament, and took his seat in the stalls of the Mermaid Theatre, [where] he is accepted as a discerning play-wright.'[60] Hoyle told the *Express* that he knew from his own children what teenagers expect from science fiction. In the play, a space-rocket and crew are launched to the constellation Ursa Major, and the rocket returns to Earth empty except for a message warning of the outbreak of a war in space. Earth is in danger of being blown up by hostile aliens. According to Hoyle:

> I am an atheist, but as far as blowing up the world in a nuclear war goes, I tell them not to worry. It will never happen – not in our generation. It is inevitable that man will kill himself if he does not sort himself out, but it will not be for 150 years.[61]

Critical reaction to the play was summed up by the *Daily Telegraph*'s headline 'Only 2 cheers for Mermaid.'[62] When the play was revived at Christmas of 1962, the same critic wrote 'slightly gayer space war; occasional thrills.'[63] *The Times*' review welcomed this addition to the Christmas festivities, but expressed reservations:

> It is this author's merit that he does not make things easy for us. The science in his fiction sounds real, and, if the play's turning point is a scientist's inspiration, its background ... is one of painstaking intellectual effort. One could wish that the dangerous plan ... to defeat mysterious invaders from the farthest reaches of Professor Hoyle's imagination had been attended by more heart-searchings, for there seems to be no recognition that moral issues are involved in the plan, although any audience capable of coming to terms with the play's physics should surely be capable of recognizing moral problems, which are far more homely and down-to-earth matters.[64]

Hoyle took four per cent of takings at the box office, which amounted to about £50 per week throughout the run,[65] and he and Barbara became life-long friends and correspondents with Bernard and Josephine Miles.

✳✳ ✳

Negotiations about an institute for theoretical astronomy continued. Sussex was still the front runner in the summer of 1962, when Hoyle told Quirk that although Hodge had 'come round' to the idea, Cambridge University was wary of institutes because a microbiology institute set up by the Medical Research Council had been 'off-loaded onto the finances of the university, and with staff which were not at all first-class.' Thus the British National Committee for Astronomy had no alternative but to choose Sussex. However, a degree of scepticism had arisen, and not just from Hoyle, about the chances of starting an institute from scratch at Sussex. Quirk told Turnbull:

> I believe Professor Hoyle when he says that he is not just influenced here by sour grapes. He has, in fact, with his large and distinguished staff in Cambridge, in effect a good, and internationally reputed, Institute there already. He feels, and so do I, that it really is rather a pity to try artificially to create one on a green-field site. One cannot, just by waving a wand, cause an internationally famous and distinguished Institute of this kind to come into existence. I suggest that we should, tactfully, discuss this question with Lord Todd [of the Advisory Council on Scientific Policy] on Friday. It is pretty impossible, I think, for the ACSP to do anything but bless the BNC's proposal, but it seems a trifle disingenuous to do so if the scheme is doomed from the start. At least, I suggest that Lord Todd, Sir W. Hodge and Professor Hoyle might have a word on the matter together, in Cambridge.[66]

Quirk did explain the situation to Todd (whose recent change of title was due to the award of a life peerage), and asked him to talk to Hodge and Hoyle. Quirk then told Turnbull that while John Fulton, the Vice-Chancellor of the University of Sussex, and Richard Woolley, Astronomer Royal and Director of the Royal Greenwich Observatory (and who was also on the council of 'the new Brighton university') had been enthusiastic about the project, Hoyle believed that it would be hard to find a suitable Director: 'there is no-one of real distinction at present who is likely to want to take it on.' Quirk stressed that 'an institute of this kind is a very ambitious venture and would obviously require a first-class Director and staff.'

In July 1962, a memo was issued from the Office of the Minister for Science. It concluded that steps should be taken to strengthen, coordinate and stimulate theoretical astronomy; that the scale of expenditure should be of the order of £100,000 annually; and that the Royal Society be invited 'to pursue with the authorities concerned the possibility of establishing an institute at the University of Sussex.'[67] Woolley quickly fired off a list of questions to Quirk. The Admiralty wanted a new computer for the work done at the Royal Greenwich Observatory on navigation and tides, but Sussex could not afford to pay half; and yet the University was appointing theoretical physicist Roger Blin-Stoyle and a professor of astrophysics, both of whom would need computers. If the Observatory shared a computer with Sussex, they could claim an educational discount and therefore get a better computer for the same money. Was it likely, Woolley asked, that the institute would be in existence by 1965, as that is when they would have to start paying for the computer if they ordered it immediately? Was there a chance of Treasury money, since the Treasury was already contributing towards the Isaac Newton Telescope? Could Quirk advise Woolley on how he might expedite the proposal?[68] Quirk replied that the matter was now in the hands of: the Royal Society, in the person of Hodge; the University Grants Committee (which funded the universities); and Fulton at Sussex. The Royal Society would inform Fulton about any decisions or advice arising from meetings of the British National Committee for Astronomy and the Advisory Council on Scientific Policy; Quirk insisted he had no part in that process. However, he assured Woolley that he would make enquiries, though he doubted there was anything he could do to help at this stage.[69] Clearly Quirk was not going out of his way to speed things along in Sussex.

Computing provision for Sussex was a complicated issue. A recent policy document on the distribution of computing facilities had not mentioned the University of Sussex, presumably, the civil servants thought, because at the time of the report, Sussex had no thoughts about needing a powerful computer. Four questions remained to be answered before Sussex's real needs could be gauged: whether the institute of theoretical astronomy would indeed be built there; whether a case for computing facilities was made by the new appointee Roger Blin-Stoyle ('a contemporary of mine at Wadham [College Oxford]', one of the civil servants noted); what Woolley's requirements might be; and lastly whether these three needs might be managed as one. The Department of Scientific and Industrial Research continued to respond with further questions to enquiries from Woolley, and told Quirk that there was very little likelihood of anything being decided about Sussex until 1965 – two and half years away – unless money was diverted from other projects.[70] In September, Quirk told Woolley that the Sussex institute had the green light from all advisory parties but these bodies did not offer funds; and so it was now up to Fulton to raise the money. Quirk recorded that: 'I suggested to Dr Woolley . . . that he ought to press Mr Fulton to get on with the scheme, if he was not already doing so.'[71]

Fulton was indeed getting on with the scheme, and the complex processes of the policy institutions were under way. Woolley was pursuing funding from the University Grants Committee for start-up money and from the Royal Society so that McCrea could be hired by the University of Sussex as Director of the Institute. At the Department of Scientific and Industrial Research, Quirk was urged to ensure that Fulton was sent copies of all relevant documentation or he might suspect that Quirk was 'not being quite open with him'; and Sir Harry Melville was to visit Sussex to discuss the Department's involvement in the project. Fulton was happy for Quirk to include a note about the new Sussex institute in the annual report of the Advisory Council on Scientific Policy, which concluded with the words: 'we hope that the plans for Sussex will come to fruition.'[72]

9
Storm clouds

Hoyle on television; the Ferranti ATLAS affair; Fifth
Planet; *what, and where, are quasars?; curing the
Los Angeles smog; a new theory of gravity (without
apples); Roger Tayler and the helium abundance*

B y the end of 1962, Hoyle's cosmology was once again to the fore.
He published a textbook, *Astronomy*, and *The Times* congratulated
him on this attempt to place the contemporary cosmological debate
in historical context. The book concluded with an exploration of the role of
radio-astronomy in deciding between big-bang and steady-state theories.[1]
He spent time during the winter in California, where he worked with Fowler
on imploding stars, which they thought could be sources of radio waves. In
the spring of 1963, around the time when the work with Fowler was being
published in *Nature*, Hoyle appeared on television, in a BBC programme
called 'The Cosmologists,' along with Geoffrey and Margaret Burbidge and
Jodrell Bank radio-astronomer Bernard Lovell. New results from Jodrell
Bank had recently been reported in the newspapers, where it was claimed
that they constituted 'a setback for the Big Bang theory.' This new work had
cast some doubt on the method radio-astronomers had been using to
determine the distance away of radio sources, thus suggesting that the
supposedly distant sources may not be so far away after all, in which case
they may not be indicating the state of the distant past – making them far
less interesting to cosmologists.[2] In the television programme, Hoyle hinted
that astronomers would soon know more about the source of the radio
signals that had been received at Jodrell Bank, and he described his
steady-state theory. The *Daily Telegraph* called the programme 'brave and
ambitious,' but said that 'even enthusiasts must have found it heavy
going': 'five scientists poured out opinions, theories, contradictions and
astronomical statistics in massive profusion.'[3]

Hoyle did not enjoy appearing on television: the process was too complicated, and it was hard to have much editorial control over the eventual broadcast. But he did appear in a television experiment called 'Dawn University,' in which he spoke straight to camera from his own script, rather as he had done on the radio. The *Daily Mail* reported that 'the full-blooded, highly concentrated flavour of the typical university lecture comes into the living room of all enthusiasts for higher learning.' The series would be 'uncompromisingly solid . . ., with no attempt to pose as entertainment.'[4] Hoyle gave the first lecture, on the mathematics of violence, and it was reviewed in the *Daily Herald* by Clement Freud under the headline '$x + y + z$ = no breakfast.' Freud thought that Hoyle was a good choice to open the series, but doubted whether the public was ready for mathematics at 7.15 a.m. He reported a tangled mass of mathematics tenuously linked to politics and sport, and that 'Hoyle's glasses glinted at us from a background of unlimited lengths of blackboard.'[5] Hoyle's lecture nevertheless made an impression on some people: it was cited in a paper on mathematical models of battle by a Cambridge psychologist that was published in *Nature* in 1964.[6]

In May, Hoyle was back in *The Times*, on this occasion to condemn the action of his colleagues and the customs of his university. Cambridge University had announced that it intended to confer an honorary doctorate on Lord Hailsham, Minister for Science and Technology, who had subsequently given a speech in which he expressed views on education and science which offended many scientists. The members of the University had therefore refused to endorse Hailsham's nomination, and it was withdrawn. Hoyle was angry because although he did not want the degree conferred on Hailsham, he found the procedure of publication and subsequent withdrawal of the award most offensive.[7] Hoyle often found university procedures baffling and irritating; but on this occasion they at least afforded him the opportunity to impress the Minister, who wrote a long letter to Hoyle to thank him for his protest: 'From yesterday's *Times* I gather you have struck a blow for decency,' Hailsham wrote; and he 'regretted the deplorable breach between public and academic life.'[8] Thus in the Minister's good books, Hoyle concluded the academic year in Cambridge and set off once again for Caltech.

Hoyle's many trips to the USA were a nuisance for his graduate students, but he kept them busy in his absence. As the astrophysicist John Faulkner recalls:

When I began to be a student of Fred's, it was in one of those typical phases of his life when he wasn't going to be around. Fred was off to the United States, so I went to see him. . . . He gave us essential reading for while he was away: three things – I had to read B^{2}FH, and then two cheek-by-jowl articles: an article by Chip [Halton] Arp on the Hertzsprung–Russell diagram [on stellar evolution] and an article by the Burbidges on stellar evolution . . . which I read religiously while he was away. When Fred returned to Cambridge, something occurred which I still regard as being really quite miraculous, which was that Fred commandeered one of the big lecture theatres, and he engaged [us] in a three-day marathon in which he introduced us to learning about computing and developing a stellar evolution and envelope code, all in the course of these three days. It was an absolutely masterly marathon performance, for something like seven or eight hours a day for three days. Fred completely laid out for us how to do automatic computation of stellar interiors.[9]

Faulkner was one of a group of graduate students working on stellar evolution. They were using EDSAC II at Cambridge, which although not the best available at the time, was still a very useful machine. One student, Sverre Aarseth, was working on the problem of interactions in systems consisting of many components, known as the n-body problem, which was important in stellar evolution and other astrophysical phenomena. He had worked though one problem by hand, and it had taken three months to complete the calculation; but once he had found its solution he used EDSAC II to compute the same calculation for other values, which it did in 16 seconds.[10]

When an IBM 7090 became available at IBM's offices in London's Wardour Street, the students had access to the most powerful computer in Europe. Using it was not a simple process: there were several machines there, the first of which, the pre-processor, was used to read in the data from punched cards. The tape produced at this stage was then fed into the 7090, 'which was treated like some sort of Queen Bee by its supplicants,' according to Faulkner, and then the job would pass back to the pre-processor which would print out the results.[11] The students had to book time on the two machines in sequence, and hope that each part of the job would run on time and let them move on to the next before their booked time ran out. Some days, they would go back to Cambridge with nothing. But on the days when everything went well, they made great progress with their research.

After a year or so of trips to Wardour Street, the students were told by the Department of Scientific and Industrial Research that money for computing time would henceforth not be available for the IBM machine, but would be available for the Ferranti ATLAS at Manchester. ATLAS had been launched with great fanfare in 1962, surrounded by government representatives and local dignitaries, as Britain's answer to the powerful IBM machines. At the launch, the assembled company watched as ATLAS absorbed a problem, and then produced the ribbons of paper tape that carried the answer. The students had to learn ATLAS's unique operating language, without recourse to much by way of manuals, and work out how to adapt their projects to the new machine. But they arrived in Manchester to find that the computer was not working at all. The staff there told them that the machine had many problems, that they were still trying to bring it into operation, and that the demonstration for the launch had been a stunt.

The students were mortified. Their projects, which had been speeding ahead on the IBM 7090 in London, were now stalled, and all thanks to a directive from the Department of Scientific and Industrial Research that they transfer their operations to a computer in Manchester that did not work. They reported back to Hoyle, who, according to Faulkner, was furious, and 'kicked up a real fuss about it.'[12] Hoyle was determined to find out why his students had been sent on such a hopeless mission. The answer to this question would lead eventually to a national scandal.

The Ferranti ATLAS was the result of a collaboration between the electronics company Ferranti Ltd and Manchester University.[13] It allowed Ferranti Ltd to carry out research and development in an academic environment, though the company's longer term aim was to make its computers commercially available. The project had enjoyed some financial support from the government, but much of the development money for ATLAS had come from Ferranti Ltd, which was still a long way from making any profit from computers. Indeed, the American IBM machines looked set to corner the market. Ferranti's chairman, Sebastian de Ferranti, was the brother of the Conservative MP Basil de Ferranti, Parliamentary Secretary at the Ministry of Aviation. Basil de Ferranti was also Director of Overseas Operations for Ferranti Ltd, and was encouraging other countries, as well as government departments at home, to 'buy British.'[14] When the ATLAS project was launched, the *Financial Times* reported that 'The Ministry of Aviation is expected to be interested in the new Ferranti Atlas computer,' which was 'unmatched in the world for speed and power.'[15] It was in this context that the Department of Scientific and Industrial Research had asked its grant-holders to transfer all operations to ATLAS.

Ferranti Ltd also produced guided missiles – a product that generated profits that were used to subsidize the computing work. It was with regard to Ferranti's Bloodhound missiles that a problem became apparent: the company had a contract to supply the government with Bloodhounds, and the Public Accounts Committee looked at this contract and raised the question of whether the government had perhaps been over-charged. The Minister of Aviation pointed out that contracts were always agreed on the basis of estimates, and that the cost of cutting-edge technology such as this was especially difficult to gauge in advance, but in the spring of 1964 an inquiry was ordered.[16] The inquiry concluded in July 1964 that Ferranti Ltd had 'unduly stressed ... the complexity and production difficulties to be expected' in order to justify excessive labour costs and overheads, and that the Ministry of Aviation should have noticed this before making the contract.[17]

By this time, Ferranti Ltd had left the computer industry: it had sold its interests to International Computers and Tabulators in September 1963. The company accepted the decision of the inquiry, and had already offered to refund £4.25 million to the government, which was now accepted. Sebastian de Ferranti did, however, ask for time to pay – the money was 'not in the bank', and he pointed out that 'gifts' such as the Ferranti ATLAS at Manchester University would not be possible if a company was prevented from making a profit.[18]

Some of Hoyle's frustrations with politics came out in a new novel in 1963: he wrote *Fifth Planet* with his son Geoffrey, who was now in his early twenties. Geoffrey had begun a Fine Arts degree at Cambridge, but he sensed some antagonism towards his father which soured the experience, and he left during his second year to join a communications firm in London.[19] The book was markedly different from the novels of the 1950s, and from the *Andromeda* books that began life as BBC family viewing. Faulkner was struck by the first two sentences: 'Hugh Conway shifted uneasily. An hour before his wife had come to him with such fervour that he knew she must have been unfaithful again.' Faulkner recalls:

> I was a little bit naughty: having read this book, I cheerily remarked to Barbara Hoyle one day that I'd really enjoyed the opening two sentences of Fred's latest science fiction. She replied: 'Oh, the co-operation between Daddy and Geoff works really well. You see, Daddy provides the science, and Geoff provides the sex. Daddy doesn't know very much about that sort of thing.'[20]

The story is set in 2087, and the 'fifth planet' is an Earth-like planet, Achilles, in the solar system of the star Helios. The Helios system passes close enough to Earth for space missions from East and West to head off to investigate. While on Achilles, some of the astronauts are possessed by the spirits of the Achilleans, and the spirit that survives the trip back to Earth passes from astronaut Mike Fawsett into the body of his lover Cathy, the wife of the discoverer of Achilles, astronomer Hugh Conway. Cathy is 'pure animal' and has no intellectual interests; she is therefore the perfect repository – an empty vessel – for the alien spirit. Thus transformed, Cathy (who is, incidentally, now much more interesting to her astronomer husband) tries to control the human race by causing mass hallucinations. Hounded by the police, she decides to go home to Achilles, and is delivered to a space shuttle by loving husband Hugh Conway.

Though set in 2087, the novel makes frequent mention of the early 1960s, particularly by referring to the origins of the space programme. Like other science fiction of the time, it explored institutions rather than ideas. Like Hoyle, Conway was sceptical about the value of space exploration, and contrasted the prestige afforded by great machines such as interplanetary rockets – impressive expensive structures, built by armies of people – with 'the systematic devaluation of achievements of comparable magnificence when they happened to come from one or two people'[21] – an echo of Hoyle's argument for people rather than gadgets that he was using in his campaign for an institute. Conway also was frustrated by committee business:

> By this time [2087] all important decisions affecting the structure of human society were taken in committee. Everybody knew that the system was wrong, but by now no one had the power to stop it. No committee was willing to vote to destroy itself. In the early days a few men had found themselves, more or less by chance, to be possessors of the power to persuade their colleagues – they were natural intellectual salesman. And like good salesmen, who can dispose of anything under the sun, they could get their way on any matter, however absurd. What had begun as a purely amateur sport had gradually developed into stark professionalism. Nowadays one did not become a good committee man by chance. One became a good committee man by sheer unremitting effort in which every working moment was spent in planning and scheming how to operate. But not all committeemen are good. There had to be some that were bad, simply because some members had to possess an adequate knowledge of the essential facts. It was by now quite impossible both to be a good committee-man and to know anything. The trouble, of course, was that those who knew what they were talking about never got their own way, although supporters of the system claimed that this was a good thing.[22]

Conway noted that in the one hundred years of space exploration, periods of public interest had helped sustain the physical sciences.[23] But he also had mixed feelings about the press. When he discovers chlorophyll on Achilles – the first evidence that there is life there – Conway decides to play his cards close to his chest:

> He had a crucial new result from his observations. If it had been almost anyone else, the news would already have flashed around the world. Since about the year 1960 scientists had announced their discoveries almost before they had made them. Not that Conway was close, but he believed in having a little time to digest his own work before it was mauled over in public. He knew what the newspapers would do with this one.[24]

Conway saw a conspiracy of information management between the government and the press to control the news:

> Of course the news could not be kept from the public if the publicity services haven't been willing to co-operate. There were certain to be scores of official leaks. This was particularly true in the West. But the plan of the Governments was really in the interests of the publicity services themselves. Responsible people soon saw to that. To prevent a break occurring it was made very clear that whatever syndicate attempted to jump the gun would have all his official privileges withdrawn. So although it would have been possible for any one group to have scooped the others, the gain – while undoubtedly large for the moment – would in the long run have been more than compensated through the long-term loss of facilities. No group was willing to run such a risk and all leaks were plugged before they could spout their delicious liquid into the mouths of the waiting public.[25]

And when problems arose in the detection of radio signals from Achilles' star Helios, the challenge Conway's colleagues faced was one that had plagued Ryle's group on Earth: how to distinguish one signal from the next. And in 2087, the government responded by throwing money at the radio-astronomers, much to Conway's indignation.

In a moment of personal reflection as he tours the world as an expert and advisor, Conway grows weary of the frantic pace of big cities and the stresses of modern life:

> Nowhere was a child to be found who ambled along alone, happy in a world of his own imagination, content to arrive late at school and not thinking it important. The psychologists said it was a good thing that there were no such children, for they would have lived unhappy lives themselves and produced tensions and unrest in others.[26]

Hoyle was still pondering how imploding stars might be radio sources. He and Fowler had published a paper in the *Monthly Notices of the Royal Astronomical Society* in which they proposed that the nuclei of evolving galaxies would be star-like objects that implode and disintegrate, transferring their angular momentum to the surrounding disk of gas. These highly energetic star-like objects would be strong sources of radio waves.[27] This idea could explain how there came to be sources of radio waves around the universe. In March of 1963, the *Sunday Telegraph* reported that Hoyle and Fowler had received support in papers published in *Nature*. The implosions were described as 'an inward explosion which may be produced by hitting a television tube with a hammer.'[28]

According to Caltech astronomer Maarten Schmidt, Hoyle and Fowler's paper gained some notoriety, 'because in the beginning they had a statement that they will not be deterred by how unlikely it is that something like that could exist. They wrote "we turn a blind eye, a deaf ear and a cold shoulder to. . . ." That was often quoted around that time.'[29] Schmidt had been observing and analysing strong radio sources, following up spectroscopically the highly luminous star-like objects located by the Cambridge radio-astronomers. So far, four of these objects had been investigated, without much success. Jesse Greenstein had written a long paper about them, but was frustrated by his lack of concrete conclusions about what they were.[30] When Schmidt studied the fifth object, known as 3C 273 (3C was the third Cambridge survey), he realized that its strange spectrum could be explained by a very large but nevertheless simple proportional shift towards the red in the frequency of its emission lines. Schmidt went along to Greenstein's office, and together they found that the spectrum of another of the objects, 3C 48, could be explained in the same way: by a very high redshift.

Schmidt then faced the challenge of explaining this redshift, and it was not something he wanted to get wrong – it would be difficult to live down. He felt it was most likely to be cosmological, and caused by the expansion of the universe. The very high redshift would then mean that the objects were receding very quickly, which would make them extremely distant, and so they must be highly energetic in order to appear so luminous from Earth. Schmidt's Caltech colleague, the physicist Richard Feynman, thought there might be other reasons – gravitational forces for example – why these strange objects had such high redshifts, and Schmidt considered Feynman's suggestions; but when he published his work he offered the cosmological explanation of the very high redshifts of the star-like radio sources.

Were these objects the hypothetical imploding galaxies described by Hoyle and Fowler? Schmidt talked to Fowler about his discovery, one day over lunch.

> [Fowler] didn't blink an eye – he hardly reacted. But I think he thought then that we had found his object. But I am still puzzled why he didn't react more strongly. If you launch these objects by suggesting cold shoulders and so on, it means that once they are found it is quite an advance. But Willy hardly reacted.[31]

But Fowler sent Hoyle a postcard that read: 'Whoopee! They have found our objects.'[32]

The press reacted with enthusiasm to the discovery. Spurred on by the ever-efficient Caltech press bureau, journalists went to some lengths to interview Schmidt. He was taken by surprise by their interest, and was not prepared for it – partly because he was ambushed at the end of a very complicated journey. He had funding from the US Navy to travel to a conference in Australia, and so was able to fly to the Philippines on a military aeroplane. He had planned to take a civil flight from Manila to Sydney, but Schmidt is Dutch, and, because he was not an American citizen, he was held at Clark Airforce Base and could not reach the civil airport. He recalls:

> So I took a plane into Vietnam, full of soldiers with rifles on their backs, and had to listen to the lieutenant's welcoming speech to all the troops. Then I managed to get out of there. I sent postcards to my wife all the time and as these came not necessarily in order she didn't know what was happening. Finally I land on a Qantas flight at six in the morning at Kingsford Smith airport in Sydney, I go down the stairs, and somebody says are you Dr Schmidt? I say 'yes', and he says 'I am from the *Sydney Morning Herald*,' and that started it, at six in the morning. I didn't know that anyone in the world knew where I was, because I had been incommunicado in this military system. But the press always knows.[33]

In December 1963, astronomers and high-energy physicists came together in Dallas for the first 'Texas Symposium'. One of the organizers was Robert Oppenheimer, who had directed the atomic bomb project during World War II. Despairing of what he saw as the inappropriate use of nuclear physics for military ends, he thought the new synthesis between astronomy and nuclear physics was a very positive step. The Symposium was one of the very first events at which scientists from these two fields, each of which had its own conferences, met each other face to face, and it marked the start of what became a very close relationship. In 1963, the new luminous objects with the very high redshifts were a hot topic. With the Symposium happening in Dallas so soon after the assassination of John F.

Kennedy there, it was suggested that they be known as Kennedy stars, but instead they were called 'quasi-stellar radio sources', which was shortened to QSRSS or the more easily pronounced 'quasars'. At the Symposium, Greenstein gave a lecture about the technical details of quasars, but before him on the programme was Hoyle, who described quasars as fulfilling his and Fowler's theoretical prediction.[34]

The next day, the *Daily Express* reported that 'Cambridge's controversial Professor Fred Hoyle' believed the immense objects to be gas masses, probably new galaxies forming. Since big-bang theory says the universe was created all at once, and steady-state theory says it is continuously created, these new objects supported his theory: they had 'driven another spike' into the theory that the universe had a beginning. 'This may persuade a lot of people to get off the fence,' Hoyle said.[35]

Radio-astronomers hailed quasars as a triumph for radio-astronomy as an observational tool. Spectroscopic analysis of the radiation emitted by quasars indicated that it had taken a very long time to reach Earth, and observing them was equivalent to looking a very long way into the past. Also they were receding at speeds very close to the speed of light, which would mean that they were in the furthest reaches of the expanding universe. However, as in the early days of radio sources, it was also suggested that quasars could be local. If they were, another explanation would have to be found for the very high redshift: one possibility was that it is caused by an intense gravitational field, as Feynman had suggested to Schmidt. Some astronomers questioned whether the standard interpretation of the redshift as related to distance could apply to an object as strange as a quasar, and the problem was reported in the press: in the *New York Times*, Walter Sullivan described:

> ... an atmosphere of uncertainty and excitement: uncertainty because inter-
> pretation of the data leads to conclusions that are hard to believe: excitement at
> the possibility that a rare event in science may be at hand – the discovery, in an
> utterly new kind of phenomenon, that supposedly immutable laws are invalid.[36]

But the majority of astronomers had settled on the standard interpretation by the mid-1960s: quasars have very high redshifts because they are very far away in an expanding universe.

In their discussions at Caltech, Fowler and Hoyle had not only been think-ing about radio sources. They had also invented a solution to the problem of the Los Angeles smog. This smog would feature in a short story Hoyle published in 1967: 'Welcome to Slippage City' tells the story of a beautiful place that becomes hellish due to 'the march of "progress" '. In Slippage City:

> Wide highways were driven through the very heart of the city, not just one highway, but an intricate complex linking the sprawling communities of the whole urban area. These were highways of rapid excess. They were crowded with furious, fast-moving vehicles throughout all daylight hours and through most of the night, too. Everybody in the City acquired the habit of driving everywhere by car. Then the leg muscles of the people atrophied, and this became the cause of the early deaths that were soon to sweep the city.
>
> The City, of all the cities of the Earth, was perhaps least suited to the use of the automobile as a primary means of transport. The very air movement, in and out over the sea, which had led to the founding of the City, was now a terrible liability. The air became a stagnant pool into which the by-products of the incomplete combustion of oil gradually accumulated. The strong sunlight induced chemical reactions, resulting in a kind of tear gas. Half-a-dozen times in a day the eyes of the people would burst into uncontrollable fits of weeping, as they vainly sought to wash themselves clean of the smear of chemicals that latched continuously onto the front surface of the eyeballs. It was difficult now for anything other than humans to live in this appalling atmospheric sewer. The oranges that grew on the few remaining trees reacted sharply to changed circumstances by suddenly becoming very small and sour to the taste.[37]

Hoyle and Fowler's solution for the real Los Angeles was a series of structures they called 'coverways'. Coverways were essentially huge tubes that ran above ground, with transparent roofs to allow light to enter. Traffic ran through these tubes, which contained the exhaust gas so that it could be pumped out via pipes that ran into the Pacific Ocean. Fowler had engaged a patent lawyer, James T. Barkelew, in Pasadena, who found a number of other inventions for ventilating tunnels and trapping smog, but he reported early in 1963 that his search had not produced any similar inventions, and the idea was indeed patentable. Fowler asked Hoyle to think about whether they would pursue the project.[38] But coverways were to fall by the wayside: even for Hoyle, there were only so many goals that could be pursued at one time.

Hoyle's latest work on the steady-state cosmology addressed a significant shortcoming of the theory: it proposed no mechanism for the creation of matter, and during this mysterious creation, mass–energy was not conserved. Hoyle and his student Jayant Narlikar knew that if matter is to be continuously created, energy would have to be continuously consumed, and as the universe expanded it would become more and more dilute – far from a steady state. The solution Hoyle and Narlikar proposed was a reservoir of negative energy which they called the C-field (C for creation). The C-field comes into effect when particles are created: an amount of negative energy equivalent to the newly created matter is radiated as a pulse or perturbation of the C-field. Thus the C-field is a manifestation of the creation of matter, and it has a repulsive gravitational effect – it will repel matter. Because the energy used to create matter is negative, the magnitude of the energy in the reservoir increases as energy is used: compared to, for example, 10 units of energy per unit volume becoming, when one unit of matter is created, 9 units of energy if the numbers were positive, −10 units would become −11 units. The expansion of the universe would spread this reservoir of −11 and return it, per unit volume, to the original −10. So the total energy is conserved: matter is created, the universe expands, and the whole achieves a balance – a steady state.[39]

Using general relativity as their framework, Hoyle and Narlikar developed the mathematics of the C-field, and produced a cosmology in which the creation of matter is embodied in the equations, as a perturbation in the C-field – not as a special one-off event beyond the reach of mathematics and physics, as in big-bang theory. They presented the first stages of this work at a Royal Society discussion meeting in February 1962, and then presented further developments in a series of papers running though the mid-1960s. At their first presentation, a member of the audience at the Royal Society suggested to Hoyle that the introduction of the perturbation destroyed the 'philosophical appeal' of steady-state theory because it violated Bondi and Gold's perfect cosmological principle – a perturbation would mean that something had changed. But Hoyle responded: 'if I can see a set of equations where a small perturbation will make one go into a steady-state solution, I like that better. That to me is more aesthetic.' Bondi was there, and he added: 'We do not all agree.'[40] Hoyle and Narlikar subsequently explicitly rejected the perfect cosmological principle, and concluded that initial conditions could be dispensed with (that is, the cosmos had no beginning) because the C-field could account fully for the observed state of the universe – there was no need to invoke a special event to bring it into being.[41] They also considered the influence of the C-field on the direction of time, and developed the gravitational aspects of the theory.[42] By the early summer of 1964, the newspapers were reporting that Hoyle had come up with a new theory of gravity.

Hoyle and Narlikar presented their work at the Royal Society when Hoyle returned from Caltech;[43] but as with Ryle's announcement in 1961, the popular newspapers pre-empted the scientific meeting. Hoyle must have cooperated with the journalists, for he was frequently quoted. The *Daily Express* reported that Hoyle's new theory explained not only why things fall down, but also why they do not fall up; Hoyle said that it was 'too complicated to be explained in simple terms.'[44] In the *Evening Standard*, under the headline 'No apples for Fred': he said 'We started by trying to tackle the old question of past and future . . . and we landed up, quite unexpectedly, with this.'[45] Hoyle had tried out this latest version of the theory on colleagues at Caltech – 'the toughest audience in the world', he told journalists – who had received it well.[46] Even Feynman had 'failed to fault it.'[47] (Feynman would win the Nobel Prize for Physics in 1965; he had told his students, with regard to continuous creation, that 'our friend Mr Hoyle injects galaxies into the universe with a hypodermic needle.'[48]). The *Daily Mail* reported that 'The driving force in Hoyle's world is his famous – and fiercely controversial – "continuous creation" of matter.' The report closed with a quote from Bernard Lovell, who said that the steady-state and big-bang theories 'are still equally balanced.'[49]

Among the audience at the Royal Society was a Cambridge postgraduate mathematics student, Stephen Hawking. Hawking had seen Hoyle on television, and had come to Cambridge hoping that Hoyle would be his supervisor; but he had instead been assigned to another member of the Department of Applied Mathematics and Theoretical Physics, Dennis Sciama. Hawking was just beginning to show symptoms of the disease that would later cripple him, and at that time he was not expected to live for very much longer. Narlikar's office was near Hawking's, and they often talked about theories of gravity; Narlikar had given Hawking a copy of the paper before the Royal Society meeting. At the meeting, Hawking interrupted Hoyle to announce that he had found a mistake in it. From the back of the room, he called out 'You're wrong!' According to Faulkner, who was sitting just in front of Hawking, 'the whole room swivelled and gasped, collectively. We all recognised this as the throwing down of a gauntlet. . . . Fred looked a little as though he'd been slapped in the face.'[50] Hoyle asked Hawking how he had come to that conclusion, and Hawking said 'I've worked it out.' Hoyle was furious:

He didn't tell us before the meeting what the objection was, and it was just sprung – deliberately sprung in the meeting. . . . Had Hawking been a normally physically fit sort of chap I would have really gone for him over that, but when he's sitting there [shaking] what do you do? And yet it was a kind of second-year undergraduate mistake that he'd made. . . . But it is very difficult. . . . had it been anyone else I would have ripped them apart on the mathematical point, but I couldn't because of Hawking's disability. Even though I'd have won the technical argument I would have lost more points on the . . . popular sympathy. So you just have to wear it.[51]

Once Hoyle had presented the work at the Royal Society, the *New York Times* and the *Guardian* both carried reports. According to the *New York Times*, this theory, which included a set of equations 'said to be more complicated than Einstein's general theory of relativity', looked at the universe as a whole, and 'points the way to the unification of conflicting views about the structure of the universe.' According to Hoyle, 'Everything . . . is tied up in totality.' Hoyle was described as 'a vigorous upholder of the "new cosmology",' and while continuous creation gained some support from the new work, 'he goes out of his way to assert that his gravitation theory is "in no sense dependent on it".'[52] The *Sunday Times* offered three explanations of three different theories of gravity – Newton's, Einstein's and Hoyle's – in the space of a paragraph each, under portraits of the three men. While Einstein's theory was about fields, Hoyle's was about particles, and 'gravity becomes a matter of relationships.' When asked if his theory could be tested, Hoyle replied: 'Unfortunately, the most interesting experiment is just not on. Perhaps it's just as well, since it would mean removing half the universe.'[53] The next day, the Newton–Einstein–Hoyle lineage gave new impetus to the story in the popular press.[54]

The *Guardian*'s report was written by its science correspondent, the former theoretical physicist John Maddox. He suggested that Hoyle's theory was 'putting the clock back 200 years', because it is more like Newton's work than Einstein's. Einstein attributed properties to space and then saw how matter behaves in that space; like Newton, Hoyle gave properties to matter, and then he inferred space from that. We would have to wait and see whether Hoyle is right or wrong, said Maddox, but in the mean time 'even wrong theories are worth having, for they are certainly better than none at all.'[55] Maddox also wrote a feature article using Hoyle and Narlikar's announcement as a peg for a survey of the recent resurgence of big theories about the Universe. Maddox said that Hoyle claimed that recent interest was a result of Einstein's death in 1955, as people were afraid to 'tinker with [the] theoretical edifice' while Einstein was still alive. Maddox gave Hoyle, Bondi and Gold some of the credit for shaking the cosmological foundations with the theory of continuous creation, a theory which 'ought to be true even if events show that it is not.'[56]

While the theory excited admiration among mathematicians – and in Cambridge, it was the hot topic among the graduate students[57] – other scientists were more critical. One member of the audience at the Royal Society told the *Guardian* that he 'thought he had been witnessing a parade of splendid mathematics "but not much of a theory of physics".'[58] Other scientists took Occam's razor to what they saw as a superfluous hypothesis: they already had a serviceable theory of gravity, and they had little incentive to try to understand a new and difficult one. According to one later description, 'Hoyle–Narlikar theory was . . . scarcely a necessity.'[59] Nevertheless, it seemed as though cosmology was once again in turmoil. The excitement was captured in *New Scientist*, a magazine that had been launched in 1956 for the growing body of people working in, or interested in, science. In an editorial in June 1964, which introduced a feature written by Narlikar,[60] *New Scientist* said that whether Hoyle and Narlikar's 'remarkable new theory' was correct was for their peers to determine, but:

> we need not doubt . . . the importance and timeliness of a bold re-examination of fundamental ideas. Nor need we suppose, even if it does prove to be as powerful a theory as it appears at first sight, that it will be more than a stage in the quest for more refined and more comprehensive descriptions of the nature of the universe and the forces that govern it.

New Scientist cast Hoyle and Narlikar as humble, heroic seekers-after-truth:

> We also have in this research an illustration of how the good scientist . . . must continuously and critically re-examine the very foundations of his belief and knowledge. He should also be conscious that he is trying to comprehend the universe from an obscure vantage point and with mental apparatus which was evolved for more mundane advantages in the struggle for survival among ancient Primates.

The new theory, *New Scientist* argued, represented the scientific spirit of the time:

> There is among physicists today a sense of expectancy, as though they were on the brink of a new revolution comparable with the formulation of quantum theory and relativity earlier in this century. Amid the new information provided by the high-energy machines and the radio-telescopes, and the computer-aided enquiries there seems to be a widespread conviction that many things are about to crystallise. The Hoyle–Narlikar theory is one manifestation of the stirring of

ideas and of the value of philosophical meditation on nature – seeking to understand even where understanding seems prohibited – in contrast with the unreflecting working-over of familiar concepts.[61]

In the face of ambivalence towards his theory of the C-field, Hoyle did score one unqualified scientific success in 1964: he and his Cambridge colleague Roger Tayler worked out where helium comes from. Tayler was, like Hoyle, a grammar school boy who had survived the Cambridge Tripos to emerge a Mayhew Prizeman. He did his doctoral work at Cambridge under Bondi in the early 1950s, on the structure of stars, while the controversy over steady-state theory was at its height. Then after a spell at Caltech, he joined the UK's Atomic Energy Authority, where he worked on nuclear fusion: the idea was to liberate nuclear energy by confining hydrogen so tightly in magnetic fields that its nuclei fused together to produce helium – a process that still taxes scientists forty years later. Armed with this experience, Tayler had returned to Cambridge in 1961 to work on the formation of the elements. He had investigated the processes described in B²FH, and calculated their outcomes for nuclei rich in neutrons.

Hoyle was still pondering the problem of the production of helium in the universe – a process barely explained either by big-bang theory or by B²FH. A helium nucleus requires two protons, which hydrogen could provide, and two neutrons, which had to be accounted for. There is a lot of helium about – around a quarter of the mass of all the elements in the universe is contributed by helium. But the nuclear processes proposed so far for the production of helium could only produce about one tenth of this abundance. Hoyle and Tayler took advantage of the recent discovery of a new sub-atomic particle, the muon neutrino, the existence of which allowed them to consider some further nuclear reactions that produce neutrons. With these reactions in mind, they concluded that there were indeed nuclear processes that could, at high enough temperatures, make helium at a rate that would account for its abundance. They wrote up their work under the title 'The mystery of the cosmic helium abundance', and it was swiftly published in *Nature*. The paper revived interest in big-bang theory, on which progress had stalled. The boost to the theory came about from the conclusion that the extremely high temperatures needed for Hoyle and Tayler's nuclear processes were those of the very early universe. However, the paper also offered an alterative scenario:

Either the Universe has had at least one high-temperature, high-density phase, or massive objects must play (or have played) a larger part in astrophysical evolution than has hitherto been supposed.[62]

While the big-bang solution – the one high-temperature, high-density phase – seemed right to Tayler, Hoyle was more interested in the ongoing role of the massive objects, such as galaxies and quasars.[63] He saw the helium being made there, formed and still being formed in explosions or implosions, as part of continuous creation in the steady-state universe. So proponents of both cosmologies could take heart from the paper; and it became a classic of the new era of nuclear astrophysics.

10

'Dear Mr Hogg'

Problems at the Department of Applied Mathematics and
Theoretical Physics; problems at the Treasury for the Sussex
institute; Hoyle lobbies the Minister; Hoyle makes a
decision and takes to the hills; an invitation from the
Minister; the Vice-Chancellor's Special Committee on the
Hoyle Problem

In the spring of 1964, and with the prospects for an institute for
theoretical astronomy still uncertain, Hoyle's work at Caltech was
interrupted by unwelcome news: he heard from his friend and colleague
Ray Lyttleton that their Department at Cambridge, Applied Mathematics
and Theoretical Physics, looked set to reappoint George Batchelor as Head.
Lyttleton had received a circular letter asking him to nominate the next
Head of Department, and he reported that Batchelor had 'himself done so
much to have his own name brought forward to the virtual exclusion of all
others' and had employed a 'stratagem of extracting votes prematurely
from inexperienced members.'[1] Hoyle wrote to Wilfred J. Sartain, Secretary
General of the Faculties, explaining that Batchelor was arranging to stay in
post despite the fact that when the Department was set up, it had been
agreed that the headship would rotate every five years.[2] The junior staff,
Hoyle suspected, were not aware this, as there had been no discussion in the
Department. Hoyle thought Batchelor had over-reached his authority during his term of office: the Head of Department was supposed to handle
practical matters such as administration and accommodation, whereas
Batchelor had determined policy: 'witness the recent very dubious Tripos
reform, Part II in my opinion has become irresponsibly difficult.' Hoyle
pointed out that the Department was formed at the instigation of Batchelor
and two colleagues, one of whom had already resigned because he was not
happy with the direction the Department had taken. The junior staff did
not know, according to Hoyle, that he and another senior member, Paul

Dirac, were opposed to the situation, nor that there were other members of staff willing to serve as Head. Hoyle was blunt: 'the only circumstance under which I would be prepared to remain in Cambridge will be if either Dirac or I become head of DAMTP.' He continued:

> The conduct of the department, while outwardly democratic, has in my view inwardly been very power conscious. Holding this opinion, I see no alternative but a reversion to the normal Cambridge practice of the senior professor being designed Head. If in this case Dirac should be unwilling, I am myself prepared to accept the responsibility.

Hoyle also told Sartain that Batchelor's proposal to found the Department within the Faculty of Mathematics had come within a few weeks of Hoyle's appointment to the Plumian chair, which he took to be no coincidence. Hoyle had supported the proposal so as to 'avoid the impression that I wished to throw my weight about.' Of the 20 original members of the Department, only eight remained, and four of them, Hoyle said, did not want Batchelor to continue as Head. Hoyle credited Batchelor with limiting the choices available to undergraduates, so restricting their education; and with creating a 'complete schism' in the Faculty between pure and applied mathematics to the extent that the pure mathematicians were now planning their own department. Hoyle told Sartain that he regretted having let things go so far without speaking out, but he had been very busy with his own research and with furthering students in research – a glance at *Physical Abstracts* would show where his energy had gone. 'A similar small piece of research will show where Dr Batchelor's energies have been during the corresponding period.' Hoyle admitted that 'to be frank, some personal animosity is probably involved'; but that 'unless Dr Batchelor is constrained now, . . . great sections of the university will have fallen under his control . . . [and] it will be impossible for me to continue in the service of the university.'

At the end of April, Lyttleton wrote to Harold Davenport, the Chair of the Faculty Board of Mathematics, urging that the five-year term for Head of Department be observed.[3] Dirac and Hoyle signed the letter along with three other colleagues, two of whom Hoyle then nominated for co-option to the Faculty Board. One of these was co-opted, and at the Board's meeting on 4 May it was decided to hold an election for Head of Department at the end of the month. Dirac nominated Hoyle, now back from his travels, to stand against Batchelor. As Hoyle had not spent much time in the Department, the staff did not know what to expect of him. Despite the cosily familiar tone of his press interviews from the living room at Clarkson Close, he was strictly 'Professor Hoyle' at the university (he lectured in gown and mortarboard), and many of his colleagues and students barely knew him. Early in May, an informal committee of staff, including Leon Mestel, Roger Tayler and Dennis Sciama, wrote to both Hoyle and Batchelor giving them two days to declare their plans for 'the long-term development of the Department; the role of the Head in the day-to-day running of the Department; and the relation of the Department to the Faculty of mathematics and other university bodies.'[4] Hoyle sat down in his armchair with his lined notepad on his knee and drafted a manifesto. He declared that the next five years – the term of office of the next head of Department – would be crucial. He urged a looking outward for funding to allow Cambridge to 'reassert its pre-eminence as a really first class international university.' On day-to-day matters, bureaucracy should be reduced, and routine business would be dealt with by a responsible secretary, while staff matters would be the responsibility of the Head. Hoyle noted that Rutherford spoke to every one of his students every day: 'perhaps this is too much to aspire to but it brings out my point.' As for the rest of the university, Hoyle thought that the only problem was that a lack of funds meant internal competition, which was destroying cooperation and morale. Finally, he felt that 'DAMTP should see itself in a larger international role than heretofore.'[5]

On 25 May 1964, the day of the election, Hoyle was featured on the cover of *Newsweek* magazine in the USA, which described him as 'No. 1 man in the tight little pecking order of cosmologists.'[6] In Cambridge, however, his colleagues elected Batchelor to a second term of office as Head of Department.

The next day Hoyle wrote again to Sartain. He pointed out that 'the unease which I had been feeling for the past two or three years, and which had several times nearly led to my accepting a post in the USA, ... really had its roots in my dislike of the atmosphere in which I have been compelled to work in DAMTP.' He continued:

> My strong inclination is to regard this [election] as a vote of 'no confidence' in me personally, and as an indication of contempt on the part of the young members of both the Department and the [Faculty] Board in those of us who have served the university for upward of twenty years. I am therefore doubtful of whether anything but an acute and disastrous split within the faculty can be avoided if I continue as Plumian professor. Before I take the inevitable step of resigning, I would be grateful to have the opportunity of a discussion with you in case you might be able to suggest some compromise that has not occurred to me. I need hardly add that because of my deep affection for Cambridge I find this whole affair agonising and painful in the extreme.[7]

Cambridge did not want to lose Hoyle. Batchelor had written to him as soon as the election result was decided, acknowledging Hoyle's personal dislike for him but expressing the hope that they might discuss their differences. Batchelor asked Hoyle for 'your candid analysis of the situation in the Department as you see it.' He noted that Hoyle did not:

> ... make much use of the Departmental facilities, and I am not clear whether this is by inclination or because they are no help to you. The Department gains a good deal of reflected glory from your presence here, and I, like everyone else, would be willing to help us make the Department a fitting centre for your work – if only we knew what you would like to see done. Can we talk about all this in a constructive way? Upheavals like the one we have been through at least provide an opportunity for some new thinking.[8]

Hoyle's reply was not conciliatory. He referred to 'a deep seated divergence as to how people should be treated.'[9] The election was reported in the newspapers some weeks later, when the *Sunday Times* reported that now that Hoyle had lost, he might leave Cambridge and join a new institute, funded with money from the Department of Scientific and Industrial Research, perhaps in Brighton.[10] Hoyle was later to blame colleagues in the Department of Applied Mathematics and Theoretical Physics for leaking the story.[11] But journalists had been reporting the general situation for some time: from the spring of 1964, the newspapers had carried headlines such as 'Astronomer–author Hoyle: UK frustrates scientists' and 'Skyman Hoyle may join Atlantic brain drain'.[12] After the *Sunday Times* report, the press arrived at Clarkson Close, where Barbara Hoyle told the *Daily Express* that her husband was too busy to talk about the Department or his possible move to the USA: 'he is in the middle of his most creative period and it is jolly hard work,' she said.[13] But Hoyle had other stories for the press: he had written, in the form of a diary, an account of the shortcomings in computing provision for himself and his students, including the problems with the Ferranti ATLAS machine at Manchester, and now he could see a use for it.[14] It read:

Aug. 1962
Ferranti prototype not ready. Programming information for ATLAS not even ready. D.S.I.R. extend period of use of IBM 7090 month by month. Projects inevitably lose momentum, however, due to uncertainties.

Jan. 1963
Programming information for ATLAS received. Research now ceases because a delay of at least three months must be accepted while programmes are rewritten. It is not pleasant to lose the drive and coherence of a research

programme that is going well, but the delay was accepted in good faith by all concerned.

But the good faith eventually expired, and Caltech provided computing facilities for the research students. The diary concludes with:

Spring 1964
After contrasting the generosity and competence of the Americans with the miserable bungling at home, F.H. decides to accept an appointment in the U.S.A.

Postscript
This story is true, except perhaps the last line. I would not have a reader believe, however, that the civil servants in the D.S.I.R. were themselves responsible for these grotesque situations. It seems that a pecking order exists among civil servants with those at the Treasury in top position. Once a bad decision has been made by the latter it is well nigh impossible to secure a reversal from below.

No scientist leaves Britain because of one experience like this. He leaves Britain because this sort of thing goes on all the time, every month, every day, incessantly. At first his patriotic responses hold him in Britain, but with too frequent use these responses at last burn themselves out. The last line may well become true in my case.

Hoyle told his son Geoffrey that he had written the diary 'with the intention of sending it to the Minister for Science.' But now he thought that 'it might be used more effectively in some other way . . . possibly one of the Sunday heavies [broadsheet newspapers] would print it . . .'. Hoyle asked Geoffrey to see if his contacts could help facilitate this. Hoyle explained to his son:

I am not interested simply in political trouble or in stirring up a lot of mud, but I am interested in starting the tremendous shake-up that must come soon if this country is to remain viable. The basic trouble is only a very few people with close contacts in the US have any real idea of how quickly we are falling behind.[15]

There would be other stories from the Department of Applied Mathematics and Theoretical Physics in the *Sunday Times*, and the publicity Hoyle sought for his computing story haunted his relationship with the civil servants for some time.

Batchelor and his colleagues in fluid mechanics were to become highly successful, both academically and as advisors to industry, and Batchelor was a popular and respected Head of Department – a post he held until 1983.

✳✳ ✳

In the discussions about an institute, Sussex still loomed large. In March 1963, notice of a Parliamentary Question had reached the Department of Scientific and Industrial Research. It fell to Roger Quirk to draft the reply, and he consulted Vice-Chancellor John Fulton at Sussex and the Astronomer Royal Richard Woolley. When, in the House of Commons, Mr William Small, Member of Parliament for Glasgow Scotstoun, asked the Parliamentary Secretary for Science what plans he had to establish an institute for theoretical astronomy, Mr Denzil Freeth replied that there was a recommendation for such an institute at the University of Sussex, which is considering the proposal.[16] *The Times*' political correspondent reported 'government backs astronomy plan.'[17] However, Quirk told his colleague Frank Turnbull that Fulton had not raised any money but was optimistic about getting it from the Department of Scientific and Industrial Research. Woolley was likely to buy a small computer that could have extra power bolted on later. In January 1964, astrophysicist Roger Blin-Stoyle, who was newly in post as Dean of the School of Physical Sciences at Sussex, sent Hoyle a copy of 'the final version of the application we have submitted to the DSIR for support for the proposed Institute of Theoretical Astronomy.'[18] On the first page of the application, a footnote explained that 'the general arguments presented in this paper are derived from an earlier document prepared by the [British] National Committee for Astronomy.' The application was indeed very similar to the proposal for an institute prepared by the British National Committee for Astronomy's sub-committee that Hoyle had chaired in 1960, although it also noted the value of the nautical work carried out by the proposed Sussex institute's intended partner, the Royal Greenwich Observatory at Herstmonceux. In April 1964, the Department of Scientific and Industrial Research received an application from Sussex for £50,000 towards the cost of the institute building, and the request was approved and forwarded to the Treasury. In July, the Treasury was asked to provide a further £637,000 for running costs over eight years.[19]

The Treasury was growing increasingly weary of its difficult position in such cases.[20] It received many requests from the Department of Scientific and Industrial Research, all of them approved by series of committees made up of people who knew much more about science than any Treasury official, and yet it was expected to make decisions about which projects should be funded from an increasingly limited budget. The scientists' requests, from many different branches of science, were often clearly self-interested and all made ambitious claims, and the Treasury found it very difficult to judge one against another. After a glance at the slender public purse, it raised a sceptical eyebrow over the Sussex proposal. Why, asked the Treasury, was the Department of Scientific and Industrial Research being asked to pay for computers when these were usually funded by the University Grants Committee? And could scientists not see that if they all stressed the national prestige that would accrue from their projects, it was no help at all in prioritizing one over another?[21] There was to be no quick answer from the Treasury to the application from Sussex.

In the final days of July 1964, while the proposal from the University of Sussex for an institute of theoretical astronomy sat on a desk in the Treasury, Hoyle began a last-ditch campaign for an institute at Cambridge. The timing of this renewed effort suggests that he knew about the hold-up at the Treasury; or perhaps the reporting of his failure to oust Batchelor had spurred him on. There was another impetus too: Roderick Redman, Director of the Cambridge University Observatory, had offered Hoyle a strip of land adjacent to the Observatory in Madingley Road, a short walk from St John's College on the northern fringes of Cambridge and just yards from Clarkson Close. Hoyle thought it would be a most suitable site for an institute.

His opening shot in this latest campaign, which would bypass the usual scientific committees and civil service bureaucracy, was both homely and presumptuous: while Barbara fended off callers, he sat in his armchair in the living room at Clarkson Close and composed a letter addressed directly to Mr Hogg, the former Lord Hailsham, now the Minister for Education and Science.

These were uncertain times for the Minister. He had had ambitions to lead the Conservative Party, and so had renounced his hereditary title in order to sit in the House of Commons. As Quintin Hogg, he had been elected to Parliament in a by-election in 1963, but he did not win the leadership of his party. The Conservative government was plagued by scandal, and was now struggling after the resignation of Prime Minister Harold Macmillan. The new Prime Minister, the aristocratic Sir Alec Douglas-Home, was under pressure to set a date for a general election, as his government was only a few months away from the maximum permitted term of five years. Hogg's Department, Science, had recently been merged with the Department of Education, and Hogg was now representing Education and Science in Cabinet, while his former opposite number in Education, Sir Edward Boyle, took on a supporting role under Hogg in the new merged Department. With the election bound to happen soon, Hogg, a barrister by profession and an active and outspoken campaigner, was busy supporting his party against strong opposition. The Labour Party, newly invigorated under grammar school boy and Yorkshireman Harold Wilson (whose election poster Hogg famously attacked with his walking stick), had an explicit manifesto for science – society was to be forged 'in the white heat of the scientific revolution'. Labour looked set to be forming the next government, and by October 1964 at the latest.

A major reorganization of the science policy structure was also under-way, in an attempt to unravel the tangle of committees, advisors and decision-makers responsible for science. The 'near chaos' identified by the Labour Party in 1961 had been apparent to the Conservatives too, and a committee under the chairmanship of Sir Burke Trend had recommended the formation of a largely autonomous funding council for science, run by scientists and housed in the Department of Education and Science. The new council would be the single agency for funding decisions about research; the influence of the Royal Society and its committees on such decisions would be much reduced. The Royal Observatories would be taken out of Admiralty control, and placed under the aegis of the council. Trend's report was delivered in September 1963, and its recommendations were accepted by the government. The new Science Research Council was to begin its work in April 1965.[22]

With the government in what seemed to be its final days, and with the funding institutions in the throes of change, it was unlikely that any decisions about a project as big as Hoyle's would be made in the immediate future. But Redman's offer of a building plot meant that there was a friendly place for Hoyle in Cambridge, and now all he needed was money to get the institute off the ground. So he decided to ask the Minister for it.

Hoyle's letter went through many drafts, with some careful editing and some more agitated deletions, as well as additions from other hands. In one version of the letter he accused the Royal Society of having changed his original idea of an institute with staff of 'active young men', into 'a home-for-the-feeble, and from then onwards all was lost.'[23] The final version of the letter was written on 1 August 1964, and read as follows:

Dear Mr Hogg,

Since I was appointed Plumian Professor of Astronomy, some six years ago, I have tried to build up a school of theoretical astronomy at Cambridge. As regards research I believe success has been achieved but as regards organisation it has not been possible to give any permanent stability to the school. Without help now, it is virtually certain that the whole endeavour will collapse very shortly.

You may recall that my proposal for an Institute of Theoretical Astronomy at Cambridge was commented on favourably some time ago by the ACSP. This proposal failed to make progress, however, mainly because it became bogged down in a Royal Society committee. The situation in the University has proved equally unsatisfactory. Although the University is on the whole well-meaning, the only clear cut policy that exists, in a very muddled situation, is directed towards the teaching of undergraduates, mainly in classical applied mathematics of a 19th century character. While our regard for undergraduates is very fine in itself, it gets Cambridge, and the country, nowhere so far as contemporary science is concerned. It is because so much of our science is flat to the taste that the United States now seems so attractive to those whose interests lie mainly in the present-day frontiers of knowledge.

Hoyle then explained to the Minister that the apparently hopeless situation in Cambridge had been turned around by Redman's offer of land. Also, Hoyle reported that he had received expressions of interest from 'fifteen people, all exciting original scientists', who would be happy to come to such an institute in Cambridge – some of them returning from the USA. Hoyle therefore urged the Minister to consider providing running costs of £75,000 per year:

The purpose of my letter is to assure you that for expenditure of this order the country could acquire, very quickly, one of the best schools of theoretical astronomy and astrophysics in the world. I hope you will not mind my writing to you but I feel I must appeal to you to help me convert an otherwise hopeless situation into one of great promise for theoretical astronomy in the UK.

Yours sincerely
Fred Hoyle[24]

It fell to Michael Proctor, Private Secretary at the Department of Scientific and Industrial Research, to decide how to respond to this letter. He was struck by the obvious omission: Hoyle's letter did not mention the University of Sussex. Proctor wrote to Elizabeth Beaven at the Department of Education and Science, on 24 August. He attached a document spelling out the history of the Sussex plan, noting that they were currently waiting for the Treasury's response. Proctor told Beaven:

Naturally an important part of the consideration of this was the advice given by the Astronomy Sub-committee of our Research Grants Committee, of which Professor Hoyle is chairman. . . . One of the reasons why it was thought preferable to establish the new institute at the University of Sussex, rather than at Cambridge where Professor Hoyle already has a small but lively research group, was the difficulty of getting the University of Cambridge to give any enthusiastic support to the proposal and to make any contribution towards its establishment, either in accommodation, permanent posts, or money. Sussex, on the other hand, was keen to have the Institute and to profit by its proximity to the Royal Observatory at Herstmonceux.

There has been some informal personal discussion about the possibility of Professor Hoyle being considered as the director of the new institute, but it would appear that he does not wish to do this and that he has lost confidence in Sussex as a home for the new institute. If he is not able to develop his researches in a way which he regards as satisfactory, it is likely that he will go to the California Institute of Technology, where he has done part of his work.

A draft letter to Professor Hoyle is attached.[25]

Proctor's draft letter from the Minister to Hoyle insisted that the Minister was keen to ensure the health and vigour of British science, but reminded Hoyle of:

> ... proposals which the Advisory Council on Scientific Policy had endorsed for the establishment of an Institute of Theoretical Astronomy in Sussex. I am therefore most puzzled that you make no reference in your letter to all this. I know that arrangements are not finally concluded with the University of Sussex, but I expect that they will be before long. . . . I am sure you, yourself, would agree that in the interests of a balanced national scientific effort we would not want to duplicate Institutes of this kind. It was after careful consideration of all the circumstances that it was decided that the Institute of Theoretical Astronomy was best placed at Sussex, and in this we were guided very largely by the advice of your committee. Would you not agree that, having reached this decision, we should now all do our best to implement it in the most effective way, by giving the fullest degree of support to the new institute and not duplicating it with another one at Cambridge?
>
> I am not sure if I understand what you mean when you say that 'our science is flat to the taste', but surely one way in which to make it lively and vigorous is for the leading scientists in any branch of science to work together to make a success of the decisions which have been taken in good faith to try to do the best for their subject.[26]

Proctor's letter did invite Hoyle to let the Minister know if the situation in Cambridge had changed in the mean time, but Beaven nevertheless thought it 'obviously unsuitable.' She pointed out that Hoyle 'had told Mr Quirk on more than one occasion that he thought that the University of Sussex was not a suitable place for the new Institute, and that it had been proposed only because Cambridge had been unwilling to accept it. This fact would have to be taken account of in the reply to Professor Hoyle.'[27] A few days later G.B. Blaker at the Department of Scientific and Industrial Research showed the documentation to Lord Todd, chairman of the Advisory Council on Scientific Policy, and then told Beaven that Todd did not think the letter suitable either: Todd, as a Cambridge Professor and Master of Christ's College, thought that there was chance that the Cambridge situation could be resolved in Hoyle's favour. Todd had told Blaker that there was a possibility of getting money 'from some other source for a programme of work which Professor Hoyle would undertake,' and that if that were possible, Hoyle might attract back to the UK:

> ... four or five scientists at present working in the field of theoretical astronomy in the United States. If this is not successful there is the opposite danger, as you mentioned, that Professor Hoyle may himself go to United States, thereby dealing a further blow to the reputation of British science in general, and astronomy in particular. Lord Todd therefore feels that it needs very careful handling.[28]

Todd also suggested to Blaker that the response to Hoyle 'should be so worded that, if it were published, no harm would result.'

Hoyle had very recently handed Todd a proposal, marked 'confidential', for an institute of theoretical astronomy in Cambridge. While much of the document repeated arguments Hoyle had made many time before, this time he emphasized the special value of Cambridge. It produces theoreticians of great skill, he claimed, but then wastes their potential by scattering them:

> The usual point of view seems to be that a general dispersion is desirable, with young men spreading themselves rather thinly over many universities (or industries/government agencies) and disseminating therefrom the things that have been learned at headquarters. This procedure is most inefficient since experience shows that only a minority continue to produce original work of real quality – the strain of combining the normal teaching schedule with the really personal beginnings of research seems too much for the majority.

Highlighting the threat of the 'brain drain', the proposal noted competition from the United States, to which losses so far had been 'moderately severe':

> ... present information suggests that a crisis point has currently been reached in which almost all our outstanding theoretical astronomers in the age group 30–50 are considering, more or less as a group, whether there is any point in continuing to work under unfavourable conditions here in Britain when excellent opportunities exist in the United States. Critical decisions, known to me personally, will be taken well within next year that will decide whether or not

there is to be a general co-operative movement to United States. It is in relation to the situation that there is urgency to set up an organisation in Britain. Such an organisation should aim at American standards and would do well to have good connections with one or more of the best American institutions. It should aim to be one of the main three or four key points in the international network of research in theoretical astronomy. It could be established in Cambridge, given appropriate financial support.

And with tacit reference to the earlier reluctance of Cambridge University to host the institute:

> Professor Redman has informed me that he would welcome an extension of activities in theoretical astronomy on the site of the Madingley Road obser-vatories. . . . Since the present proposal is very largely concerned with research at the post-doctorate level, I see no reason for requesting the creation of posts from the University, and therefore do not think that this proposal conflicts in any way with current University policy.[29]

Todd discussed Hoyle's document with Blaker at the Department of Scientific and Industrial Research. Todd said that although Hoyle would prefer to stay in Cambridge, 'he may feel compelled to go to the USA if his proposals for a school of theoretical astronomy in Cambridge are not accepted (he would need a decision by end of the year, although the money would not be needed before the summer).' Todd thought that Hoyle could muster 'about 50 young research scientists in theoretical astronomy, none of whom would consider going to Sussex.' He also thought that Sussex was only likely to attract 'older men', and would not be of interest to American scientists, as Cambridge would be. Nevertheless, Todd suggested that Hoyle be reminded, in the reply from the Minister, that he had been convenor of the sub-committee that had recommended siting the institute at Sussex. But Beaven, at the Department of Education and Science, noted, after reading a fat file on the issue, that Hoyle had resigned once Sussex had been agreed upon, and that the British National Committee for Astronomy had invited Hoyle to pursue the matter in Cambridge. This meant that Hoyle, in her view, no longer had any formal responsibilities to the Sussex proposal. In any case, the British National Committee for Astronomy had then come to think favourably of Cambridge. Hoyle had, however, approved the grant application from Sussex when it came before him as Chairman of the Astronomy Sub-Committee of the Research Grants Committee at the Department of Scientific and Industrial Research, which Beaven thought should be mentioned in subsequent dealings with Hoyle. But, she concluded:

It seems clear from the papers . . . that Professor Hoyle was at no time in favour of setting up the Institute at Sussex, although it is not clear whether he expressed these doubts officially as well as privately. This office was quite clear about his views.

In the summary memo attached to Beaven's letter to Blaker, the position was set out as follows:

> Professor Hoyle has . . . created a very difficult situation. If his proposal is turned down, the Government would be vulnerable to the criticism that, after due warning about the consequences of refusal, they had done nothing to prevent Prof. Hoyle and his followers from going to USA. If, on the other hand, it is accepted, the Government would be open to criticism because they had duplicated the provision for research in this particular field at the expense of other fields of research.

The memo also recorded that Todd felt that Hoyle could be supported, and the brain drain averted, without compromising the plans at Sussex, if money for an institute in Cambridge could be found from an independent source, such as the Nuffield Foundation, of which Todd was a Trustee. However:

> . . . [it] would be a temporary solution only, as the grant is unlikely to be continued indefinitely. A new source of finance would have to be found when the grant came to an end, unless the authorities at Cambridge had a change of heart, the burden would presumably fall on public funds. Lord Todd undertook to explore the possibility of a grant for Professor Hoyle's new school, but before this proposal could be taken further, more would have to be known about the effect on the new research institute at Sussex. It is suggested that this is the main point which the Secretary of State should take up with Professor Hoyle at that stage.

Todd was consulted over the new draft of the letter from the Minister, which he advised should be expressed 'in a form suitable for publication.' If Hoyle were to accept the Minister's invitation to talk to him about his new proposal, it might be suggested that Hoyle could be Director of the Sussex institute, and that if he refused, his reasons for so doing should be discussed.[30] The new draft letter from the Minister to Hoyle read as follows:

> Thank you for your letter of 1st August about your proposals for a school of theoretical astronomy at Cambridge, which I have read with great interest. You do not mention the proposal for setting up a theoretical astronomy Unit at Sussex University. As Chairman of the Astronomy Sub-committee which considered this and made recommendations to the Research Grants Committee

of the DSIR, you will know that this proposal is going forward. Although I am greatly interested by your suggestion for a school of theoretical astronomy at Cambridge, I am sure you will agree that it would be unfortunate if support for your present proposals were to jeopardise the success of the scheme agreed for the University of Sussex. Before going any further, therefore, I should like to know in more detail what you propose, and to have your assessment of the effect that your new proposals are likely to have on the unit to be set up at Sussex University. I should like to have your view of what the relationship would be between the Institute in Sussex and your Unit of Cambridge, and in particular how far you would expect them to be mutually competitive. Would you care to come and talk to me about this?

By now, Frank Turnbull at the Department of Scientific and Industrial Research was very worried. The second draft was not right either. 'What are we going to do about this?', he asked Blaker. Blaker was running out of patience with Hoyle, and thought it 'astonishing that Professor Hoyle should again make the proposal . . . without any reference to the proposed establishment of an Institute for theoretical astronomy in Sussex,' despite having approved the grant application from Sussex – an application that had already progressed as far as the Treasury which 'is currently raising various queries in connection with it.' Blaker explained to Turnbull Todd's views on the importance of staunching the flow of scientists to the USA, on the possibility of money from an independent foundation, and on the potential of Cambridge; but he also stressed the danger of appearing to duplicate resources. Again, the problem of publicity was noted:

> In view of the political capital to be made out of this it is unreasonable to expect any government decision on the issues involved until after the Election, and Professor Hoyle is already aware of this. . . . the reply to be sent to him meanwhile by the Secretary of State can obviously not be on the lines suggested by DSIR, but must be considered, among other things, from the point of view of its suitability for eventual publication in case of need. A draft reply intended to meet these conditions is attached. . . .[31]

The third draft letter from the Minister to Hoyle was dated 11 September 1964, and read as follows:

> The Secretary of State has asked me to reply to your letter of 1st August about your proposals for a school of theoretical astronomy at Cambridge, which I have read with great interest. There are, however, some aspects of the matter which are not entirely clear. The proposal for an Institute of Theoretical Astronomy at Sussex University has been supported by the ACSP, and DSIR, and this Department, and we are now seeking funds for it. Would there, in your view, be

room both for the school you envisage in Cambridge, and the Institute? Secondly, we were advised, at an earlier stage, that Cambridge University was not willing to have an Institute of Theoretical Astronomy in Cambridge. Has this situation altered in any way? Would you care to come and talk about this? If so, I shall invite Sir Harry Melville and Lord Todd to come to the discussion.[32]

With Hogg busy campaigning, Turnbull sent the draft to the Minister for Education, Sir Edward Boyle, on 11 September 1964, along with an explanatory note:

> I have had another talk with Lord Todd about this this morning. We all recognise that Professor Hoyle is being pretty difficult. He was actually chairman of the DSIR sub-committee which recommended that the Institute of theoretical astronomy be placed in Sussex. One of the reasons for this decision was that Cambridge University were not willing to entertain the proposal to have it in Cambridge. But Lord Todd says that there is a lot of internal politics in Cambridge behind the subject. There are considerable strains within the faculty of mathematics, in which Professor Hoyle is at present located, and if it is true, as Professor Hoyle says, that the Professor of Astronomy in Cambridge could be willing to house the proposed school of theoretical astronomy, Lord Todd thinks it's just possible that Cambridge University would take a different view of this matter now to that which they took a year ago. They too may wish to avoid a general exodus of Professor Hoyle and his followers to the United States.

Turnbull then suggested that they consult Sir Harry Melville, Permanent Secretary at the Department of Scientific and Industrial Research, because 'this subject is his affair now, and will also be the responsibility of the Science Research Council when it comes into being', which Melville was to chair.

> Also, it is quite exceptional for a minister to see a Professor about a matter of this kind which is part of the current business of a Research Council. On the other hand, I do not see how you could allow this matter to proceed to a rather disastrous conclusion without intervening. If you approve, therefore, I would propose to send this draft to Sir Harry Melville, if you are disposed to reply on these lines, and to see Professor Hoyle if he is willing to come. . . . Although, no doubt, a decision would have to wait until after the Election, I think it is pretty important to keep Professor Hoyle under the impression that the government is taking notice of his representation.[33]

On 16 September, Boyle wrote to Hoyle following exactly the third draft letter. Shortly thereafter, Blaker suggested to Turnbull that they should push Sussex harder to find a larger share of the money for an institute. 'But,' he added, 'we should be wary of pressing Sussex too hard, at least until we are sure of Professor Hoyle.'[34]

The next day, Turnbull wrote a 'Confidential note for the record' which contained a revelation: according to Todd, Redman's offer of hospitality on the Observatory site had been prompted by the University in an attempt to extricate Hoyle from the 'considerable turmoil' in the Department of Applied Mathematics and Theoretical Physics:

> The university authorities consider that no permanent solution to the problem of the mathematics faculty can be found which will leave a place there for Professor Hoyle and his group of theoretical astronomers. They have therefore suggested that Professor Redman, the Professor of optical astronomy, should incorporate Professor Hoyle and the theoretical astronomers in his faculty, and Professor Redman is willing to do this.

Todd also told Turnbull that the proposal Cambridge had rejected was on a rather grander scale than Hoyle now wanted. Todd thought that there had been some confusion at the University with regard to Hoyle's status in the negotiations: when Hoyle had made his proposal as chair of the sub-committee of the British National Committee for Astronomy, speaking, in effect, for the Royal Society, the University had seen it as simply yet another demand from one of its professors. Also, the University's support had been 'lukewarm' because 'they were afraid of setting up a grandiose Institute which might not be a success and which would then become a burden to them.' Todd considered that the likelihood of Sussex attracting any 'young, top-flight astronomers' was slim: an institute there would be 'led by some competent, elderly theoretical astronomer, but it would not attract the brilliant young men from the Cambridge mathematical faculty (which is way out ahead of anything else in the country) whom Hoyle wants to keep in this country, and who will otherwise go to America.' In conclusion, Turnbull noted:

> We agreed that a pretty delicate operation was in prospect. If Hoyle sees the Minister, the best line will probably be to say to him that he really must work out a reasonably modest costed scheme and sell this to the University and also give it to the Government. The [University Grants Committee] or DSIR must then ask Cambridge point blank whether they are willing to support the scheme and how much of the cost they will bear. If it then became apparent that Cambridge would foot a larger proportion of the cost than Sussex are prepared to pay for, there might be good justification for withdrawing from the Sussex scheme. The difficulty would be to carry out this operation without it becoming known in Sussex that the Government was turning away from the Sussex scheme, in which case there might be embarrassing allegations of bad faith.[35]

At the beginning of October, Turnbull approached Melville, setting out the problem and asking that the matter be treated as confidential. The tide was turning:

> It would appear that there is a possibility now of Hoyle and the University agreeing to set up a much more modest school in Cambridge and that such an agreement may be the only way of avoiding the departure to the US of Hoyle and his followers. It would also be much cheaper than the Sussex project. The existence of both a Cambridge School and Sussex Institute would seem to be absurd and wasteful. It therefore seems to us that we ought to consider, pending the arrival of the new government, whether we could not change course on this subject. If Hoyle and Cambridge put up a firm proposition acceptable to them both, would it perhaps be possible to use Treasury objections to the present Sussex plans to release us from the Sussex project?

Hoyle had not yet replied to the Minister's letter, and Turnbull thought that in any case the proposed meeting could not be held until the next government had taken office. But he asked Melville to 'give us your views on how the matter should be handled.'[36]

Hoyle had not replied to the Minister's letter because he had gone on holiday. Before he left, he considered the situation. He thought that resistance in Cambridge to the institute idea was too strong, and he could not face another five years of Batchelor as Head of Department.[37] So in September 1964, he submitted his resignation from his post at Cambridge, and he and Barbara hitched up their caravan and headed for the Lake District.

The Hoyles returned to Cambridge in time for the start of the academic year on 1 October. Among their mail, they found the letter from the Minister for Education, Sir Edward Boyle, inviting Hoyle to discuss the proposed institute. Hoyle replied on 2 October:

> Thanks for your letter. [. . .] the situation at Cambridge over the summer has been rather critical. In my negotiations with the University I set myself a target date of Oct. 1 for clearing the various points of difference. Since this date has now gone, and since my points have not been met, I have decided to resign my post. This naturally means that I will not be available to take part in the development I outlined in my letter to Mr Hogg. I am sorry to have caused you trouble over the matter, but as I frankly said, my appeal was very much in the nature of a straw to a drowning man![38]

By now, a General Election had been called, for 15 October, and campaigning was intense. Boyle was defending his seat as MP for Handsworth, Birmingham, and was busy in his constituency. His Private Secretary Christopher Herzig sent a note acknowledging receipt of Hoyle's letter, and Barbara used the back of it for a 'to do' list, which included 'sign papers', 'note for milkman' and 'wash socks.'[39] Herzig sought advice from Turnbull, who had spoken to Todd. Todd noted that Hoyle's letter said not that he had resigned, but that he 'had decided to resign'. Thus they took the letter to be 'designed to try to cause the minister to put pressure on Cambridge University to meet Professor Hoyle's wishes.' They would have to take care that the University did not consider any approach from government as one designed to oblige them to stop Hoyle leaving the country. If the University then asked for government money, and this were not forthcoming, the government could appear responsible for Hoyle's emigration. Todd advised that the Ministry should do nothing except issue a statement to the effect that the Government had for some time been involved in plans for an Institute at Sussex, after recommendations from a committee of which Hoyle had been chair. Todd had reminded Turnbull that 'Professor Hoyle had had the opportunity of joining this Institute if he had wished to do so; and that his departure to the United States was not, therefore, due to any failure by the government to provide on a substantial scale for an Institute of theoretical astronomy.' Todd said he would try talking to people in Cambridge once more, and that 'if a settlement was reached between the University and Professor Hoyle, and Cambridge made some reasonable request for [Department of Scientific and Industrial Research] assistance, no doubt DSIR could be persuaded to consider the case sympathetically.'[40]

In Cambridge, the Vice Chancellor, the Revd. John Boys-Smith, who was also Master of St John's College, had received Hoyle's letter of resignation, but had decided not to act upon it immediately. He invited Hoyle to suggest how matters might be put right for him. According to Geoffrey Burbidge, who was in Cambridge that summer along with Fowler, 'there were enough people within his friends who were senior members of the university who were very disturbed about this and didn't want Fred to leave, and they wanted the [institute] done.'[41] Indeed, the University had set up a committee of three to meet Hoyle and talk about his problems: they were the Vice Chancellor; physicist John Cockcroft, who had supported Hoyle since his Fellowship application in 1939; and mathematician Mary Cartwright, the Mistress of Girton College – Turnbull called them 'the Vice Chancellor's Special Committee on the Hoyle Problem'.[42] Metallurgist Alan Cottrell, a mutual friend of Hoyle and Todd, also offered to help, but he warned Todd that 'there is always trouble in the Maths Dept.' and that it would be 'very difficult to find out the real problem.' Hoyle was due to leave shortly for the USA – he had arranged unpaid leave for one term, starting on 16 October – so time was tight, and Todd was worried that if Hoyle left while the prospects in Cambridge were poor, he might not come back. Permanent Undersecretary Sir Maurice Dean at the Department of Scientific and Industrial Research was worried that the Government could be embarrassed if 'something blew up in the Press'; but Todd thought that Hoyle would not want to publicize the matter, although 'if he did it would certainly cause a great amount of harm.'[43] Todd was about to make a short visit to Israel, but he dashed off a note to Dean before he left on 9 October:

> I managed to get hold of Fred Hoyle last night and & talked to him like a father. He has now agreed to see the Vice Chancellor, Cockcroft and Miss Cartwright some time next week & will not do anything final about resigning at least until he has seen them. I got him to promise he would do nothing rash like giving out any statements (or indeed resigning at all) without first talking to me on my return from Israel when I could also discuss with him what help the Foundations might give him. So I don't think ministers need worry for the moment. I also got John Cockcroft & asked him to let you know by phone if anything of note transpires at the meeting with Hoyle in my absence. I hope, therefore, that all this is more or less in order. Pardon the scribble but I'm just about to depart for the airport.[44]

According to Burbidge, Hoyle's friends – Cockcroft, Cottrell and Cartwright – called round and talked to him 'night after night', asking him to withdraw his resignation if the institute could be brought about in Cambridge. After 'enough gins and tonics', Hoyle agreed to give it another try.

Although Hoyle had been tempted by the USA, Barbara had already made it clear that she did not want to emigrate.[45] Burbidge was sure that Barbara's influence had been crucial in this decision too:

> I suspect that Barbara had a lot to say. Barbara's always had a lot to say. Barbara has been very dominant in Fred's life, she always looked after Fred, she looked after most of the family. Fred, of course, was very thankful for all that. For example, one of the reasons why they were in Cambridge was that Barbara's mother was living with them and she ran the Society for the Blind in Cambridge and she didn't want to leave. I don't know what Barbara will tell you but that was the truth.[46]

Pacing Hoyle's garden one frustrating evening, Todd told Burbidge: 'You know, we're doing this and it's absolutely right that it should be done, but we can't let everything be determined by Fred's mother-in-law.'[47]

Fred Hoyle, approaching ten years old, strikes a pose for the camera in his home
town, Bingley.

Fred Hoyle is first

"exceedingly clever . . . more popular than Pickles."

WHO is Britain's most popular broadcaster of 1950?

Pickles? Richard Dimbleby? Gladys Young?

The answer—believe it or not—is Fred Hoyle, Esq., M.A., astronomer, mathematician, Fellow of St John's College, Cambridge, and University lecturer in mathematics.

He has been rated by radio listeners as a better broadcaster than anyone else whose popularity has been investigated by the B.B.C. Audience Research Service. B.B.C. pollsters have found him more popular than Bertrand Russell, Dr. Joad, Tommy Handley or even Wilfred Pickles.

He is the star of the year—a fact which is reported this morning in the B.B.C. 1951 Year Book; a fact, incidentally, which staggers even the chiefs of Broadcasting House.

Fred Hoyle, 35 - year - old Yorkshireman who looks like a · benevolent owl, was something of an unknown quantity when he was first brought to the microphone this year.

Blunt answer

His success even in the field of "highbrow radio" was far from certain. For Hoyle's subject was cosmology—the study of the Universe. With another maths. lecturer, R a y m o n d Lyttelton, he had developed a new theory of the universe.

Astronomers who boggled at his theory of "continuous creation" w e r e answered bluntly. They did not know enough physics or mathematics to follow him, he said.

And this was the man and the subject that the B.B.C. launched · on the Third Programme to give a series of talks about "The New Cosmology."

The Third's 300,000 listeners voted him a "hit" · and the B.B.C. promptly repeated his talks in the Home Service.

Basil Blackwell collected his lectures in a five-shilling book called "The Nature of the Universe" and has sold 77,000 copies in six months—which makes it one of the biggest scientific best sellers ever.

The B.B.C., in the Year Book, reserve for Hoyle the sort of effusions that are usually devoted only to variety stars.

'A triumph'

"He is exceedingly clever," they say. "Hoyle's broadcasts have been a triumph in the use of radio for the communication of new, difficult and highly important ideas.

"This stocky Yorkshireman, with his · curly black hair and genial smile, seems certain to become one of the important broadcasting figures of o u r generation."

Hoyle's challenging ideas have been helped by his homely accent, his dogmatism, his confidence. But they have not avoided criticism. He has been attacked in pulpits up and down the country.

Hoyle welcomes the attacks. And meanwhile, as maths. lecturer, astronomer, · physicist, Fred Hoyle, the lad from Yorkshire, pursues his important way.

He has proved many things, not the least important the fact that British radio listeners are not exclusively interested in variety and dance music.

There are millions who don't mind using their minds in the seat by the fire.

Walter Hayes

On 22 November 1950, in the wake of Hoyle's radio broadcasts *The Nature of the Universe*, the *Daily Graphic* announced that Hoyle had been voted Broadcaster of the Year.

St John's College

DESCRIPTION-SIGNALEMENT

Bearer—Titulaire		★Wife—Femme
Profession) Profession)	University Lecturer	
Place and date of birth Lieu et date de naissance	Bingley Yorks. June 24, 1915	
Residence) Résidence)	St. Johns College, Cambridge.	
Height) Taille)	5 ft 9½ in	ft in
Colour of eyes Couleur des yeux)	Brown	
Colour of hair Couleur des cheveux)	Brown	
Special peculiarities Signes particuliers)		

★CHILDREN-ENFANTS

Name—Nom	Date of birth-Date de naissance	Sex-Sexe

Signature of Bearer Signature du Titulaire	F. Hoyle.
Signature of Wife Signature de sa Femme	

3

TADA H · 3
PHOTOGRAPH OF BEARER

WIFE   FEMME

(photo)

Hoyle acquired this passport for his first trip to Caltech in 1951.

Hermann Bondi
in 1954

Tommy Gold
in 1968

George Gamow
in 1958

In 1957, Fred Hoyle and Willy Fowler celebrated the completion of B²FH with a trip to Rome for a conference on the evolution of the stars. Their wives joined a programme of social events and day trips designed to amuse them while the men talked astronomy. Here Barbara Hoyle takes centre stage, and looks much more suitably dressed for summer sightseeing in Rome than the other ladies in this group. Fowler is wearing the bow-tie, immediately to the left of Barbara; Hoyle's head is just visible at the back on the right; and Alan Sandage is to the rear on the left.

FRED
HOYLE

A SCIENCE
FICTION NOVEL
BY A FAMOUS
ASTRONOMER

**The
BLACK
CLOUD**

The first US edition of *The Black Cloud* carried a picture of the Mount Palomar telescope on its dust jacket.

Bernard Lovell at work at Jodrell Bank in 1959, in the shadow of the radiotelescope.

Evening Standard

42,505   FRIDAY, FEBRUARY 10, 1961   ● 3d.

Gay look for the new Piccadilly—Page Twelve

SCIENTISTS GRASP CLUE TO RIDDLE OF UNIVERSE

'How it all began' fits in with Bible story

By PETER FAIRLEY

Six British scientists will announce to the world tonight that they have proof to explain one of man's greatest mysteries — how the Universe began.

It is the first proof. It follows years of speculation. It is sure to cause a furore among scientists, who have been split into two opposing camps on the subject. But it may delight churchmen.

For the scientific facts fit in nicely with what the Bible says — *"In the beginning God created the heaven and the earth."*

THE SCIENTISTS — five men, one woman—are a group of radio astronomers at the Mullard Observatory, Cambridge, led by 42-year-old Professor Martin Ryle.

They have taken nine years to reach their conclusions after probing deep into the heavens with the world's most powerful radio-telescope.

Expanding

Now they say confidently:

1—The Universe, of which we in our solar system are but a tiny part, is expanding.

2—All matter in it—planets, suns, stars, huge galaxies—is rushing out into space at fantastic speed, leaving a hole in the middle.

3—There was a definite beginning. Some might interpret this as the biblical Creation.

4—The Universe will not last for ever. It is changing with time. But it has existed so far for at least 10,000 million years.

Everlasting?

THE NEW EVIDENCE points to the Universe having originated from a colossal explosion. One massive lump of matter — or several lumps parked turnidy together—may

● Page Thirteen, Col. One

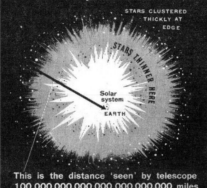

STARS CLUSTERED THICKLY AT EDGE

STARS THINNER HERE

Solar system

EARTH

This is the distance 'seen' by telescope
100,000,000,000,000,000,000,000 miles

Stars and planets rush out into space like fragments scattered by an explosion.

'Lumumba escapes from jail'

ELISABETHVILLE, Friday. — Patrice Lumumba, ex-Premier of the Congo and the new republic's biggest headline-maker, has escaped, according to an official statement today.

The news was announced by Mr. Godefroid Munongo, Minister of the Interior of the breakaway Katanga Government.

The Katanga Government of Moise Tshombe, one of Lumumba's most bitter enemies, offered a £2100 reward for his recapture.

Two of Lumumba's lieutenants, Maurice Mpolo and Colonel Okito, escaped with him.

Rewards of £350 were offered for their recapture.

Way out

BUT, according to Associated Press, one diplomat said it was possible that this would be an excellent way for the Katanga Government eventually to announce that Mr. Lumumba and the two others had died—or been killed—while trying to make their way out of the country.

Last month Mr. Lumumba and his aides were flown from a Leopoldville jail to Katanga in the belief they were more secure there.

Reports said they were beaten up as they were taken off the airplane.

Mr. and Mr. Munongo, is how the three escaped:

Mr. Lumumba and his aides were being kept in a lonely farmhouse south of Katanga.

▲ Back Page, Col. One

PROFESSOR RYLE
He led the team.

Pies and lorry haul

A lorry loaded with meat pies, pasties and sausages was stolen from outside Telfers factory in Lillie Road, Fulham.

INSIDE NEWS IN THE Evening Standard

Ford strikers go back on Tuesday
PAGE TWELVE

Queen Mother's cousin buys Belgravia house
BACK PAGE

WEST INDIES 252 FOR 8
PAGE TWENTY-TWO

Monday debate for Touche censure motion
BACK PAGE

Amusements Guide
PAGE TEN

The Queen's cold 'almost better'
PAGE THIRTEEN

£8000 LONDON SAFE RAID
BACK PAGE

Tory MPs tell Mac: Slow up in Africa
BACK PAGE

Paris newsletter
PAGE SEVEN

Youth, who posed as a spy
BACK PAGE

LIMELIGHT
PAGE EIGHT

PERSONAL SERVICE!

A trained receptionist will take your small-ad.

TILL 9 TONIGHT

for insertion in the Evening Standard. Ring
FLEet Street 3000

WEATHER — Dry — See Page THIRTEEN

This superb rum makes the most delectable long drinks with:
BITTER ORANGE
BITTER LEMON
BITTER LIME

Good judges always say ... MAKE MINE MYERS

The dark and mellow rum · Aged in the wood · Bottled in Jamaica

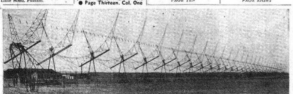

Part of the radio telescope's giant aerial which straddles several fields.

The *Evening Standard's* report of Ryle's press conference in 1961, which broke the press embargo and earned its writer a blast of Ryle's temper.

The 1962 US edition of *A for Andromeda*.

The newly built Institute of Theoretical Astronomy, in 1967.

The staff at the Institute of Theoretical Astronomy in its first summer in 1967 included (front row, from right): John Faulkner, Jamal Islam, Willy Fowler (with tennis racket) and Geoffrey Burbidge. Behind Burbidge, on Hoyle's right, is Frank Westwater, and to Hoyle's left are Margaret Burbidge and Donald Clayton. The building behind them housed IOTA's IBM computer.

Astronomer Royal Sir Richard Woolley stands before the dome of the Isaac Newton Telescope, which, when this photograph was taken in 1967, had only recently come into operation after 20 years under construction.

Hoyle's friends from the USA dominated the summers at IOTA. In 1970, visitors included (from right) Wal Sargent, Willy Fowler, Geoffrey Burbidge and Donald Clayton.

Chandra Wickramasinghe,
in 1970.

The great dome of the Anglo-Australian Telescope tops the ridge of Siding
Spring Mountain. The smaller dome houses the UK Schmidt Telescope, which
was constructed shortly after its larger neighbour.

Martin Ryle in 1972, the year he became Astronomer Royal.

Jocelyn Bell Burnell and Antony Hewish, among the poles and wires that constituted the radiotelescope with which they first detected pulsars.

At the height of the controversy over Hoyle's charge that Jocelyn Bell had been
unfairly excluded from the Nobel Prize to Ryle and Hewish, Hoyle visited Don
Clayton in Texas. They rented a beach-house for a weekend and composed a
letter to *The Times*.

In 1975, a conference for Hoyle's sixtieth birthday was held on San Giorgio Island. Here he is joined by (from left) Geoffrey Burbidge, Don Clayton, Margaret Burbidge, Martin Rees, his daughter Elizabeth Butler, Barbara Hoyle and Willy Fowler.

Jayant Narlikar

Halton 'Chip' Arp

Having persuaded Hoyle that the institute could still happen in Cambridge, Todd and Cockcroft set about raising money for it from charitable foundations. Conveniently, Todd was a Trustee of the Nuffield Foundation, and Cockcroft was a Trustee of the Wolfson Foundation. Hoyle took up the Vice-Chancellor's invitation to suggest how matters might be put right, and on 11 October sent him a proposal entitled: 'An Institute for Advanced Studies in Theoretical Astronomy'. This time, Hoyle left no room for ambiguity: the institute was to be in Cambridge, and it was to be the reason why he would stay there. His proposal began by lamenting the great expense of research at the forefront of the physical sciences; but:

> . . . in theoretical physics and astronomy it is still well within the scope of a major university to support a self-contained centre of international standing, because the expense involved is one or two orders of magnitude smaller. I believe that this is the direction in which the university should go in order to maintain its long-standing pre-eminence in the physical sciences.

Hoyle then pointed out the value of theoretical astronomy as a discipline that capitalized on a new synthesis – one that Hoyle had helped to popularize among scientists: 'it is bringing together in the most surprising way large parts of elementary particle physics, relativity, cosmology, and astronomy itself.' Hoyle stressed his loyalty to Cambridge, but:

> In order to collaborate with leading people in this field, and to make use of the modern facilities necessary to keep up with the present state of development of the subject, it has become unavoidable that I must spend large amounts of time away from Cambridge at other centres, particularly at the California Institute of Technology. Through their size, organisation, policy, and concentration of outstanding people, such places are now the real forcing points for advance in theoretical astronomy. I want to see a similar development here in Cambridge.

Hoyle explained that 'Cambridge has a long and outstanding tradition in this field', and was 'a prolific producer of first class young theoreticians.' But these young men were forced to work elsewhere. If they could be retained in Cambridge, for example at an institute, theoretical astronomy could flourish there, and it would quickly become the leading centre in the subject – and all for much less money than a similar development in an experimental field. Hoyle concluded his case by saying that he had 'discussed the proposal in an informal way with government representatives and have reason to believe that they are broadly sympathetic to it.' He then spelt out his needs: staff (but 'I would not expect more than one or perhaps two of the present members of [the Department of Applied Mathematics and

Theoretical Physics] to be involved, and the institute would not overlap the activities of DAMTP'); a building; and 'adequate and convenient computing facilities'. Computers could be hired, since they were developing so rapidly, so as to avoid being stuck with an obsolete machine. With tacit reference, perhaps, to EDSAC, Hoyle stressed that 'This computer must be entirely reliable in operation and must therefore be a standard commercial machine. The day has long since passed when home-made computers or semi-home-made ones can provide an adequate routine service.' He then spelt out the costs, the site promised by Redman at the Observatory, and the urgency with which he felt a decision should be reached. He also repeated his position on the 'relation to the University':

> To start with at least, I think the Institute should lie outside the formal activities of the University, although I would very much hope that the university would recognise and assist such a group, and maintain a good informal contact similar to that which now exists for example with the MRC unit.[48]

Mentioning the Medical Research Council Unit at this point was a clever move. The Council was, in the medical sciences, the longstanding equivalent of the forthcoming Science Research Council: it distributed government money for medical research. The Unit, which was in a new building in Hills Road in the south of Cambridge, undertook interdisciplinary work on the structure and composition of biological molecules. The University had expressed scepticism about the quality and value of such units, but five scientists at this one had recently won Nobel Prizes: Fred Sanger in 1958 for the structure of insulin; Jim Watson and Francis Crick in 1962 for the structure of the nucleic acids DNA and RNA; and Max Perutz and John Kendrew, also in 1962, for the structures of proteins. A parallel development in theoretical astronomy could surely only be welcomed.

When he received Hoyle's proposal, the Vice-Chancellor wrote to the Department of Scientific and Industrial Research asking about the Government's interest in the project. Sir Maurice Dean replied that while, because of the election, it was an awkward time for the Minister to make any detailed response, the Government would be interested in any proposals in this area that the university might care to submit.[49] Dean had recently been in Cambridge, and he had seen Cockcroft at Churchill College. Cockcroft, with his connection with the Wolfson Foundation, had told Dean that he thought that a charitable foundation could probably find the money, and Dean told Cockcroft that if the foundations could not find the money, 'the Government could probably help.' But he also warned Cockcroft that 'if there was publicity about this threatened resignation before the Election immense damage would result to the University cause.'[50] They agreed that they needed to keep the matter open – that is, that Hoyle should be prevented from actually resigning – until the ministers could give it some attention. Back at the Department of Scientific and Industrial Research, however, Dean told Turnbull that 'we should not push forward with Government money if the matter could be fixed up through the Foundations, not withstanding the possibility of criticism from Hoyle,' and he suggested that they should 'drop hints to Cambridge of the same kind as he has already given to Cockcroft.'[51]

Two weeks after returning from his holiday in the Lake District, Hoyle left Cambridge once again, headed first for Columbia University in New York. He would be away until the New Year. Ministers and journalists were busy with the election. For the civil servants, though, it was business as usual, and Cockcroft arranged an appointment at the Department of Scientific and Industrial Research in London.

11
His institute

A change of Government; Of Men and Galaxies; *more trouble for the institute; a blow for steady-state theory; Hoyle despairs of Cambridge; the institute comes to Cambridge*

On 15 October 1964, Harold Wilson's Labour Party ousted Sir Alec Douglas-Home's Conservatives, after 13 years in Opposition. The Labour victory was seen by many as a triumph for the working man, for the North over the South, and for the grammar-school meritocracy over the aristocrats. The pronouncements of the new Prime Minister would be delivered from Downing Street in a Yorkshire accent.

Hogg and Boyle both held their seats in Parliament, but with the change of government they lost their ministerial responsibilities for Education and Science. The civil servants, however, spent election day preoccupied with Hoyle. Cockcroft saw Dean at the Department of Scientific and Industrial Research, and word of their discussion quickly spread. Cockcroft had told Dean that Hoyle had said that he was 'not prepared to go to Sussex and two other names which had also been considered had also turned it down.'[1] Cockcroft had brought a copy of the document of 11 October that Hoyle had prepared for the Vice-Chancellor. The key points, according to Cockcroft, were that an institute could be both in Cambridge but 'outside the formal activities of the University, like the MRC Unit of Molecular Biology'; and that the charitable foundations were likely to be willing to pay for a building. He and Dean agreed that 'neither this Government nor the next could envisage the departure of Hoyle to America, because the Government had failed to accept a proposition of this sort.' Turnbull noted that the Department of Scientific and Industrial Research had wanted to 'contribute to a solution', and that while 'it is a novel idea for the DSIR to maintain a Unit at a university, . . . I imagine that the S[cience] R[esearch] Council will do this kind of thing, and there would be no great harm in making a start now.'[2]

By now, the civil servants were starting to worry about the University of Sussex, which was still awaiting news of funding for its institute of theoretical astronomy. They could still see the advantages of Sussex,[3] and did not want to be seen as having influenced any move to Cambridge. It had long been an important principle of science policy in Britain that scientists themselves should determine the shape of scientific research, and universities tended to be nervous about government intervention. A statement placed on record at the Department of Scientific and Industrial Research affirmed that:

> At no time has DSIR tried to force the Institute on to Cambridge or Sussex. The idea came from the astronomers themselves after consulting the two universities. This is in line with DSIR's views on the freedom which Universities should have to decide their own research programme.[4]

It fell to Sir Harry Melville, Permanent Secretary at the Department of Scientific and Industrial Research, to write to John Fulton, the Vice-Chancellor at Sussex. Melville reminded Fulton that Hoyle had declined the post of Director of a new institute at Sussex, and explained to him that the earlier circumstance, of Cambridge being unwilling to host an institute, had now changed:

> ... matters have come to a head in the last few months and both Cambridge University and the Government feel, naturally, that it would be unfortunate if Hoyle were to have to go to the States, and that we ought to make a reasonable effort to prevent this. There, therefore, seems a possibility now of a proposal being put forward on quite a considerable scale for an Institute at Cambridge and, as usual, DSIR is being looked to for support.
>
> We do not know yet what might come of this, but if it does happen we shall be in the position of being asked to help to develop its existing school of theoretical astronomy, which is of high repute, and we shall be in the embarrassing position of having to consider this in competition with your Sussex proposals, which were nearly approved. I think it is inevitable that they will be in competition, as I think it is unlikely that we can, or should, afford two Institutes of Theoretical Astronomy in this country, when there is such a great need for more effort to be put into applied science and technology. I think, however, that you would probably agree that we ought to make quite an effort to keep a man of Hoyle's reputation and achievement in this country. It may well be, therefore, that we shall have to change from the idea of an Institute in Sussex to an Institute in Cambridge, and I can appreciate what a disappointment this would be to you, and to Sussex.

> I am writing to you in confidence about this immediately, even though nothing is yet decided, because these things have a way of leaking out, and I should not like you to learn of the possibility of this change except from me. . . .[5]

Turnbull and his colleagues were taken aback by Melville's letter. They had planned on using the fact that the Treasury 'were finding difficulty in swallowing the Sussex scheme,'[6] and now they were nervous of how Sussex might react to Melville's confession that the motive for the change of course had been to keep Hoyle from leaving the UK. They feared 'embarrassing allegations of bad faith,' and hastily put together a note for the record about the history of the issue 'in case of awkward publicity.'[7]

Fulton was, as expected, disturbed by the letter. He went to see Melville, and expressed concern that he could have done more to attract Hoyle to the scheme: perhaps Richard Woolley at Herstmonceux had deterred Hoyle by saying that the Directorship ought to go to someone who would stay in Sussex all year round; whereas Fulton would have been happy to see Hoyle for six months of the year, when he was not in California.[8]

Having alerted Sussex to the existence of the rival bid, the civil servants grew nervous when no news came of progress in Cambridge. In December, Francis wrote to Todd, asking for 'any news of the Hoyle affair.'[9] Todd replied some weeks later to the effect that he was hoping to see Hoyle soon, but that 'Meanwhile I gather that the university has decided to agree to the idea put forward a couple of months ago so that when I have seen Hoyle I may be able to give you some hard information.'[10] The civil servants noted that Cambridge was looking for money from the Foundations, but this was unlikely to be enough, in which case the Science Research Council might expect an approach. If such were to be made the matter would have to be considered by the Council's Research Grants Committee, where Bernard Lovell was now chairing the Astronomy sub-committee. They also noted that the 'exact relationship of proposed institute to Cambridge University is not clear.'[11] Given the inaction from Cambridge, Alan Cottrell asked the Department of Scientific and Industrial Research if it could make enquiries of the University.[12] Francis's colleague Joliffe said they could not. His view was that the University should demonstrate its commitment to the institute, otherwise the Department would feel that 'the University would give [the institute] no support and would leave others holding the baby'. Joliffe saw Cambridge's silence as 'a typical university move to get something for nothing.'

Hogg's successor at the Department of Education and Science was Herbert Bowden, who had known Hoyle for some time. Bowden met Hoyle when Hoyle returned from the USA, and they had a long and difficult conversation. Hoyle told Bowden that the situation with the mathematicians 'made his position untenable unless the organisation can in some way be changed.' He intended to leave Cambridge in June 1965 unless some help could be promised for him by then. Bowden told Hoyle that he was unhappy with an institute that would take some of the brightest mathematicians in Cambridge away from undergraduate teaching, but Hoyle countered that the row had been about undergraduate teaching – he was referring to the changes Batchelor had made to the Mathematical Tripos. Bowden wrote to Melville about the meeting, and while he acknowledged that it was 'inappropriate for me to get involved in a matter of this kind . . . I have known Hoyle for a long time and have a high regard for him . . . my sympathies are with him and I would like to feel that we could do something to help him.'[13]

Cambridge broke its silence shortly thereafter: an application was sent to the Department of Scientific and Industrial Research for money for running costs for the Institute, to complement funds being raised at the Foundations. The application would be submitted to the Science Research Council when it came into being in April.[14] Staff at the Department prepared a briefing paper for the new Council, which duly opened a file. The paper concluded that the application from Cambridge:

> . . . is now being considered in consultation with the University of Cambridge and the Wolfson and Nuffield Foundations. The sum asked for is a considerable one; and a number of other bodies are involved and the final decision taken will have a significant effect on the future of British Astronomy. [Because of] these factors we must expect this to take some time to settle. Of course, the Government hopes that Professor Hoyle will stay in this country and wants his work to prosper. This is why the Government and the University are considering together the best way to ensure this.[15]

Hoyle was still not confident of victory. In February 1965 he wrote to Fowler at Caltech, and marked his letter 'confidential':

> . . . my negotiations concerning an Institute have now gained considerable momentum. The Foundations I have approached are strongly supporting it. On the government side the present argument is that, while the project is excellent in itself, it should not be sited at Cambridge. Since I know you have views on this it would be a great help, when this struggle gets to ministerial level, if I had statements from you, Geoff and Margaret, as to how you view Cambridge

compared, say to Brighton. My guess is that another two months will see the critical stages past.[16]

The Burbidges had spent five years at the University of Chicago since B²FH, and then had both joined the University of California at San Diego. Geoffrey Burbidge, responding from their new base in La Jolla on behalf of himself and Margaret, told Hoyle that they were rather too distant from the British university scene to be able to offer much by way of advice, but as to the advantages of Cambridge:

> I have always had the feeling . . . that . . . there will always be some underlying latent hostility to this development there, and after all your two chief backers [Cockcroft and Todd] are both mortal. On the other hand I do not know if it would be a good idea to have the Institute associated with the University of Sussex either. . . . I feel that it is entirely possible that the Institute could be run in conjunction with say the University of London. . . . Another possibility is Bristol. . . .[17]

Burbidge did, however, conclude with some more concrete advice:

> . . . I feel it is exceedingly important that your true attitudes about this matter be fully displayed to whoever is arguing for or against Cambridge or for or against Sussex. Otherwise we shall get into a situation in which they believe that the Institute is being set up largely to supplement the university facilities and just to keep you in England. Those of us involved in it will know that other factors are much more important.[18]

Fowler's response must have been more what Hoyle was hoping for: after stating that 'the work of your group in theoretical astronomy and nuclear astrophysics will be of the utmost importance in analysing and interpreting our observational and experimental results' and endorsing Hoyle's arguments for 'extensive computer facilities', Fowler congratulated him on his choice of a Cambridge location:

> This seems very wise to us. . . . I hardly need remind you that Cambridge is a great attraction, second to none in England, to all Americans with scientific and cultural interests. The amenities are most important for those who wish to spend a sabbatical year abroad. . . . You have my best wishes and prayers for the success of your venture – it will means so much to science in our two countries.[19]

Then a bombshell exploded over Cambridge: the dreaded publicity was at large. On 7 March, *Sunday Times* journalist Nicholas Tomalin, who had reported the election for the Head of the Department of Applied Mathematics and Theoretical Physics the previous May, now spelt out the situation, including details of the row with Batchelor – and the outcome of a university enquiry which had not found any improper behaviour during the election. Tomalin reported Hoyle's resignation, which was 'in a state of diplomatic suspension', the proposal from Sussex, and the possible funding arrangements for an institute in Cambridge.[20] According to the article, there were three possible outcomes of this difficult situation, which the Department of Scientific and Industrial Research would discuss at the meeting of its Research Grants Committee on 19 March:

> Hoyle may leave, with wife and mother-in-law, to work permanently at one of the many American universities or institutions which have offered him research facilities.
>
> The government may put up between £50,000 and £150,000 a year to create a new Institute of theoretical astronomy, with Hoyle in charge and insulated from abrasive contacts with rival academics.
>
> If the government refuses, or attaches too many conditions to their offer, the Nuffield Trust or the Wolfson Foundation might put up money instead.
>
> Nothing is definitely decided and there is furious lobbying going on. But it looks as though the DSIR will provide the money. No government, particularly one committed to the new technological society, could afford the outcry Hoyle's departure would raise. . . .
>
> Fred Hoyle will not talk to anyone about these negotiations. He is very sensitive about any suggestion that he is threatening either the government or the university with any personal ultimatum.
>
> But despite his *amour propre*, all those involved in the search for a solution realise that the whole institute project is basically a device to keep Fred Hoyle happy, productive, and here. Either he is given the opportunity to think and work as he wants, where he wants – or we will lose him. . . . If he does decide to emigrate to America, whatever the effect on the advancement of Science, it will be the most dramatic and painful brain drain we have yet suffered.

Unlike his protest about computers, Hoyle had been silent on the matter, but now he was furious that the story was out in the open, and he wanted his say. Whether he went to the USA or stayed in Britain, he told the *Guardian* in time for the Monday edition, would depend on whether or not his institute were funded. He had, according to *The Times*, 'for some time been chafing against the conditions under which he has to work,' and had been offered several jobs at high salaries by US universities.[21] With the new Science Research Council due in just a month, Hoyle told the *Daily Mirror* that he would prefer to stay in Britain: 'For personal reasons, yes. For technical reasons, no. . . . Britain is spending a tenth of what she should be spending on scientific research.'[22] The *Guardian* concluded:

Professor Hoyle has seldom shown much patience in his dealings with politicians, and he has constantly emphasised that all he wants to do is get on with the
work in the best possible way. . . . A head-on collision seems very possible.[23]

Hoyle was even angrier when he realized that colleagues thought that he
had been the source for the story in the *Sunday Times*. Within a few days he
had drafted a letter to Joliffe at the Department of Scientific and Industrial
Research, reporting that that 'the article has been reproduced in the *New
York Times* and the following comment from a shrewd, very scrupulous,
friend must be typical of what is thought in the United States – ". . . the
story in today's *Times* came as something of a surprise. I had thought that
this was all being held tightly under wraps. Did you decide that this publicity might be the final bit of pressure needed to bring about acceptance
of your plan?" ' Hoyle thought Tomalin had a source in the Department of
Applied Mathematics and Theoretical Physics, and that this latest leak, as
well as Tomalin's report about the election the previous summer, were part
of a strategy by his colleagues to ensure that 'Cambridge guarantee the
creation of a large number of senior positions [at the new institute] – clearly
with the intention of acquiring them for themselves.' According to Hoyle,
the General Board of the Faculties – the body responsible for the academic
and educational policy of the University – had been suspicious of this
and was resisting the institute idea for exactly that reason. Hoyle also told
Joliffe that Tomalin's source had also briefed the politician and Oxford
philosopher Richard Crossman, formerly Labour's spokesman on science
and now Housing Minister in the new government, who had spoken against
the Cambridge plan in the House of Commons. Hoyle concluded:

It seems only right that I should state my position quite frankly before the
[Research Grants Committee of the Science Research Council] meets on the
19th. From . . . last October I have been equally happy either to go ahead with
the Institute project or to go my own way. As I said in our recent telephone
conversation, the *Times* leak has swung me on balance away from the Institute,
for I can see possibilities of similar incidents to extending for many years into the
future.[24]

The civil servants swung into action, drafting and circulating a press release
explaining the state of play, and then abandoning it. In the mean time,
Bowden had seen Melville, who had reported that the Vice-Chancellor of
Cambridge University had told him that 'the university would be willing to
make some contribution towards the running costs in the next quinquen-
nium.' Melville had also noted that Sussex would need to be 'placated':
Woolley had already been to see Francis, and he had pointed out that an

Institute was no guarantee that Hoyle would stay in Britain.[25] But Bowden suggested to Melville that Sussex make Woolley a visiting professor and 'develop work in practical rather than theoretical astronomy.'[26]

With the Science Research Council's Research Grants Committee meeting looming, Hoyle once again set out his plans on paper.[27] He observed that the increasing student numbers during the 1960s had meant less time for research in universities, which was one reason why institutes were becoming a common way of organizing research. He acknowledged that institutes raised problems in their relationship with the universities, but thought that such problems in Cambridge could be avoided by a simple arrangement that was modified in the light of experience. He did not want staff members of the institute to be in the usual university hierarchy and to have the usual titles: he suggested that everyone be called 'staff member', and that 'prospective members who felt this to be inadequate to their status aspirations are probably not the right people for an institution weighted on the research side.' Despite Bowden's advice, Hoyle insisted that teaching would be at postgraduate level only. Staff members should not expect jobs for life, which, while it was 'right and proper for those who carry the load of undergraduate teaching', was not appropriate for researchers, who 'must be prepared to live dangerously' on short-term contracts. In any case, with the initial funding sought only for the years to 1972, there was no guarantee of continuity; this would have to be 'openly explained to the young men.' Hoyle felt that the overlap with the activities of the Department of Applied Mathematics and Theoretical Physics would be small and 'there should be no mass transfer of staff.'

The decision of the Research Grants Committee on 19 March 1965 was not unanimous, but a recommendation was submitted in confidence to the Science Research Council on 12 April that it provide two-fifths of the recurrent costs of an institute in Cambridge up to a maximum of £375,000 for the period until 31 July 1972, on the understanding that the Wolfson Foundation pay for the building, the Nuffield Foundation two-fifths of recurrent costs, and the University of Cambridge not less than one-fifth of recurrent costs after 31 July 1967. So Cambridge would get some money from the government – £375,000 – but it would have to find some itself. The recommendation noted that 'an Institute with Professor Hoyle would be better than one without', and 'if he left the country it would be a great loss.' However, 'in the long term it would probably be unwise to base a decision which would have repercussions on British Astronomy for the next decade on the presence of one man, no matter how brilliant. Also it cannot be assumed that the siting of the Institute in Cambridge would, in itself, keep Professor Hoyle in this country.' As for the University, the recommendation noted that it had changed its attitude towards 'large projects started with outside funds' since Hoyle's earlier bid, and that:

> the University is keen to have the Institute at Cambridge with a close academic and scientific association which it would expect to strengthen as time went on so that the Institute would eventually be merged closely into the University. As a first step towards this, the Vice-Chancellor would put to the General Board of the Faculties the proposition that the University should meet one fifth of the running costs from 1967–1972.[28]

The Research Grants Committee had also suggested that Hoyle be required to be at the institute for at least six months each year. The recommendation concluded that on the exact relationship between the institute and the university, 'detail[ed] proposals for organising and financing the Institute acceptable to both Professor Hoyle and the University would have to be worked out.' Francis told Turnbull that in view of the recent publicity and the university calendar, it would like to be in discussion with Cambridge in early May. Francis's gloomy but prescient conclusion was that the final decision about a grant 'will not be an easy one if full agreement can not be reached between Hoyle and the University, particularly as we shall probably not escape criticism whichever way the decision goes.'[29] Despite the new policy structure for science, the Science Research Council still had to secure approval from the Treasury for such a large sum of money – approval that came only after battles over the amount (despite it being considerably cheaper than the Sussex bid, as the Council had pointed out) and over the timescale for the project, which was too long.

On 24 May, Francis wrote to the Vice-Chancellor asking him whether the University would accept the money. In return, Hoyle would be required to be present at the Institute for six months of the year, 'an administrative officer of standing would be appointed' to manage the institute, and alongside money from the Wolfson and Nuffield Foundations, Cambridge would contribute one fifth of running costs.[30] Now, if Hoyle left Britain, it would be Cambridge's fault, not the Government's. In London, the civil servants sat back and waited. In Cambridge, intense discussions began across the University.

In the spring of 1965, Hoyle also had a book to promote. In 1963 he had given the John Danz lectures at the University of Washington, and his texts were published, in 1964 in the USA and 1965 in the UK, under the title *Of Men and Galaxies*. The book's US publisher announced that Hoyle, 'this "twentieth-century Leonardo" reports to the layman on the present and future role of man as a citizen of the Galaxy'. In one lecture, 'An Astronomer's View of Life,' Hoyle speculated about the possibility of systems analogous to the terrestrial biological systems elsewhere in the universe:

> In growing up a child receives many rude shocks in its encounters with the world. Similarly, the human species as a whole must expect many shocks as it grows up. One of these I am convinced is that the phenomena of consciousness, of intelligence, independence, aesthetics, are going to come in ways that may seem strange to us. We must be prepared to find in the larger universe outside the Earth not only creatures very much like ourselves but widely different ways of doing things, even 'inorganic' collections of matter endowed with a sense of 'justice,' for example. It is, of course, much easier to think in terms of creatures like ourselves. Our imaginations are hardly adequate to go much beyond this. Fantasies soon reach the realm of science fiction.[31]

These ideas about intelligence in the cosmos made headlines in Britain among the coverage of negotiations about the institute. An article by Hoyle for the *Daily Mail* – it was verbatim extracts from *Of Men and Galaxies* – was headlined 'But can you be sure?', and subtitled 'Published soon – by one of the liveliest minds of our age . . . the most staggering speculation of all . . . many write it off as cosmic doodling.'[32] Reviews of the book were positive: the *Observer*, for example, said that '[Hoyle's] ideas rest on a sound basis of possibility, and his book makes stimulating and enjoyable reading.'[33]

Hoyle was looking to the skies for answers not just about astronomy, but also about the future of the human race. In the third lecture in *Of Men and Galaxies*, 'Extrapolations into the Future,' Hoyle concluded that because of the high probability of intelligent life elsewhere in the universe, 'We are following an inevitable path, one that must have been followed many, many times on other planets':

> If in other places other species have already followed the difficult route ahead of us, then plainly it would be an enormous advantage to know exactly where the dangers lie. My suspicion is that ample information exists in what I might call a 'galactic library' to show exactly what is going to happen to us if the world persists in following current policies. It will be known, for example, what policies lead to nuclear war and what policies avoid it. Acquisition of such knowledge would lead to perhaps the most revolutionary step in human thinking. From time immemorial man has looked up at the sky in wonder. He has invariably placed his gods there. Instinctively he may well have been correct. To paraphrase the

well known psalm I will end by saying, 'I lift up mine eyes to the sky from whence cometh my help. . . .' It may prove to be so.

While the newspapers were enjoying Hoyle's speculations, *Of Men and Galaxies* was quickly trumped by events: on 6 June, the General Board of the Faculties 'agreed in principle to accept the offer' from the Science Research Council,[34] and government and charitable funding for the institute was announced on 29 June. Francis had prepared a press release in which he had said that 'Cambridge university has signified informally its willingness in principle to accept such offers,' but Turnbull chided him for making Cambridge sound like a 'reluctant debutante . . . I wonder if you can not word your statement to the Press in a way that covers up this rather unusual situation a little more . . . [it] conveys an impression of detachment that is bound to excite comment.' Francis revised his statement to read that 'consultations are now in progress between Cambridge University, the Science Research Council and the Foundations on the detailed arrangements.'[35] The *Daily Express* reported 'Hoyle triumph: 'Britain will give him his new institute.'[36]

However, government backing, sufficient funds and press acclaim were not enough: Cambridge had still not approved the plan. When the General Board had agreed in principle to accept the grant, it had also set up a sub-committee, consisting of physicist Brian Pippard and geologist William A. Deer, to explore the practical implications of setting up an institute. Other colleagues raised questions: Harold Davenport, Chair of the Faculty Board for Mathematics, wrote to Hoyle to ask for clarification on 'the proposed Institute of Theoretical Astronomy.' Davenport wanted to know more about 'relations between the institute and the theoretical astronomers already in Cambridge', that is, those in the Department of Applied Mathematics and Theoretical Physics. Was Hoyle thinking that they would be appointed to posts at the institute? If not, would they have to bear the full burden of undergraduate teaching, while the institute staff only did research? Davenport considered that scenario 'anomalous.' On the other hand, if the institute's staff were to teach, why should it not be constituted as a University department? And if it were to be a Department, why would it be funded by the government and the foundations? Davenport's 'own guess is that you are thinking of the Institute as a research institute with no teaching commitments below the Part 3 level, and that you are not thinking of taking into it the theoretical astronomers already in Cambridge, at least not "en bloc". If this is so, what do you think about the points I mentioned earlier?' Hoyle was asked to consider these questions in advance of an imminent meeting of the Faculty Board, to which he was invited.[37]

Other colleagues watched the situation in despair. One young member of the Department of Applied Mathematics and Theoretical Physics, John Dougherty, who was well aware of the tensions between Hoyle and Batchelor, was going to miss the Faculty Board meeting, but felt strongly enough to write to express his disappointment that:

> . . . difficulties derived from purely personal relationships would stand in the way of the success of such a promising project. One hears of this sort of thing elsewhere, but most of us have always been on such good terms with our colleagues that (like motor accidents) we imagine 'it will never happen to us.' Last year's experience [the election for Head] has however destroyed this idealism (or perhaps naivety) . . .

Dougherty also recorded on behalf of the junior members: 'how disappointed we have been at the behaviour of the senior members of the Faculty', who had 'brushed aside' attempts at reconciliation.[38]

At a meeting in July 1965 of the University's Council of the School of the Physical Sciences, which included Hoyle as usual, and Martin Ryle by invitation, Hoyle expressed disappointment that the Science Research Council had made it a condition of the funding arrangement that the University contribute some money to the institute – Hoyle felt that was unnecessary. He also argued that the institute should be exempt from the statutes and ordnances of the University and instead should be governed by a board that would include representatives of the organizations that were funding it. Several members raised the problem mentioned by Davenport: that this would create two groups of scientists working in the same area but subject to different rules – those in the Department of Applied Mathematics and Theoretical Physics working within the university, and those at the institute. Others felt there could be many kind of associations between the university and the institute – some sort of working association could surely be found. Ryle protested that once the Science Research Council had given money to the institute it might consider its duty to Cambridge discharged, and there would be no further money for other astronomers. The Faculty of Mathematics raised similar questions about the extent to which its activities might be compromised by the new institute.[39] But other members of the Physical Sciences council countered that surely neither the Science Research Council nor the institute would behave in such a way as to privilege one group of researchers over another. Similar reassurances were offered on the issue of computers.[40] In any case, a national review of researchers' computing needs was under way, and everyone would have their say. But not everyone at the meeting was convinced, and afterwards

Professor Sir John Baker, the Head of the Department of Engineering, wrote to Wilfred Sartain, the Secretary General of the Faculties, to express the concerns of the Physical Sciences council that this welcome contribution to astrophysics would 'handicap' other 'equally desirable scientific developments', particularly with regard to computing provision.[41] The formal minutes of this meeting were not circulated until December, five months later – just after Hoyle had left for a trip to the USA. When the General Board met on 21 July, the sub-committee of Pippard and Deer was expanded to include Baker and the physicist Sir Nevill Mott, as well as Hoyle's old adversary William Hodge representing the Faculty Board of Mathematics. Pippard told the General Board that the support of the Physical Sciences council would depend upon 'satisfactory resolution of the difficulties which they foresaw might arise' in relation to theoretical astronomy being undertaken in the Department of Applied Mathematics and Theoretical Physics.[42] The Board itself was split over the sort of departmental status that might be appropriate for the institute, and decided to defer formal acceptance of the Science Research Council's offer until further discussions had taken place.

A few days later, as if to remind the University of the temptations elsewhere, the BBC repeated a television programme Hoyle had made the previous autumn, originally for the Christmas schedules. The programme, *Fred Hoyle's Universe*, was not filmed in Cambridge: instead it proudly claimed to be 'the first ever television visit to the world's largest optical telescope on Mount Palomar in California.'[43]

By the autumn of 1965, Hoyle was despondent. Not only was the institute still not approved, but the steady-state cosmology had suffered a serious set-back.

In 1948, George Gamow and his colleagues Ralph Alpher and Robert Herman had concluded that the big explosion at the start of their evolving universe would have left behind it traces of radiation that could potentially be detected. The radiation was likely to have a temperature of just a few degrees above absolute zero, and to fall in the microwave region of the spectrum. So it was a very cold and rather feeble glimmer, but it was required by the physics of the situation nevertheless. It was, however, soon forgotten, and no one tried to find it. During the 1950s, a number of scientists stumbled across radiation of this kind at very low temperatures in space, but they did not connect it with the earlier prediction and the big bang.[44] Hoyle and Gamow discussed the problem of this mystery radiation – they disagreed over its likely temperature – when they met in California in 1956, but even they did not connect it to cosmology. Then in 1963, two physicists at the Bell Telephone Company's Laboratories in New Jersey, Arno Penzias and Robert Wilson, found what they thought was noise in an antenna they were developing into a very sensitive radio-telescope. They spent some time trying to discover the source of the radiation that was causing this noise, and decided eventually that it did not come from Earth, nor from the solar system, nor even from the galaxy. Nor could it be attributed to the pigeons nesting on the antenna.[45] They decided that the radiation, whatever it was, had to be cosmic. In the mean time, at the nearby Princeton University, physicists Robert H. Dicke and James Peebles were developing cosmological models that required that some kind of cold, left-over radiation would still be at large in the cosmos. In the spring of 1965, when they heard about Penzias and Wilson's mystery radiation, they felt they had found the missing piece of their puzzle. As indeed did Penzias and Wilson, who thereby found an explanation for their baffling observation.

To the chagrin of Gamow, Alpher and Herman, none of these young physicists connected the discovery with their prediction of 1948 – the reason was, Peebles later said, that as physicists, they did not know the literature in cosmology.[46] However, the general cosmological significance of the radiation was immediately noted and widely celebrated: *New York Times* science editor Walter Sullivan won the front page for his report, in which he said that several cosmological theories could accommodate the newly identified radiation, but if it had indeed come from an explosion then steady-state theory could not be among them.[47]

Robert Wilson was rather disappointed by that conclusion: he had learnt cosmology from Hoyle as a student at Caltech, and 'very much liked the steady-state universe.'[48] He held on to the theory for a while longer, until further data along the same lines convinced him to abandon it. The radiation was eventually to tell scientists much more about astrophysics than just cosmology, and Penzias and Wilson were awarded a half-share of the Nobel Prize for physics in 1978 when the achievements being celebrated were all in low-temperature physics. By then, they were able to give full credit to their many antecedents in physics and cosmology who had contributed to the saga of the cosmic microwave background radiation.[49]

In September 1965, Hoyle gave a lecture, 'Recent Developments in Cosmology', at the annual meeting of the British Association for the Advancement of Science, which was held that year in Cambridge. He discussed the idea that Penzias and Wilson had found the left-over radiation from the explosion at the start of the evolving cosmos, and acknowledged the problems presented for steady-state theory by the receding quasars and the radio source counts. In conclusion, Hoyle accepted that 'the universe must have been denser in the past from what it is today,' and therefore was not in a steady state. This conclusion was qualified, however: Hoyle proposed some new models for the universe: in one, the universe is oscillating between expanding (such as observed now) and contracting modes, and would eventually settle down in a steady state; and in another the universe contracts in some regions and expands in others (such as in the region of our observation) and thereby maintains a steady state overall.[50]

The lecture was printed in *Nature*, where it drew wide attention from journalists.[51] 'Hoyle says: I was wrong about creation', announced the *Daily Mirror*;[52] the more sober *Daily Telegraph* reported: 'Hoyle retracts 20-yr-old theory of universe: "steady state" abandoned'.[53] The *Sunday Telegraph*'s science correspondent noted that radio-astronomers were jubilant at Hoyle's retraction: the *Nature* paper marked the end of 'the century's most brilliant academic controversy.' One unnamed astronomer, however, pointed out that all the evidence against steady-state theory that Hoyle had accepted was his own, and he had not actually taken note of his critics at all.[54]

Many proponents of big-bang theory believed they had won the war; and steady-state supporters began to admit defeat. According to Felix Pirani, a steady-state supporter who was now working on relativity at King's College London:

> The discovery of the background radiation was a heavy blow, I think to everybody except Hoyle. That was pretty much it. I remember thinking at the time, perhaps you could get round this by some clever idea . . . there must be some integral you could do that would give the background radiation as coming from somewhere, but I wasn't motivated enough to go back to it. . . .[55]

Hoyle was sure he had only lost a battle, and continued to work on steady-state cosmology. But the tide had turned, and Hoyle's retraction, though qualified, was taken as a final declaration of surrender by his rivals. But rather than this undermining his reputation, Hoyle seems to have emerged in a stronger position: according to the *Sunday Telegraph*, Hoyle's colleagues were pleased to welcome Hoyle 'back into the fold' and were sure that his ideas would continue to be of great value.[56] And to his public, Hoyle was valiant in defeat. According to the *Daily Express*:

> Professor Fred Hoyle is my hero. I understand only about one word in ten of what he is saying, but . . . he is a man to be admired extravagantly. . . . 'Half the troubles of the world' [said Hoyle] 'stem from people who never admit

mistakes. . . . It is necessary for everyone to open his mouth to give progress the best chance.'[57]

Hoyle's retraction of the steady-state cosmology struck a chord in the popular press, and may have convinced his colleagues that he was settling into a more conventional position. But his institute was still in jeopardy. By the end of October 1965, Cambridge University had not responded to the Science Research Council's request that it contribute to the funding of the Institute. Hoyle wrote to both Cockcroft and Todd, suggesting that Cambridge wanted a rather different sort of institute from the one he had proposed:

> Since the Spring I have had the strong conviction that things were going wrong with the Institute project, but since nothing was ever put on paper . . . it was difficult to know exactly what to do. As one might have guessed, the University has avoided answering the SRC proposal, by making a counter proposal of its own. The ultimate effect, in my view, of accepting this new proposal would be to downgrade the Institute into a minor university department.[58]

Indeed, one of the first reports of the sub-committee had described the institute as being a Department of the Faculty of Mathematics. Hoyle also wrote to Francis at the Science Research Council, telling him that he could 'see no clear course other than to call off the project', because the University's delay in answering the Council's proposal most likely indicated a negative response. Hoyle apologized for having been 'overoptimistic about the changed attitudes of the University', but he also laid some blame for the situation with the Council itself, for making the condition that the University contribute to the project, 'which has had the effect of bringing on exactly the state of affairs we all wished to avoid. In exchange for this 10 per cent [it was in fact 20 per cent in the Science Research Council's recommendation], the university seems to expect to have what amounts to 100 per cent of the control.' Hoyle told Francis that he was not giving up just yet, but since he had other responsibilities – family, students, colleagues – he was setting the end of the Michaelmas term, in December, as his deadline for abandoning the project. Francis replied by return, expressing hope that the difficulties could be resolved, but also pointing out that the University's contribution had been required not to test its enthusiasm but to ensure its interest.

However, the Cambridge bureaucracy had made some progress: draft regulations for a new institute were agreed with a sub-committee of the General Board in November, and Hoyle accepted them.[59] He then left for a short trip to the USA, and while he was away, the minutes of the Physical Sciences council's July meeting were circulated, in which the Faculty of Mathematics spelt out its earlier concerns. It was 'not satisfied' with the situation, for several reasons. It felt that the Institute would be an extra drain on funds for studentships, which could have 'serious repercussions elsewhere.' It also felt that the Science Research Council would not fund theoretical astronomy that was in Cambridge but not in the institute. It argued that the case for an institute was considerably weakened by the considerable strength of the Department of Applied Mathematics and Theoretical Physics in theoretical astronomy, which came from its 26 members in relevant areas, including six lecturers (this was considerably more than Hoyle's estimate). It expressed concern that external funding for the institute was available only until 1972 – would it become a drain on the University thereafter? The document also pointed out that the Faculty of Mathematics would lose Professor Hoyle, and with him a senior teaching appointment, and regretted that a replacement for him had not been mentioned. The mathematicians also wanted the institute staff to be involved with undergraduate teaching, and members of the Department of Applied Mathematics and Theoretical Physics to have associate status at the institute, and to be represented on its governing body. They also wanted a thorough assessment of the impact of the proposed institute building on the operation of the Observatories.[60]

At the General Board meeting of 1 December 1965, the sub-committee submitted a report noting that even if the institute were not a Department within a Faculty, and had external funding, it would still be subject to the decisions of the General Board. It suggested that the institute staff, while not being involved in routine undergraduate teaching, might still contribute lectures on more advanced topics. It also reported assurances from Francis at the Science Research Council that while the number and amount of grants available would not increase should the institute come into being, the institute would not be privileged over other parts of the university in the distribution of funding. The sub-committee noted, however, that 'it was difficult to legislate for cooperation' but that Hoyle had expressed his intention to cooperate in the resolution of outstanding points of difficulty.[61] On 14 December 1965, Professor of Geophysics Edward Bullard wrote to Hoyle at Caltech to tell him that the Physical Sciences council had decided to recommend to the General Board that to prevent one group of theoretical astronomers being privileged over another, all of the relevant staff in the Department of Applied Mathematics and Theoretical Physics should be transferred into the institute, and that the General Board was likely to accept this advice (though Batchelor might resist it, Bullard suspected).[62] This was not at all what Hoyle had in mind for his institute.

On 12 January 1966, the sub-committee reported to the General Board that there were many matters still unresolved, and it was decided to inform the Science Research Council that 'the University had not yet considered proposals for the establishment of the Institute.'[63] But the sub-committee also reminded the Board that, contrary to the suspicions of the mathematicians and the physical scientists, there was no reason to doubt that the Science Research Council would consider all theoretical astronomers in Cambridge fairly, irrespective of whether they were connected with the institute or not.[64] It recommended that the General Board should advise the University that the institute be established; and in the mean time it should continue to address the practical problems of the relationship of the institute to other parts of the university.

When Hoyle returned from the USA, he set out his thoughts on the outstanding problems. He could not see how any member the Department of Applied Mathematics and Theoretical Physics could be part of the governing body of the institute, as the only two (according to his estimate) relevant staff were far too junior. If sections of the university could insist on being represented, what sort of composition might the board achieve? One 'quite unfitted to the spending of £800,000 of public and charitable money', suggested Hoyle.[65] He thought that the alleged undermining of the Department of Applied Mathematics and Theoretical Physics by the institute had been exaggerated by swelling the numbers of relevant lecturers to six – 'an untruth'; and that he had taken measures to ensure that the Department was not undermined. Hoyle noted that his friend and colleague Ray Lyttleton 'should not be considered a member of DAMPT for this purpose since he is opposed to the politics of that Department and does not want to be considered a member of it.'[66]

In Hoyle's notes about the possible relationships between the Department of Applied Mathematics and Theoretical Physics and an institute, he accused Batchelor, 'himself not an astronomer', of 'demand[ing] parity with myself in theoretical astronomy. He has even asked the University to fund a new chair in theoretical astronomy to be attached to his department. The whole case is built on the presence of two university lecturers on his staff.'[67] Hoyle quoted Hermann Bondi, who was now at King's College London, who had written to him in May 1965 to warn him that if the institute were to become subservient to the General Board of the Faculties, it 'may be out of your hands very soon.' Bondi had also told Hoyle that 'you are unlikely to get any Cambridge Institute of the kind you want and where you can do your work without constant and excessive annoyance.'[68] Hoyle concluded his notes thus:

> It remains for me to end with an apology – for not listening to the clear warnings of last March. What was clear to people on the outside, like Bondi, was not so clear from the inside. It has taken me six bad and unpleasant months to learn my lesson.[69]

His thoughts thus in order, on 20 January 1966 Hoyle wrote in strong terms to the Vice-Chancellor, insisting that 'it has become necessary for me to stop waiting patiently and dumbly for an answer from the University. The sensible procedure I am now convinced, is for me to state the conditions under which it is possible to continue this project and for the University to give a plain yes or no to these conditions.' Hoyle argued that his proposal was being treated as 'a rag-bag thrown from one committee to another.' This 'must cease': it was an insult to those who were taking it seriously, and was damaging the University's reputation. Next, 'the Institute is to be governed by a body composed of six trustees,' representing the funding bodies in proportion to their contribution. Last, 'the deliberately whipped-up fuss over theoretical astronomers in DAMPT ceases.' Hoyle insisted that only two people, Roger Tayler and Dennis Sciama, had relevant interests, and they could choose between the Department and the institute. Hoyle wanted a response to these conditions by the beginning of April. If this were not forthcoming:

> I suggest that any disposition on the part of the University to fall back on the well-known ineptness of the Cambridge 'system' be taken as equivalent to no. I trust that in saying this you will excuse my bluntness. Please ascribe it to my Yorkshire background.[70]

Hoyle copied the letter to Cockcroft, Todd, the Science Research Council and the Wolfson and Nuffield Foundations. A day or so later, he saw Sartain's deputy and found that he was too late: many of the recommendations from other parts of the university that Hoyle had found objectionable – in particular the representation on the institute's governing body – had been adopted by the General Board on 12 January. On 24 January, Hoyle wrote a furious letter to the Vice-Chancellor, reminding him that the Science Research Council's offer of a grant, and those from the Foundations, 'would be specific to Professor Hoyle and Cambridge University', and that 'The only interpretation to be put on this statement is that the conditions under which the Institute operates must be acceptable both to the university and to myself.' Hoyle insisted that he had offered to discuss any outstanding issues before leaving for the USA, but that he had been assured that there were none. 'Yet in my brief absence this latest development, going exactly contrary to the agreement of 1 November, was rushed through without reference to me.' Hoyle saw a conspiracy at work: 'particular persons were involved both in the agreement of 1 November and in this latest development.' He suspected that a fait accompli had been engineered behind his back, and he had had enough:

> What this manoeuvre has now convinced me is that Cambridge is not the right place in which to set up an institute. The aim of the Institute was to deal in problems of some magnitude and grandeur. This cannot be done in an atmosphere of petty intrigue. Were I to accept a bad compromise, which would prejudice the success of the Institute, I would be in breach of faith to the donors. I feel it then my duty to report back to the donors that I no longer see Cambridge as a suitable place in which to set up the Institute.[71]

Hoyle again copied his letter to Cockcroft, Todd, the Science Research Council and the Wolfson and Nuffield Foundations. On the same day, he drafted a letter to Francis at the Science Research Council, claiming that the Council's condition of a contribution from the University had led to 'a mounting criticism of the project which has now exponentiated to a point where the atmosphere of goodwill, so necessary for the Institute to flourish, does not exist.'[72] This draft was moderated, and Francis received a letter spelling out the problem, as Hoyle saw it, of having junior staff from the Department of Applied Mathematics and Theoretical Physics on the proposed governing body of the institute and their being involved in appointing staff. Now, Hoyle wrote,

> I find the whole project, the whole of 15 months' effort, has been destroyed in one morning by a group of men who were incapable of understanding that one cannot have junior people appointing senior people. This is the inner reason why I feel I must now exercise my option of withdrawal.[73]

But the General Board and its sub-committee were now close to resolution. In its report dated 8 February 1966, the sub-committee recommended that the institute be established from 1 August 1966, and it had found a solution to the problem of the relationship with the Faculty of Mathematics: it recommended that the Institute of Theoretical Astronomy be added to the formal list of 'Departments not assigned to any Faculty,' a classification that more usually applied to administrative than academic departments.[74] The head of this new Department was to be known as the Director of the Institute of Theoretical Astronomy, which 'shall be the Plumian Professor of Astronomy and Experimental Philosophy for so long as that Chair is held by Professor Hoyle.' However, the report also listed a management committee that included members nominated by the Faculty Board of Mathematics and the Physical Sciences council. The report was circulated to the funding bodies and submitted to the General Board on 9 February, where the Board noted that the relationship with existing Departments and Faculties was still problematic. It also pointed out that if the staff were all to be called 'staff member', it would be impossible to know what to pay them, since staff pay was based on categories such as 'lecturer' and 'professor'. The Board was also concerned about recurrent costs to the University; and in the light of these issues it resolved to defer further consideration until the funding bodies had responded to the sub-committee's report.

There was no discussion about the institute at the next meeting of the General Board at the end of February. The Board had plenty of other matters to deal with: the University was in the throes of setting up clinical teaching for its medical students; and all of the sciences were under review: compared to the pre-war years, student numbers in physics and biochemistry had tripled, while botany and zoology were shrinking.[75] Provision of teaching staff in the Department of Applied Mathematics and Theoretical Physics had been under review for some time. Hoyle was busy too: his latest work with Narlikar on gravity, which they had submitted, like the earlier papers, to the *Proceedings of the Royal Society of London*, had been criticized by referees who had said, much to Hoyle's irritation, that 'there is something wrong somewhere' with the work but that they 'can't tie it down at the moment' or 'give chapter and verse.' One of the referees also mentioned Stephen Hawking's earlier criticisms, which irritated Hoyle even more.[76]

However, by the General Board's meeting of 9 March 1966, the sub-committee's report recommending the establishment of the institute had been sufficiently accommodating of enough of the interested parties' views that 'after some discussion the Board amended, approved and signed the report.' The report charged the institute with the duty 'to cooperate with other bodies in the encouragement of research in theoretical astronomy.'[77] When the report was published on 16 March, it confirmed that the institute was to be established and that Hoyle would be appointed to head it. Relieved of that long-running uncertainty, Hoyle settled down to compose a letter of complaint to the President of the Royal Society about its referees' failure to articulate their criticisms of his gravity paper,[78] while journalists composed their news stories about the new institute. The *Daily Telegraph* emphasized that it 'will in no way jeopardize or conflict with the work being done in this field within the University,'[79] and *The Times* said that Hoyle would get the computer he had needed for so long.[80] When the University ratified the decision in June, *The Times* reported that assurances had been given to Batchelor and the Faculty of Mathematics about their continuing responsibility for undergraduate teaching in theoretical astronomy, and announced that 'any danger that Professor Hoyle would be lost to America is thus averted.'[81]

12
The end of the beginning

Establishing the institute; October the First is Too Late; *more on the quasar redshift; steady-state theory revisited; Stonehenge is an observatory; Australia will host a telescope for the southern hemisphere; organic polymers in space*

A few days after the newspapers had announced that Hoyle would head the Institute in Cambridge, he wrote to Fowler about his new project:

> Truth to tell I have become psychologically exhausted with the Institute business. Now that the struggle is all over I am recovering rapidly. Already there is a big change here in Cambridge. At the last moment there was a sort of cooperative effect in which the project received nearly unanimous support. I trust plans stand – that you, Geoff and Margaret will be coming next summer. . . . the builders have promised to finish next summer.[1]

Hoyle had strong views about the design and organization of the Institute. He had seen a building he liked in La Jolla: the Institute of Geophysics and Planetary Physics, which was built of redwood and overlooked the Pacific. That institute was one storey high, and arranged around a single straight corridor. The offices all had glass doors that opened onto a common veranda along the length of the building, facing the ocean. Hoyle intended to copy the layout around a single corridor: it would avoid 'the physical segregation of people [that] leads inevitably to a psychological segregation.'[2] Along the corridor were to be offices: 21 small rooms on one side, for the scientific staff, and seven larger shared rooms on the other side for the postgraduate students and secretaries. All rooms were to have blackboards. The computer room and library would be at one end, along with a lecture room and a room for making coffee and sandwiches.

The single corridor would have other benefits: a long straight route 'permits of a brisk walking gait and has the exhilarating effect of the old walking galleries.' However, the exhilarated walkers were not to disturb the workers in their offices: there was to be good sound insulation in the walls, and, unusually for a public building at that time, it was to have carpet. Since the offices were going to be quiet, the library did not need to be: that was for storing books and chatting. Hoyle wanted wood to feature in his Institute: he noted that 'internal walls [should be] of wood, not breeze blocks. I also prefer as much wood as possible in the external structure. I have a strong objection to concrete in any place where it can be seen.' Although he could not offer his staff a view of the ocean, they would have big windows, so Hoyle wanted a garden started among the trees and meadows around the building 'as soon as possible.'

Hoyle had started working with a company of surveyors, Bernard Thorpe and Partners, in January 1965, long before the Institute was officially approved. In March 1965, the *Sunday Times*' 'leak' and subsequent coverage of the project prompted one of the surveyors to suggest that the architect might offer some services free of charge, since he looked likely to benefit from all the publicity.[3] Hoyle set in train the building work in August 1966, and it was to be completed within a year.

Hoyle had also recruited staff before the Institute was approved. His early notes about staffing mentioned Willy Fowler, Geoffrey and Margaret Burbidge, Ray Lyttleton, Dennis Sciama, Roger Tayler and the radio-astronomer Cyril Hazard. Fowler's wife did not want to leave California and Fowler suggested that Hoyle might offer him some sort of permanent visiting post.[4] From among his graduate students Hoyle wanted to recruit Jayant Narlikar, Sverre Aarseth, John Faulkner and Jamal Islam, who was working on relativity theory, Peter Strittmatter, who was working on quasars and stellar evolution, and Chandra Wickramasinghe who was working on interstellar dust.

Aarseth was now a research assistant in the Department of Applied Mathematics and Theoretical Physics, and was happy to make the move to the Institute. Faulkner and Strittmatter were working in California with Fowler, but were enticed back to Cambridge: on an earlier trip to La Jolla, Hoyle had invited them to take a walk on the beach, and had outlined his vision for the Institute. At the time, Fowler had advised caution, knowing then, as Faulkner and Strittmatter did not, that the Institute was still not approved.[5] But they were prepared to take the risk: 'This was a wonderful opportunity, which we seized.'[6]

Although there were some reservations among Hoyle's colleagues about his administrative capabilities – one remark, attributed to a close and loyal friend, that was circulating at the time was that Hoyle should not be trusted to run a fish and chip shop[7] – his scientific vision for the Institute was very attractive. Many of his young staff had had other offers: both Faulkner and Strittmatter had been approached by William McCrea, who was now at Sussex, about jobs there, and Strittmatter had been offered a post at King's College London by Hermann Bondi, and despite their enthusiasm for the Institute both men told Hoyle that he had competition for their services.[8] Eventually they received written offers – written by hand from Clarkson Close, with salaries that appeared to them to reflect Hoyle's vague recollections of what junior staff might expect to be paid.[9] The jobs were to start in September 1966, but when they arrived for their first day at work, Faulkner and Strittmatter, along with Wickramasinghe who had also been recruited, found that '(a) the Institute was five courses of bricks high, and (b) little practical details like salaries for the intending staff had not yet been arranged.'[10] Faulkner and Strittmatter both had fellowships at Peterhouse College which gave them room and board and a small allowance (plus wine to the value of £300 per year), but they did not see a salary cheque until January 1967. Faulkner recalls that Hoyle explained the situation to him, and 'not with the embarrassment I should have thought he ought to have had', as follows:

> . . . the way the university works is that certain decisions have to be made in the order A-B-C, and these decisions are made by a sequence of committees that meet once a quarter in the order C-B-A. It's a policy deliberately arranged so that nothing can be done in less than a year.[11]

One unexpected drawback of the gap between arriving to take up the post, and receiving a salary, was to do with travel expenses. Faulkner and Strittmatter knew that new appointees to the University of Cambridge could claim relocation expenses, so they went along to the Old Schools, the administrative centre of the University, and asked how they should go about applying for the cost of their move from California. The administrator there could not see how they qualified: they had arrived in Cambridge in September, when the Institute did not exist. It would come into existence on 1 January, which is when they would take up their appointments. By this time they would already be in Cambridge, and so would not be entitled to relocation expenses. Faulkner recalls:

> Neither of us was sufficiently confident about our position. We didn't go to Fred – we were sufficiently intimidated by the aura of Fred that we couldn't go to him and say, excuse me Professor Hoyle, we have gone without salary for the last

three months having been brought back here by you, now we are being told that the travel expenses aren't payable. So we just let it slide. But it rankled a bit.[12]

Aarseth had also been told that his job at the Institute would start as soon as it opened in the autumn: in his case, the stated date was 1 October. Aarseth had to some research to publish, and:

> I wanted to be the first. So I wrote a Letter to *Nature*, and it was accepted, and it was published on 1 October 1966, which happened to be the publication date, and the address: Institute of Theoretical Astronomy. I am very proud of that. 1 October 1966 was supposed to be the starting date and nothing happened, there were delays, delays. The paycheques started from 1 January.[13]

Hoyle was not oblivious to the situation: like Aarseth, Narlikar and Islam were also on grants that would run out before they could draw a salary, and he arranged for their money to be extended until the Institute was functional. The University also contacted the Science Research Council on the matter of Faulkner, Wickramasinghe and Strittmatter, and asked permission to use money from another grant to give them each £200 – about six weeks' wages – to keep them going.[14]

Hoyle had hoped to attract Dennis Sciama, Roger Tayler and Ray Lyttleton from the Department of Applied Mathematics and Theoretical Physics. Lyttleton agreed to move to the Institute rather than stay in the Department, but wanted professorial salary and status, so that he could 'bypass the areas controlled by "you-know-whom" ' – George Batchelor had recently been appointed to a newly created Chair of Applied Mathematics, and Lyttleton did not want to be out-ranked.[15] Sciama stayed in the Department, but several of his PhD students subsequently joined, including Martin Rees and Stephen Hawking; but Tayler accepted a post at Sussex before the Institute was up and running. Cyril Hazard joined the Institute: he had worked out a method of pinpointing radio sources from rough positional data by recording the blocking of radio emission when the Moon passed in front of the source. This had been used with great success at the Parkes Observatory in Australia to determine the exact position of 3C 273, the radio source that became the first quasar. To this first cohort of staff, the Institute became known by its initials, IOTA.

Hoyle accepted the Science Research Council's recommendation that a manager be appointed, and he asked his old friend Frank Westwater to take on the role of Secretary. Westwater was the naval captain who had travelled with Hoyle to the radar conference in Washington in 1944. He was a highly competent mathematician, and he had developed a strong interest in computing, so when he retired from the Navy he had joined ICT, International

Computers and Tabulators. He had risen to a senior management position by the time Hoyle approached him, but Westwater resigned from ICT and took the job; and despite, at Brian Pippard's urging, his appointment being at professorial level,[16] he took a substantial pay-cut in the process.[17]

Barbara Hoyle became influential in the running of IOTA. When Hoyle was working from Clarkson Close, she had been closely involved in the organization and administration of his diary and working environment, as well as being a sounding board and advisor, press liaison and editor. Now that Hoyle had a base away from home, Barbara saw her role continuing there. She was involved in the appointment of the secretaries (who were often from Yorkshire[18]), and the staff did not consider this unusual: according to Aarseth, 'You wouldn't expect Fred to interview a secretary! The point is, in those days the secretary would be for Fred, so [Barbara] would then say this person will be suitable for him. She would be a better person to decide that than he would.'[19] Barbara made sure that the appointees would protect and manage his time just as she had done at home. She also took an interest in the interior decoration at IOTA, and shopped for furniture. She organized parties for the staff, and helped the younger members to settle in: Aarseth, for example, was buying a house in Cambridge – a process complicated by the late appearance of his salary – and Barbara went with him to check it over.[20] So although she never had any official role in IOTA, she was deeply involved in her husband's project.

An important strand of Hoyle's case for the Institute had rested on its need for a powerful computer. Hoyle wanted an IBM machine, so he had to make a case for not buying British. He enrolled Aarseth to see what was available from the three British companies, in order to spell out why the American computer was a better buy. Aarseth recalls:

> Fred said very prophetically 'there are three companies, and if they merged into one it could be a good company, and I would back it.' But the three he didn't want. Later there were problems with the British computer industry and it became just one company, ICL. This was one occasion where he actually had some commercial instincts – that was really quite perceptive. As a result he was allowed to buy the IBM computer.[21]

In a departure from Hoyle's original floor-plan, the IBM computer was installed in an adjacent building in February 1967. There were no operators for it: the research staff had to work it themselves (at one point, the General Board had thought that the absence of technicians in Hoyle's proposal was an oversight). The computer used a tape which presented a fire risk, so it could not be left to run unattended: Aarseth recalls spending 50 hours there over a long weekend with a camp-bed and provisions, running a long calculation. Soon a manager was appointed, and, fulfilling a long-standing arrangement with John Kendrew, the computer earned some of its keep from a leasing arrangement with the molecular biologists from the MRC Unit.[22]

By the summer of 1968, IOTA had a permanent staff of 35, and it was soon attracting a similar number of visitors every summer. Conferences brought in other researchers for shorter visits. Visitors were crucial to Hoyle's vision for IOTA, and they came largely from the USA, and particularly from Caltech, against the flow of the 'brain drain'. Initially they were drawn from among Hoyle's friends and their students;[23] but these people – the Burbidges and Fowler in particular – were a great attraction for other visitors whose applications would be reviewed every spring. Where Hoyle and his students had been travelling to the USA to find decent facilities, now the Americans came to Cambridge for the same reason. Researchers from one part of the USA would come to the Institute to meet researchers from elsewhere in the USA, and, free of the demands of their usual work, they found the atmosphere stimulating and productive.

The hospitality was always warm and generous. Although Hoyle was unable to fulfil his plan to buy a house for the visitors, there was plenty of other accommodation, either in the ancient, imposing Colleges or in the more bucolic environs of the Observatory (Geoffrey Burbidge remembers sheep cropping the grass there). One astronomer visiting from the USA was astonished to find a black limousine waiting for him at London Airport: Barbara had arranged for a chauffeuring company in Cambridge to collect him.[24] Hoyle would drive his American friends to Scotland or the Lake District for hill-walking trips, and invite them to Clarkson Close for gin and tonic or to watch Test Match cricket on television (with explanatory commentary by Hoyle).

Hoyle would still often work from home at Clarkson Close, just a few minutes away from IOTA. When he walked over to IOTA – keeping to his wartime resolution, he never rode a bicycle – he would typically be dressed not in a jacket and tie but in a sweater, as though for open country.[25] He would chat with the other staff members about his interests beyond astronomy, such as sports and mountains, and he was still passionate about chess. He tackled its problems as he would a scientific challenge, and, more than a decade ahead of commercial chess computers (he was to spend £420 on one in 1982), he encouraged his staff in vain to write programmes that could play chess. He also encouraged them in their research with his quick scientific mind and his great facility for mathematics: in a field that relies on numbers and formulae he carried these in his head and produced them in an instant where others would resort to the textbooks; and to the irritation of those who would have preferred him to cite their original papers he would rather work from first principles than from other people's research, deriving from scratch the tools he needed in less time than it would take someone else to find them in the library.

The changing mix of staff and visitors was a powerful stimulant, and was an innovation compared to traditional patterns of scientific research. According to Aarseth, 'Fred should be given a great credit for that. That was new, and it was his achievement and his vision that brought this about.' Soon, IOTA staff were producing over a hundred papers each year, and in 1972, an alphabetical review of publications would run neatly from Aarseth's paper of 1 October 1966 to very recent work by Sataoshi Yabushita, a visitor from Kyoto University in Japan.[26]

Despite the pressures in the run-up to the founding of IOTA, Hoyle had found time to write another science fiction novel. In 1966 he published *October the First is Too Late*. Reviews were mixed: the *Sunday Times* said:

> Fred Hoyle is the Buchan of Science Fiction. His fantasies are not only firmly rooted in scientific possibility, but are told at a galloping pace and with an appealing no-nonsense authority. Having said this, I have to admit I far from fully understood the central thesis of *October the First is too Late*. . . .[27]

The novel tells of two men, one a Nobel laureate in mathematics (there is no real Nobel Prize for mathematics), and the other, the narrator, a musician, who are caught up in the management of a global crisis that occurs when time jumps to different periods of the past and future in different places around the world. While the UK remains in 1966, western Europe is in the throes of World War I, ancient Greece is flourishing and the USA is thousands of years into the future. The mathematician and the musician (who, like many of Hoyle's fictional characters, enjoy walking holidays, caravanning, and food) carry the story through a dialogue in which the mathematician, John, explains what is going on to the musician, Dick. John believes that a strong, information-rich infrared beam that astronomers have detected coming from the Sun is being used by a superior intelligence to construct parallel worlds that co-exist despite apparently being in different moments in time. At certain moments it is possible to travel from one country to another, even though one country is in 1966 and the other in the fifth century BC; and at other times these boundaries in time are closed – hence *October the First is too Late*, a date that was also the start of the Cambridge academic year, and a crucial marker in Hoyle's relationship with his university.

In the novel, John and Dick meet some people from the fourth millennium who show them a film history of the human race:

> The record was relentless. . . . Added to the horror of intimate detail, I had the feeling of a whole species in some monstrous, unclean cycle from which it could never escape. Each cycle was occupying a little less than a thousand years. Always during the reconstruction phase we could see the same bland confidence that this time it would be different. Because these phases were reasonably long drawn out, over three centuries or so, it always seemed as if the disease had been cured. Then quite suddenly, almost in a flash, the monstrous expansion started again. It was a kind of shocking social cancer. Then came the major surgery of flame and death, and so back to endeavour, to a temporary happiness, and to unrequited hope.[28]

As the novel ends (the chapters have musical titles, so the last is 'Coda'), John predicts that all time boundaries will soon close, and he decides to return to Britain in 1966, while Dick and the narration stay in the future, from which perspective Dick tries to explain what has happened:

> Although much more science is known here than was known in the world of 1966, the detailed operation of the singular mixing of epochs is not well understood. As I make it out, issues involving time-reversal were involved, but the physics of the matter is not within my competence. What is quite certain is that the affair was brought about from a higher level of perception than our own. That such levels exist seems reasonable. That we ourselves are unable to comprehend the thoughts, the actions, the technology perhaps, of an intelligence of a higher order also seems reasonable. Disturb a stone and watch ants scurrying hither and thither underneath it. Can those ants comprehend what it is that has suddenly turned their tight little world upside down? I think not. It emerges very clearly that humanity can also be stirred up at any time, just like ants under a stone.[29]

In 1965 Hoyle had published a book in the USA entitled *Galaxies, Nuclei and Quasars*. The book raised some hackles in America, not least for a perceived anti-American slant to its criticism of the space programme,[30] and it made clear that for Hoyle, the cosmological debate was far from over. In 1966 the book was reprinted by a UK publisher, and Hoyle used the opportunity to re-open the debate about the redshift of quasars.[31] In the early days of quasars, there had been some discussion about whether they were local objects, or very far away. The majority of astronomers had decided by the mid-1960s that the very high redshift of quasars means that they are very far away, and this became the standard explanation. But Hoyle was not convinced, and in 1966 in a letter to *Nature*, he and Geoffrey Burbidge proposed that quasars were closer than generally thought, which, since measures of distance are related to the rate at which the universe is expanding, had implications for the expansion of the universe.[32] According to the *Daily Telegraph*, this put 'space theories in the melting pot.'[33] Other scientists too saw the matter as still open: Jesse Greenstein, a key player in the discovery of quasars in 1963, was still scratching his head about them in 1967, when he wrote:

> Horrid quasar
> Near or far,
> This truth to you I must confess:
> My heart for you is full of hate
> O super star,
> Imploded gas,
> Exploded trash,
> You glowing speck upon a plate,
> Of Einstein's world you've made a mess![34]

Gamow too was nervous: in 1968 he would tell Greenstein that despite the boost to evolutionary cosmology of the background radiation, 'the problem of [quasars] . . . presents a challenge to our understanding of the evolution of the universe.' He was looking forward to visiting Ryle, who was 'getting very amusing results'.[35]

Another astronomer who was sceptical of the standard explanation of the quasar redshift was Halton 'Chip' Arp. He had been in the first cohort to graduate with a PhD from the new Caltech astronomy department, and he had worked there since 1957. He discussed the quasar redshift in a talk he gave at Caltech in 1966, which Hoyle attended. He had been observing quasars – he was compiling his celebrated *Atlas of Peculiar Galaxies* – and he had found them to be nearer to galaxies rather more often than chance would suggest, and particularly nearer to what he called 'disturbed galaxies.' If quasars were more likely to be found around galaxies, then perhaps their association had some physical basis. But this was a puzzling observation, as the galaxies were sometimes much closer than the quasars seemed to be: they had much smaller redshifts. How could two associated objects have different redshifts – redshifts that implied that they were vast distances apart? Perhaps the association was merely one of apparent location: it was possible that the quasars just appeared to be in the same region of sky as the galaxies, but that quasars were way off in the distance and the galaxies were in the foreground. But Arp thought the association between these galaxies and quasars was real, and that it had profound implications for our understanding of the meaning of the redshift, and thus for cosmology. But his colleagues challenged his statistics: they were not persuaded that the distribution of quasars near galaxies was anything other than random, and they rejected the idea of any physical connection between them.[36] Arp nevertheless published his findings in the *Astrophysical Journal*, where he concluded that 'It is with reluctance that I come to the conclusion that the redshifts of some extragalactic objects are not due entirely to velocity causes.'[37] Some supporting evidence was forthcoming, including a paper by the Royal Greenwich Observatory astrophysicist Donald Lynden-Bell and colleagues, published in *Nature*.[38] Subsequent papers tended to contradict Arp's position, and he was to become more and more isolated at Caltech. But after the talk in 1966, Hoyle told Arp that he and Burbidge had been studying quasars and had concluded that they did not fit into an expanding universe. They thought that quasars originated in nearby galaxies. According to a Caltech colleague at the time, Hoyle 'tended to agree with . . . anything that would work against a cosmological interpretation of these redshifts which would imply an expanding universe';[39] but in Hoyle's view:

> . . . what Burbidge and I felt . . . was that there were too many quasars with positions close to bright galaxies. . . . Our statistics were not very good, but we felt instinctively that there were too many, and yet the redshifts of the quasars and the galaxies were quite different. And so we reported it and we got a lot of unpopularity out of that – people said we were mad. . . . they said your statistics are [flawed] . . . you've just got a coincidence – that [quasar] is very distant and in the background, and this [galaxy] is in the foreground. But there was one case that was one of our best statistical cases – a galaxy and a quasar – and . . . the redshifts are quite different.[40]

Burbidge felt that theory was being protected from the challenge of these inconvenient observations: 'So we began to say you've got to take these [observations] into account and that made us very sceptical of the way things were being taken.'[41] Burbidge was considering possible alternative explanations of the redshift: he thought that gravitational effects, and not the expanding universe, were the better candidate. Burbidge and Hoyle kept in touch with Arp, and, separately and together, they were to return to the problem many times.

Hoyle's *Galaxies, Nuclei and Quasars* also included a chapter on steady-state theory, which was followed by a chapter entitled 'A Radical Departure from the Steady State Concept,' in which Hoyle set out a scheme for an oscillating cosmos. While the universe may be expanding now, it may well have been contracting in the past, and expansion followed by contraction means that a balance and constancy is achieved over time. So this cosmos manages to be both dynamic, and steady.

So despite his apparent retraction of steady-state theory after the discovery of the cosmic microwave background radiation in 1965, Hoyle was persevering with the idea. When he gave the prestigious Bakerian Lecture to the Royal Society in 1968, he was defiant:

> . . . over the past 20 years the steady-state theory has had to face six different crises. Five in my view have been wholly or largely surmounted, the sixth – the microwave background – remains to be resolved. I think it fair to say that the theory has demonstrated strong survival qualities, which is what one should properly look for in a theory. There is a close parallel between theory and observation on the one hand and mutations and natural selection on the other. Theory supplies the mutations, observation provides the natural selection. Theories are never proved right. The best a theory can do is survive.[42]

The background radiation was a major nuisance for steady-state theory. Hoyle told Fowler in June 1966 that 'I keep worrying about the Penzias and Wilson temperature,' and suggested that the radiation could be a relic of starlight. Fowler replied with encouraging news about his own thoughts on the many ways ('*n* ways, where $n \gg 1$') in which big bang failed to produce heavy elements ('what a wash-out!').[43] In 1969 Hoyle told an audience in the USA that if the universe had always been expanding, matter could never have come together to form galaxies, and that steady-state theory could provide an answer to this conundrum.[44] In August 1970, he told the Uppsala Symposium in Sweden about his oscillating universe, and argued that the commonly accepted age for the universe, calculated on the basis of its apparent expansion, was not actually its age, but the period of one cycle of oscillations. Then:

> All our astrophysical evidence relating to cosmology would refer to the current cycle. Evolutionary effects if they exist, for example in the space density of radio sources, would be related to the phase of the cycle. It might be supposed, for example, that the space density of the radio sources would be greater when the average mass density was increasing than when it was falling. If these considerations are correct the microwave radiation cannot be attributed to the origin of the universe. The microwaves must be astrophysical in their nature.[45]

Thus the background radiation might not be ancient and cosmological, and a reason might be found for it in the physics of the present-day universe.

Another of Hoyle's preoccupations in the late 1960s was Stonehenge, the prehistoric circle of standing stones in Wiltshire, in the west of England. In 1966, Hoyle proposed that Stonehenge, rather than being simply aligned with events, such as the solstices, in the astronomical calendar, is instead a much more sophisticated observatory. The idea was not originally Hoyle's: he had responded to a review in *Nature* of a book by the American astronomer Gerald Hawkins, who had argued that the stones were aligned with significant astronomical events such as eclipses, and could be used as an astronomical computer. The *Nature* review, by archaeologist Professor R.J.C. Atkinson of Cardiff University, had condemned Hawkins' work as 'tendentious, arrogant, slipshod and unconvincing.'[46] Atkinson had also charged Hawkins with 'a disregard for accuracy,' 'invalid reasoning,' and 'a taste for bizarre interpretation.'[47] Hoyle had repeated Hawkins' calculations, and had come to the same conclusions; he published his results in the journal *Antiquity*.[48] Hoyle told the *Daily Telegraph* that 'a veritable Newton or Einstein must have been at work,' for his theory:

> . . . not only requires Stonehenge to have been constructed as an astronomical instrument, but it demands a high level of intellectual attainment in orders of magnitude higher than the standard to be expected from a community of primitive farmers.[49]

The ancient observatory theory attracted considerable press attention, and Hoyle would continue to write and lecture on Stonehenge in the 1970s.[50]

Administrative responsibilities outside IOTA took up much of Hoyle's time in the late 1960s. He was one of 16 members of the Science Research Council from April 1968, and was chairman of the Astronomy, Space and Radio Division, which was now run by civil servant Jim Hosie, a former classmate of Hoyle. Much of their work in the late 1960s concerned provision in radio-astronomy, and they would soon turn their attention to the Anglo-Australian Telescope project: a long-running and highly complicated saga that was already piling up problems for them on the other side of the world. The haphazard history of this telescope, and the rancorous relationships within the astronomical community that clouded its early years, show that Hoyle's experiences in Cambridge may have had their own distinct colour, but in character were far from unique.

For the time being though, the Anglo-Australian Telescope was other people's responsibility. It had come about as the solution to the problem that had been nagging at the astronomical establishment in Britain for some years: it was to provide access for British astronomers to the southern hemisphere sky. Much of the impetus for it was due to the Astronomer Royal Richard Woolley. Woolley had been appointed to the Mount Stromlo Observatory, just outside Canberra in Australia, in 1939, and his plan was to turn the Observatory from solar to stellar astronomy. It was put to munitions work during the war, making optical components such as gun sights, but by the time Woolley left Australia in 1955, Mount Stromlo had a 74-inch telescope that would serve Australian astronomy well.

In the early 1950s, Woolley had begun to canvas opinion among Australian astronomers about a much larger telescope – he wanted a 200-inch instrument to rival Mount Palomar at Caltech. He discussed the idea with Marcus Oliphant, Head of the School of Research in Physical Sciences at the Australian National University and Founding President of the Australian Academy of Science, and they decided that a Commonwealth collaboration with Canada and the UK might generate the necessary funds.[51] The Canadians declined to join in, but Woolley, by now in the UK as Astronomer Royal, and his successor at Mount Stromlo, the Dutch-American astronomer Bart Bok, pursued the idea of a collaboration between the Australians and the British.[52] The UK's commitment to the Radcliffe Observatory in Pretoria was becoming less useful because of the political problems in South Africa, which reduced the competition for funding; but the British were negotiating new relationships in Europe. Five European nations were looking for further members for a consortium to run a European Southern Observatory in Chile, and with European trade increasingly closely organized exclusively within a group of the continental nations – Britain was not included in the prototype European economic community that was forming through the 1950s – the political mood was to favour friendly links with Europe. Woolley was opposed to the idea of a European Southern Observatory as he felt that the UK, as one of many partners, would get only a small share of the facilities; but he was also concerned that any commitment to a European project would necessarily mean less money for a Commonwealth project. In 1955, the Royal Society supported membership of the European project, but the Treasury rejected the application for funds.

In 1959, the British National Committee for Astronomy at the Royal Society, which Woolley chaired, had endorsed the idea of a Commonwealth project in Australia, and recommended that: 'the Chairman [Woolley] be empowered to enter into discussion with appropriate Australian authorities.'[53] At the same time, the Royal Society approached the Australian Academy of Science. The Academy was just five years old, and the prospect of a large sum of money going into an astronomy project revealed internal factions: Fellows from other disciplines, and especially from biology, saw the proposed telescope as diverting resources away from them. The Academy decided it would have to investigate the financial implications for other branches of science before committing to the project.[54] In the UK, Woolley, supported by his assistant Olin Eggen, continued to raise objections to a European collaboration, and to push for a Commonwealth project. By 1961, when the British National Committee for Astronomy was also considering Hoyle's proposal for an Institute, events at Sharpeville had pushed the balance in the Commonwealth options away from South Africa in favour of Australia.

At this meeting Hoyle had suggested a third option: cooperation with the Americans to run a telescope in South America. Woolley contacted C.D. Shane, formerly Director of the Lick Observatory in California and now President of the research consortium that was developing the new National Observatory at Kitt Peak, Arizona, whose reaction to the idea was positive. Then South Africa withdrew from the Commonwealth, and the UK's application to join the European Economic Community was rejected, taking the South African and European schemes out of the running. The focus now was on the Americans and the Australians.

In Australia, the Academy was keeping its options open. Bok was President of the Australia and New Zealand Association for the Advancement of Science in 1962, and, pessimistic about prospects for the collaboration with the UK, he used his Presidential Address to set out the case for the telescope, and to chide the Academy for its lack of enthusiasm. Other champions continued to lobby hard. On a private visit to Australia in 1962, Harrie Massey visited friends in astronomy and in politics, and reported much interest, and early in 1963 Bok and Hermann Bondi, in his role as Secretary of the Royal Astronomical Society, talked to the Prime Minister's Department in Australia. In March 1963, Woolley mustered an impressive line-up in Australia, including the physicist John Cockcroft, Master of Churchill College Cambridge, who was Chancellor of the Australian

National University, Oliphant and Bok, and the radio-astronomers Bernard Mills from Sydney and Edward 'Taffy' Bowen, Chair of the Radiophysics Division of the Australian national research organization CSIRO, who had built the Parkes radio-telescope and gained considerable experience of working with government during his wartime work on radar. This group endorsed Woolley's proposal and agreed to lobby for a large optical telescope in Australia. In April 1963, Prime Minister Robert Menzies put a bill before the Australian House of Representatives to remove restrictions on the geographical locations available to the Australian National University, so that a field station could be set up at Siding Spring, in the mountainous region 500 kilometres north-west of Canberra. The lobbyists were able to interest enough members of the House in their cause for the debate to spend little time on the wider issue of the University's rights and become almost entirely concerned with the telescope and the collaboration with the UK.[55]

The Australian Academy and the Royal Society had both set up committees to investigate the idea, and they agreed to bid for a 150-inch telescope to be sited at Siding Spring, and to make representations to their respective governments. Woolley went to Australia with members of the Royal Society committee to meet with Australian astronomers, compile a shortlist of possible positions at Siding Spring for the telescope, and draw up a technical specification for it. These plans went before the Royal Society and the Australian Academy of Science, and were submitted for funding.

In Australia, it was usual for big funding bids for special projects to be handled by the Australian Academy of Sciences which would make a representation to government. In the UK, funding would be decided by the Science Research Council, which was just a few months old. Its astronomy section was still struggling with the negotiations over IOTA. The Royal Society could no longer go directly to the government to ask for funds for astronomy, and control of the Royal Observatories had passed from the Admiralty to the Science Research Council. So all branches of astronomy were now competing for funds from the Science Research Council, where they were also competing with other disciplines. Apart from IOTA, the Australian telescope was also competing with bids for new radio-telescopes in Cambridge and Manchester, and for new optical telescopes for the Royal Observatories at Herstmonceux and Edinburgh. Despite Prime Minister Harold Wilson's stated commitment to science, the pot of money for which they were competing was small: the major portion of Science Research Council funds was already committed to keeping existing projects running, and there was less than £5 million left over for all areas of science, until 1971. The Astronomy, Space and Radio Board of the Science Research

Council had calculated the cost of the Australian telescope over that period at £11 million.[56]

Nevertheless, the project was put forward to the Science Research Council in the spring of 1966, which passed it to the Treasury where money was authorized for a design study. The Department of Education and Science, now under Minister Anthony Crosland, informed the Australian government that the UK would share in the costs of the project, and suggested that the two governments set up a joint preparatory committee. But the reply from the Australian Prime Minister, now Harold Holt after Menzies' resignation, was that the Australians were expecting the offer of a joint project with the USA. Senator John Gorton, minister for Commonwealth activities in education and research, had wanted a wholly Australian telescope, but he knew that Prime Minister Holt would not pay the full costs, and that Holt did not have the UK at the top of his list of potential partners.[57] The Australians also thought that the British idea of a joint preparatory committee was designed to stall the project. So Gorton decided to muster two acceptable potential partners and let the Prime Minister choose one, to allow the Cabinet to maintain its sense of control over the project. Gorton's first choice of partner was the USA, and he and Bowen went there in the summer of 1966 to drum up support. Gorton was asking for half of the costs of the full project, rather than just a share of the design study as had been promised by the UK. They found much interest from the Lick Observatory in California, but the process there would be slow: while the University of California could invest some funds, it would have to apply to NASA for the rest. The British knew that the Australians had approached Lick; in any case, Woolley was still talking to the Americans too. But Massey went to Australia and assured Gorton that the idea of a joint preparatory committee, far from being a stalling tactic, was actually a statement of intent and commitment. In September, the Australian government asked the Science Research Council to make a formal statement about the availability of funds, and it fell to William Francis, now Secretary of the Science Research Council, to handle the negotiations. He went to Australia in October and assured the Australians that British money was ready and waiting in expectation of input from the Australians. Australian newspapers reported that British money was ready, and that the Australians were dithering. By now, hopes in Australia of a link with the USA had dwindled, but the British were losing confidence in the Australians: Francis reported in November that 'there was no purpose in any further discussions between Anglo-Australian experts before the Australian Government had decided on their next move.'[58] Woolley was still working on various other plans for access to the southern skies; and other bids that would compete for funds were emerging, including one for a northern hemisphere telescope in the Mediterranean by the Cambridge consortium of Hoyle, Redman and Ryle. But the profusion of alternatives did not help speed matters along: in response to an enquiry from the Royal Society, Bernard Lovell, who preceded Hoyle as Chair of the Astronomy, Space and Radio Board of the Science Research Council, reported that 'the trouble of course is the lack of unanimity and procrastination amongst the professional optical astronomers. Indeed, it has occasionally seemed during the last few months that only the theorists and radio-astronomers are interested in providing the optical people with a decent facility.'[59]

But on 14 April 1967, the UK government received a formal commitment from the Australian government. It came as a surprise to the British astronomers when they heard that their project had, as Redman put it, 'suddenly risen from the dead.'[60] Redman walked across from the Observatory to IOTA to tell Hoyle the news that the Australian project would go ahead and their Mediterranean idea would not. Although Redman was, according to Hoyle, rather depressed at this outcome, he nevertheless applied himself diligently to the Australian project and was to make important contributions to the optics.[61]

Among the conditions of Australian participation were that the design used for the Kitt Peak telescope should be used in Australia too. The explicit reason for this was one of efficiency: it would shorten the planning and design stage. However, Eggen, who was well known for his forthright style, told Hoyle that the Australians had thereby hoped, given the protracted history of the Royal Observatory's Isaac Newton Telescope – it had taken 20 years to bring it into operation – to prevent Woolley 'foisting another "Isaac Newton" on them.' Also the radio-astronomer John Bolton, who had returned to Australia from Caltech in 1961, had been campaigning in Australia for a different design, and Hoyle suspected that the specification in the agreement was to end that debate.[62] Another condition was that the observatory should be should be at Siding Spring, as had been planned, and that facilities should be shared with the Australian National University which already had an observatory there. The controlling body should be jointly managed, with the two countries equally represented and having equal status, with a single director being chosen by the controlling body.

A few months later, the astronomers met in London. Bok had tired of the political struggle, and had taken a job in the USA; his post at Mount Stromlo belonged now to Eggen. Eggen spent a couple of days with Hoyle

when he came to Britain for the meeting, and told Hoyle that his negotiations with the University of California had foundered, and that he had
therefore urged the Australian government to commit to the British. At the
meeting, the delegations from the two countries decided that the conditions
were acceptable and that a Joint Policy Committee should be established to
start the construction phase. In advance of this committee starting work, a
technical committee was convened to work on the design. Its first meeting
was held in Tucson, Arizona, so that the committee could study the Kitt
Peak telescope at first-hand. Soon, a Project Office was set up at Siding
Spring, and the first manager was appointed in 1968.

The Joint Policy Committee started work in August 1967. The Australian
members included Bowen and Eggen, and Ken Jones from the Department of Education and Science. From the UK came Woolley, Bondi and
Hosie. Bondi was soon appointed Head of the European Space Research
Organisation and he resigned from the committee, to be replaced by Hoyle,
who by now was Chair of the Astronomy, Space and Radio Board of the
Science Research Council. The Joint Policy Committee established an
office in Canberra, and had made great progress by the time the bureaucracy caught up with it – the Committee needed a particular status under
Australian law if it was to spend money and hire staff there. Thus the
Anglo-Australian Telescope project was formally constituted in September
1969. But there was plenty of work still for Hoyle to do, and some of it was
to be very fraught indeed.

In 1939, Hoyle had written a paper with Ray Lyttleton on the effect of
interstellar dust on the climate.[63] This dust was known to occupy the spaces
between the stars, and to amount to a huge quantity of material throughout
the universe, but its composition remained a mystery. In the early 1960s,
Hoyle revived his interest in this dust, and began a new line of enquiry
that was to occupy him for thirty years. Though the work was eventually to
take him into highly controversial territory, it had started simply enough
when he set his student Chandra Wickramasinghe the task of investigating
the dust. Wickramasinghe was from Ceylon, and the son of an eminent
mathematician who had won a double-first at Cambridge. Wickramasinghe
too came to Cambridge on a Commonwealth scholarship after gaining a
first-class degree in mathematics at the University of Ceylon in Colombo.
He became one of Hoyle's many postgraduate students in 1960, and
published his PhD thesis in 1963.[64]

The results of Hoyle and Wickramasinghe's work had started to appear in 1962.[65] They had considered the possibility that the dust grains were frozen ammonia or methane, but proposed instead that they were graphite particles covered in a film of ice.[66] A year later Hoyle and Wickramasinghe published spectroscopic data that corroborated the graphite–ice proposal,[67] but in 1964 new data from researchers in Edinburgh gave no indication of the presence of ice. Hoyle and Wickramasinghe pressed on with the graphite theory – this had been the subject of Wickramasinghe's PhD thesis – and by 1965 the best fit to their data was pure graphite.

Now at IOTA, Hoyle and Wickramasinghe were considering the role of these graphite dust grains in the scattering of light and heat in space. In 1967 they published a paper in *Nature* suggesting that graphite interstellar grains might act as thermalizers, absorbing and emitting infrared radiation and dispersing it throughout the universe, and thus accounting for the uniform background radiation that others were explaining as a relic of the big bang.[68] The thermalizer idea did not impress Hoyle's old rival George Gamow, who commented on it in a letter to Jesse Greenstein, adding: 'It really seems that the world of science is getting crazy.'[69]

Hoyle and Wickramasinghe then turned their attention to other possible ingredients in the composition of the dust grains: in 1968 and 1969 they published papers on the iron and silicate content of the interstellar medium.[70] By now, the data 'seemed in 1969 to provide more or less direct spectroscopic evidence of both graphite and silicate particles in interstellar space.'[71] The next paper examined 'the astrophysical consequences' of a mixture of graphite and silicate grains.[72] But Hoyle and Wickramasinghe soon became dissatisfied with silicate, and looked for other explanations of the infrared spectrum. Their method was to study spectra produced by observers studying the interstellar medium, and then to compare these with reference spectra, made in the laboratory, of various materials. Rather than thinking only in terms of the kinds of materials that constitute the rocks on Earth (what they called the 'mineral hypothesis'), Hoyle and Wickramasinghe looked instead at two possible types of grain: simple inorganic ices (such as water) and organic polymers, the large carbon-based molecules that constitute oils and plastics as well as much of living matter. In the early 1970s, while Hoyle was away in the USA, Wickramasinghe looked for 'a polymer that would aid us in explaining the astronomical data, especially of the finer details which caused difficulties for the mineral grain hypothesis.'[73] The polymer he found was polyformaldehyde, which is also called polyoxymethylene or POM.[74] This is a polymer made from repeated units of a simple molecule, formaldehyde, CH_2O. The formaldehyde molecules join together to form POM when they condense onto silicate grains. The POM reference spectrum showed a particularly good match with a spectrum of light coming from the Trapezium nebula. Hoyle and Wickramasinghe decided that the POM grains explained the spectroscopic data as well as, if not better than, silicate. Their conclusion was that the interstellar grains were a mixture of graphite and silicate particles, with a coating of the organic polymer POM.

POM is a tough material, as strong as steel and much used by engineers for joints and seals in extreme conditions. In that regard, it would be appropriate for the inhospitable interstellar environment. But it is also organic, and organic materials are associated with living things. For Hoyle and Wickramasinghe, the organic aspect of POM was to provide the basis for a challenge to the fundamentals of biological science.

13
The Astronomer Hoyle

Problems at the institute; Northern Hemisphere Review; an attack on the Royal Observatories; the new Astronomer Royal; the merger at Cambridge; Sir Fred

Now that Hoyle had his institute, he seemed to have joined the establishment after all. He was a frequent if combative correspondent to *The Times*, not only on steady-state theory and quasars but also on matters such as economic inflation, foot-and-mouth disease in cattle, and the Common Market of Europe, which he opposed both on grounds both of nationalism, and because it seemed to be tempting the UK into what Hoyle saw as fruitless pan-European collaborations on minor but expensive projects in space technology. He managed to annoy former Science Minister Quintin Hogg with this protest.[1] In 1968 both his popularizing and his science were recognized: Hoyle won UNESCO's Kalinga prize for popularization, after a nomination by the American Association for the Advancement of Science. The AAAS had tried before to nominate Hoyle, in 1965, but the nomination required the nominee's cooperation and, amid the stresses of the Institute campaign, Hoyle had declined to give it.[2] He was also awarded the Gold Medal of the Royal Astronomical Society, which is given for outstanding personal research or leadership in astronomy. When the news arrived of this award, Hoyle shocked some younger colleagues by grumbling, over coffee at IOTA, about having to 'go to London to collect a bloody medal';[3] but he enjoyed his trip to New Delhi for the Kalinga Prize ceremony,[4] despite being shocked by the frantic, noisy traffic. Hoyle told the audience there that for a scientist 'not engaged in urgent practical problems', the popular science for which he had won the prize is what 'earns one's keep.' And in a rare positive comment on the space programme, he remarked that photographs taken by the recent

Apollo 8 mission, showing Earth as a 'glowing blue jewel', should help us to think in the longer term about our responsibilities as trustees of the planet.[5] Other honours followed: in 1969 Hoyle became a Foreign Associate of the US National Academy of Science, and Professor of Astronomy at the Royal Institution in London. In 1970 he became Vice-President of the Royal Society, and in 1971 he was President of the Royal Astronomical Society.

Hoyle published another novel in 1969, *Rockets in Ursa Major*, based on the earlier stage play about a war in space, and, in 1970, yet another: *Seven Steps to the Sun*, a tale of time travel in which the future reveals the problems of the present. It centred on two of Hoyle's long-standing concerns: the problem of over-population; and the inability of politicians to run the world. In the novel, Mike, a writer, is researching a science-fiction story about using future knowledge to solve present-day problems, but he finds himself repeatedly flung forward in time by a decade or so, so that he is living his own story. On each leg of the trip, in a series of daring all-action adventures around the world, he searches for the musician friend he has left behind, the now much older Pete. In the future, Mike witnesses the disintegration of a human society that is expanding to the point where the Earth's resources can no longer sustain it, and where warring groups clutch at the last chances for survival. At the beginning of the novel, while outlining his story idea to Pete, Mike says that the general public have been so conditioned to worry about nuclear war that they have not thought about the more pressing problem of over-population. A decade later, Pete tells Mike that the politicians have been unable to control the population and are blaming the scientists, who are struggling to find new ways to produce food. A decade later, nations have isolated themselves to keep their resources for their own people, and life is organized around the protection and distribution of scarce food. In the final future scene, Mike meets a man who has witnessed a demonstration in the streets of Washington DC that was the start of a popular revolution. The man tells Mike that 'the whole world's gone mad':

> It's all to do with the total lack of authority over many years from the politicians. They've always been so interested in their own personality politics that they've conned the public into believing that they know everything. This time they've really come unstuck and so have their advisers. . . . They should have listened to the cries from the wilderness from the great minds of the last fifty years. The emotional section of society nowadays is too bloody scared of science, as they have been for more than 60 years. I bet people 60 years ago wouldn't have believed that the breakdown of civilisation would have started with a simple demonstration. They probably thought the end would come in fire and

brimstone from a nuclear war. You realise that if the politicians of the [nineteen] sixties had acted on symptoms that were being pointed out about overpopulation and food production, by scientists and philosophers, we wouldn't be in this situation today. But the politicians were so sure of themselves. Fools.[6]

Alongside the novels, Hoyle's opera, *Alchemy of Love*, had at last found a backer, and the newspapers reported that it was in production in New York. According to the *Daily Mail*, 'the professor's love lyrics are as inspired as most things he turns his hands to.' The *Mail* continued:

It will surprise no-one if this year's top of the pops turns out to have been written by the Director of the Institute of Theoretical Astronomy at Cambridge University. Fred Hoyle's imagination spends most of its time on quasi-stellar objects on the rim of our universe which are burning away at a rate that makes the H-bomb look like a bedroom candle. It's natural territory for a man widely held to be touched with genius. . . . At 54, Hoyle is already in the history books as a protagonist in one of the century's most brilliant academic controversies – whether the universe began with a bang (a clap of God's hands, as it were), or whether there exists a state of continuous creation. Perhaps the greatest cosmologist of the day, part of his genius lies in his gift for making abstruse problems intelligible to the layman. When Hoyle broadcast his theories in 1950, working men stopped him in the streets of Cambridge to argue – and got his arguments in the uncompromising accents of the West Riding [of Yorkshire].[7]

Behind the public face of a triumphant career, all was not well at IOTA. Hoyle's presence there was sporadic: travel and frequent meetings kept him away from Cambridge. Replies to correspondence were often very short, handwritten as ever, and sometimes months late. According to Aarseth:

. . . he kept going to Caltech and in term time he was supposed to be here – I don't know how he got away with certain things. He would give regular lectures, and one day he didn't show up. Two days later we came back for the lecture, he just walked in and said 'Oh, I had to go to Washington'. And then he carried on with the lecture. What a great statement – as if the President needed him. Forget about us, just go.[8]

Funding was secure only until 1972, a date that drew ever closer. Secretary Frank Westwater, as well as other staff members, could see that Hoyle was investing little time or energy in generating income for the future. Westwater was in charge of IOTA's budget, and knew exactly how bad the financial prospects were. He made every effort to deal with trivial and routine matters himself, knowing that Hoyle would not be interested in them – one staff member recalls Hoyle's attitude as 'that's just business, that'll take care of itself.'[9] But Hoyle was needed for the important issues such as future funding, and Westwater became increasingly frustrated at Hoyle's lack of engagement with the problem.[10] Westwater also felt that some members of the administrative staff, who usually reported to Hoyle, were unwilling to cooperate with him when Hoyle was away. Hoyle accepted responsibility for this problem, and undertook to discuss arrangements for future absences with Westwater before leaving.[11]

Just as the family home at Clarkson Close had been Hoyle's office for so long, so the boundaries between work and home were blurred at IOTA. Not only did Hoyle continue to work from home so that he could be free from interruptions, but also his family would call at IOTA; and there was some competition between Barbara Hoyle and Westwater to manage his time.[12] In the autumn of 1968, tensions arose when Westwater asked for a meeting with Hoyle to discuss a matter of 'staff unrest': there had been some interpersonal friction and unpleasant gossip. Also, there had been suggestions that stationery had been handed to Hoyle's son Geoffrey, and that postage stamps destined for Barbara had been given to IOTA's computer officer Nick Butler, who was courting Hoyle's daughter Elizabeth. Hoyle and Westwater had a very fraught meeting discussing staff behaviour and practices, and they were able to resolve some matters – they agreed that some accounting procedures needed to be tightened up, for example. But the implication of Hoyle's family in these charges was, Hoyle felt, 'astonishing and outrageous'.[13] He told Westwater he would be dismissed if there were any further 'smearing innuendoes', and advised that 'accusations against my wife should not be made under the heading "staff unrest".'[14] Geoffrey Hoyle sought legal advice and came to IOTA to tell Westwater that he would take the matter seriously if 'innuendoes spread outside the Institute.' Hoyle wrote a detailed diary of events in the style of a police witness statement, accounting for the stamps, stationery and a stapler which were used for work done from Clarkson Close. Barbara later labelled these records with the Yorkshire phrase that had become a euphemism for poor labour relations: 'Trouble a't'mill.'[15]

John Faulkner, who had been called to give evidence to Hoyle's meeting with Westwater, felt that the episode was symptomatic of broader problems at IOTA. While he had been a loyal supporter of Hoyle's project and had worked hard to get IOTA up and running, it was a reason to move on: along with the uncertain financial future, it was 'the sort of thing that was beginning to make me feel that no matter whether the place was scientifically exciting or not, I didn't like this sort of social, unprofessional background setup.'[16] Faulkner decided to quit after a trip to the University of California at Santa Cruz at the end of 1968, where he was offered the chance to set up a theory group to complement the observational work of the nearby Lick Observatory.

And all the while, Westwater was concerned to find ways of retaining staff, and was struggling with the ever-present problem of finding funding for Institute to continue beyond 1972.[17] The personal tensions were resolved in an unexpected and unhappy way in April 1969, when Westwater died from complications following routine surgery. By this time, many of the staff were aware of the precarious position of IOTA, managerially and financially, and several of them talked to Hoyle about the implications of Westwater's death. He appeared:

> . . . quite blasé about it, and said 'I don't think we really need to have a secretary, Frank wasn't doing very much recently.' So some people said of Fred that once he had won the battle of having the Institute founded, that he really lost interest in having it continue – the battle had been won, he had proved his point.[18]

Yet some embarrassing irregularities would shortly appear: in the annual audit in July 1969, £9000 could not be accounted for – the equivalent of four salaries for junior staff.[19] And in August, the Director of the Wolfson Foundation, Major General A.R. Leakey, wrote to Hoyle after a visit there: 'I failed to notice if there is a plaque or other such device indicating that the Wolfson's grant made possible the building of the Institute. Perhaps you would be kind enough to enlighten me.'[20] There was no plaque, but Hoyle assured Leakey that arrangements were being made – they had been, he said, in the hands of Captain Westwater. Lord Todd, as Chair of the governing body of IOTA, could see that the Secretary's post could not be left unfilled. However, he could also see that it would be difficult to appoint anyone of any seniority when so few promises could be made to them about the future of IOTA. Eventually a deal was done whereby the appointee would be transferred to a post elsewhere in the University should IOTA cease to exist in 1972, and Westwater's post was filled by Peter Vaugon early in 1970. Vaugon was a young Cambridge graduate with civil service and university registry experience, and he went on to a long career on the General Board of the University, where he became Deputy General Secretary.

✳*✳

In 1969, Hoyle, in his role as chair of the Astronomy, Space and Radio Division of the Science Research Council, was charged with reviewing 'policy for support of optical astronomy in the Northern Hemisphere.'[21] Hoyle's committee for this Northern Hemisphere Review included Jim Hosie from the Science Research Council as the senior administrator; Bernard Lovell; James Cassels, a physicist from the University of Liverpool; space scientist Robert Boyd from University College London; the two astronomers Royal: Richard Woolley for England and Wales and Hermann Brück for Scotland; and the expatriate British astronomers Geoffrey Burbidge and Wallace Sargent representing British astronomers overseas. Sargent was a 15-year-old schoolboy in Scunthorpe, Barbara Hoyle's home turf, when he first heard Hoyle's voice, in 1950, on a radio set he had built himself. The radio broadcasts affected him in two ways: first, he became determined in his atheism; and second, he identified with Hoyle's regional accent: he realized that 'someone like me can do science.'[22] Sargent studied physics at Manchester University where he became rather taken with steady-state theory, which he found very pleasing aesthetically. Unlike their Cambridge colleagues, the physicists and astronomers in Manchester did not find steady-state theory particularly controversial. Sargent wrote his PhD thesis on the expansion of supernova explosions into the interstellar medium. He then started looking for a job, and wondered if he could at the same time indulge his passion for baseball: while scanning the airwaves on his home-made radio set, he had become hooked on commentaries from American baseball games. So when his American supervisor Franz Kahn told him that only Oxbridge graduates got jobs in British astronomy, Sargent thought that since his team, the Brooklyn Dodgers, had recently moved to Los Angeles, he might try for a job at Caltech. Kahn wrote to recommend him to Jesse Greenstein and a post was arranged, and Sargent first met Hoyle there in 1960, at a Friday lunch for astronomers at the Athenaeum, Caltech's faculty club. Hoyle was with the Burbidges, and Sargent was too shy to say anything.

Woolley visited Mount Wilson in 1961 and met Sargent there, and subsequently hired him to work the Royal Greenwich Observatory at Herstmonceux. Sargent thought Woolley was very old-fashioned: the Observatory was still in the charge of the Admiralty then, and working for Woolley, the son of an admiral, felt like being in Navy. There was not much discussion about issues in cosmology, as the Observatory's interests were mostly in positional astronomy. Sargent found the intellectual environment extremely staid. By 1964 he had decided to go back to California, and having heard of Hoyle's plans for an institute at that time he had written to Hoyle explaining his decision:

> I am leaving the RGO more because I cannot stand the prevailing spirit of continental amateurism than through any love of California. Therefore I would be pleased to hear of any further developments in your expansion plans.[23]

Sargent moved to La Jolla, California, and he was soon working with Hoyle and Burbidge on the nature of quasars. Although Sargent had not joined IOTA when it started, Hoyle did have plans for him. For now, though, he wanted him and Burbidge on the Northern Hemisphere Review committee to represent the 'brain drain' community, by which Hoyle meant that they were a good example of people who had fled Britain because of the poor state of its observational astronomy.

Hoyle's view was that Britain needed a national centre for observational astronomy. He was a great enthusiast for the Kitt Peak National Observatory in Arizona, which had opened in 1958 and was providing the technical bench-mark for the Anglo-Australian Telescope. More importantly in this context, it was run by a consortium of university astronomers and funded by the National Science Foundation – the nearest US equivalent to the Science Research Council. Hoyle had written to *The Times* in 1966 to argue that one could have something like Kitt Peak in Britain for only three per cent of what Britain was spending on a minor satellite project in Europe.[24]

Hoyle wanted the new national centre to be entirely separate from the Royal Greenwich Observatory. He had a very low opinion of the quality of the work done at the Observatory, and felt it was a drain on resources that might otherwise support more competent institutions. He was also unhappy with the post of Astronomer Royal: this senior and influential figure was also Director of the Royal Observatory, and so was always an observational astronomer, giving undue influence to observational over other branches of astronomy. Hoyle also felt that 'modern astronomy does not mix well with older pursuits',[25] by which he was comparing the challenges of cosmology to the compiling of tide tables. Hoyle thought that the nautical work ought to be done at a separate site.

Since the UK's national observatories were also Royal, they were on sovereign territory, and therefore subject to British weather. Hoyle thought it ridiculous to expect to do decent observational work under those conditions. In a document he prepared for the committee, Hoyle wrote that astronomers need a large telescope and a good sky: transparent, cloudless and full of stars:

Such a sky does not exist in Britain and it is therefore the first recommendation of this report that the pretence that modern astronomy can be done from an observatory sited in Britain should finally be abandoned.[26]

Just in case anyone should think that Hoyle's plan for what he called a 'national centre for optical astronomy' could be fulfilled by reorganizing the Royal Observatories, Hoyle wrote a paper for the Review entitled 'Why the Royal Observatories are not suitable for adaptation as the National Centre for Optical Astronomy.' Hoyle argued that the Royal Greenwich Observatory had shown itself incompetent to set up a new telescope by bungling the construction of the new Isaac Newton Telescope at Herstmonceux, a process that had taken two decades. He also thought scientific standards at the Royal Observatories were shamefully low. Another charge he brought against the Royal Observatories was that they had named a road after George Airy, an Astronomer Royal who in 1845 had ignored a representation from the Cambridge astronomer John Couch Adams who had predicted the existence of a new planet. Airy did not search for the predicted planet, and Neptune was discovered by German and French astronomers rather than by British. The details of this story are still contested, but to Hoyle, 'we see little point if one wishes to be serious about doing astronomy in commemorating an arrogant Astronomer Royal who did considerable harm to the science he was supposed to serve.' Hoyle also spelt out the many ways in which the Royal Observatories compared unfavourably to those in the USA. Hoyle presented another document, co-authored with Burbidge and Sargent, along similar lines.

Burbidge and Sargent presented a paper entitled 'The Reorganisation of Astronomy in the UK'. It described a national centre where facilities would be collectively owned and open to all on merit. For example, 'Bright young people should not be denied the opportunity to do radio astronomy because . . . one powerful man does not approve of them.' They acknowledged that some interesting astrophysics was being done at the Royal Observatories, including the work of theorists there such as Donald Lynden-Bell, and of Woolley himself; but they felt that the routine work, such as the nautical almanac and the time signal, could have a dedicated laboratory and be kept out of the way of the astrophysics, which was better suited to a university environment.[27]

Worse was to come for the Royal Observatories: along with Margaret Burbidge, Burbidge and Sargent described an on-going fiasco at the Isaac Newton Telescope, which by then had been working for just under two years. Their paper was a catalogue of delays, inefficiency and poor performance:

For most aspects of modern astronomy, and all of those involving the study of faint objects, the INT in its present location is not merely ineffective but useless. . . . The leadership of the RGO is erratic and amateurish and the administrative structure is totally inadequate. . . . By world standards the Observatory is considerably overstaffed – particularly in view of the modest amount of data that the telescopes produce. . . . The RGO has a history of complacency and low productivity. It is full of relatively well paid people who have been trained by the Observatory, who have made no impact on world astronomy . . . we consider that it would be folly on the part of the SRC to go ahead with a Northern Hemisphere large telescope if any part of its development or operation is to be put in the hands of the present RGO.[28]

Some Royal Observatory staff sympathized with this view. Vincent Reddish wrote to Hoyle from the Royal Observatory in Edinburgh, noting the shift in astronomy towards astrophysics, and suggesting a closer cooperation between the Observatories and the universities. Reddish proposed reducing the numbers of salaried staff at the Observatories and replacing them with new blood in the form of postgraduate students on research grants.[29] But the Astronomers Royal on the Review committee were aghast. As Hoyle wrote to Hosie in July 1969, 'The discussions so far . . . leave me with the impression that different members have different ideas on this topic.' At the end of July, Woolley and Brück asked for the position of Burbidge and Sargent to be clarified – Burbidge felt that the two Astronomers Royal were trying to stop him and Sargent having a vote, on the grounds that they were no longer concerned with British astronomy[30] – and Hoyle confirmed their status as assessors, who did not count towards the quorum.[31] In September Hosie wrote to Hoyle assuring him that he would make sure that meetings were held on dates when he knew Burbidge and Sargent could attend.[32]

Hoyle's commitment to observational astronomy was simple: theoreticians could not do good work without good observational data, and lack of good instruments was holding them back. There was new physics to be found in space, if only the telescopes could reach it. In the summer of 1970, Hoyle expressed his frustrations in a reply to a letter from G.A. Winbow, a postdoctoral researcher at CERN, the European particle physics laboratory in Geneva, who was concerned about his prospects for a career in particle physics and was asking whether there might be openings in astronomy for someone of his education and experience.[33] Hoyle replied that he expected a steady transfer of physicists into astronomy over the next ten years. 'In the UK we are trying to set up an organisation to assist this process, but it is uncertain as to whether this can be done in the face of opposition from the traditional astronomical establishment. Until this question is resolved, it is not easy to determine what the prospects of future employment will be.' However, Hoyle encouraged Winbow to consider his knowledge of quantum theory as a major asset, and added that 'undoubtedly there is a connection between cosmology and high energy physics. This would seem the best point of attack for a theoretical physicist.'[34]

The Committee could still not find a consensus even as the report was being written. With Hoyle and his expatriate friends on one side and the Astronomers Royal on the other, the rest of the committee struggled to make themselves heard. Lovell, for example, wanted the report to focus on the one important point on which they were all agreed: that British astronomers needed a good telescope under clear skies. Lovell was President of the Royal Astronomical Society at this time, and was not prepared to endorse any proposal that would mean a decrease in funding for the Royal Observatories; he thought it should be possible to establish a national centre without compromising the Observatories.[35] But Hoyle was committed to the dissolution of the Royal Observatories, and saw the Review as his means to destroy them. Hoyle had an isolated location in mind for the new centre, and Sargent called it 'Marston Moor', after the open countryside in Yorkshire where Oliver Cromwell defeated the Royalists during the English Civil War.[36]

The result was that two reports were produced, after much editorial and diplomatic intervention from Hosie: the majority report, from the Hoyle camp, outlining the plan for the national centre, and the minority report – Appendix 1 to the majority report – from the Astronomers Royal objecting to it. The majority report, was, according to Lovell, 'iconoclastic to say the least. In essence, this report . . . recommended the dissolution of the Royal Greenwich Observatory.'[37] Its main recommendations were for the development and funding of exchange and cooperation between physics and astronomy; for the establishment of a good telescope in a suitable climate, i.e. outside the UK; that the disproportionately well-funded Royal Observatories should compete alongside the universities for Science Research Council funding, with allocation to be decided on scientific merit; and that a management committee for the new national centre be set up immediately.

The minority report questioned the wisdom of placing astronomy in a national centre which would be in effect outside the university system as a 'monolithic Government institution.' It also objected to the characterization of the Royal Observatories as aloof from university research, and questioned the right of the Committee to make decisions about the future of the Observatories without reference to their Council (Woolley was due to retire, and was uncomfortable about making radical changes at the end of his term). The minority report concluded that 'We are strongly opposed to the creation of a new [Science Research Council] establishment which we consider wholly unnecessary and against the best interests of British astronomers.'[38] Hosie encouraged Hoyle to submit the reports swiftly, as a press conference was scheduled at the Science Research Council in a matter of days at which its chairman, the theoretical physicist Brian Flowers, would present the annual update. But the report and its appendix were never published. The issue did, however, become a matter of public discussion. According to Burbidge:

> In the middle of the shambles that summer, the refusal of the Brits to pay any attention to these radical recommendations and so on, I wrote a Letter to *The Times* in which I attacked the system head on, and they published the letter of course. I wrote several letters to *The Times*, but the thing that really upset [people] was the thing I wrote for *Nature* in which I attacked the system and told them that there was no future in the way they were going. To give you some flavour of it I said the astronomical climate and the people were basically third rate. Wal Sargent was called by the *Telegraph* or somebody to comment on this and he said no he didn't agree with Geoffrey Burbidge, they weren't third rate, they were fourth rate. It was that kind of a level.[39]

Journalist John Maddox, now the editor of *Nature*, asked Burbidge for a copy of the report so that its recommendations could be published there, but Burbidge did not provide one.[40]

One of Hoyle's aims for the Northern Hemisphere review had been the abolition of the post of Astronomer Royal. He was not alone in this ambition: the chairman of the Science Research Council felt the same way. Lovell had approached Brian Flowers, who was a colleague in the Physics Department at Manchester, on behalf of the Science Research Council when the previous chairman, Sir Harry Melville, was about to retire, to sound him out about taking on the job. According to Lovell:

> I went into Brian's office and to my amazement he neither said yes nor no: what he said was unforgettable. He said, 'well, if I did accept that position, my first job would be to get rid of the Astronomer Royal.' There was a jealousy of the physicists against the astronomers who had had this close link with the Astronomer Royal, whereas the physicists had no such higher echelon to aspire to.

Flowers did take on the chairmanship of the Science Research Council, and some years later he explained his remarks about the Astronomer Royal to Lovell:

> What concerned me was firstly, that astronomy was wider than the optical branch alone, and was becoming more so every day; secondly, that the leading astronomers seemed unlikely to be in the optical branch for some time to come; and thirdly, that we had to have someone to lead and manage the RGO in its optical work who did not necessarily need or deserve any special title other than Director. If there was to be an Astronomer Royal, he or she had to be someone the astronomical community, the scientific community more widely, and the scientifically informed public could feel was outstanding among that generation of British astronomers generally.[41]

As Woolley's retirement drew closer, the Science Research Council, now under Flowers' chairmanship, undertook to nominate the next Astronomer Royal. Since taking over responsibility for the post from the Admiralty in 1965, the civil servants at the Science Research Council had been trying to find out exactly how the post was constituted, even at one point referring back to records of the purchase of the manor of Greenwich by Humphrey, Duke of Gloucester, in 1423.[42] Now, with the outcome of the Northern Hemisphere Review in limbo, the future seemed as murky as the past, and it was difficult to see what sort of job the new person might be expected to do. According to Hosie, they faced a 'chicken and egg' situation. Eventually they decided to find Woolley's successor, and then see what course of action the new appointee favoured. Flowers circulated a letter to observatories and university science departments, asking for advice as to who should succeed Woolley. But the letter mentioned only the post of Director of the Royal Greenwich Observatory – it contained no reference to the post of Astronomer Royal. Many of the replies nevertheless assumed that the vacancy was for an Astronomer Royal who would run the Observatory, and made recommendations accordingly. However, some respondents took the opportunity to argue for a discontinuation of the connection between the two roles, or for the abolition of the post of Astronomer Royal itself. Donald Blackwell of the University of Oxford Observatory said that the younger generation of astronomers in particular thought that the post of Astronomer Royal should be abolished. Hoyle and Tommy Gold both wrote to Flowers to argue that the next Director should not also be Astronomer Royal.

Flowers' survey resulted in a short-list of ten candidates, and Flowers undertook to visit interested parties in America and Australia. The result was that the Science Research Council recommended the appointment of the distinguished Caltech observer and quasar expert, Maarten Schmidt. Schmidt was a popular choice: genial and respected, and, according to Lovell, 'an outstanding astronomer ... a very good appointment.'[43] Schmidt came to Britain and saw that proper arrangements could be made for his family, and then, as Lovell recalled, 'the bombshell dropped.' The confusion over the relationship between the Directorship and the post of Astronomer Royal was brought into sharp focus by the fact that Maarten Schmidt was a Dutch citizen:

> ... quite reasonably, as every director of the RGO for 300 years had been Astronomer Royal he naturally assumed that he would become Astronomer Royal. The question then got as far as the Palace – the appointment of Maarten Schmidt as director of the RGO to succeed Woolley was passed to the Prime Minister who then sought by the usual process to get permission of the Queen. I don't know which politician got involved in this ... but it was pointed out ... that there was an Act dating from 1700-and-something which forbade the appointment of any foreigner to hold an office of trust under the sovereign. [The appointment] was overruled. And Maarten Schmidt said right, if I can't be Astronomer Royal, I'm not going to come, and that was that.[44]

Schmidt did not want to step into a role that was being undermined from the start. As he wrote to Lovell:

> ... the separation made the offer less attractive. Coming in as an outsider to take a post as important as that of Director of the RGO is a difficult and risky venture, and it seemed unfortunate to have the stature of the directorship diminished by the separation at this particular time.[45]

That left the other nine candidates for the Directorship. Some were crossed out as being so unlikely to accept that it was not worth meeting with them. That reduced the list to four: Gold, Hoyle, the Australian Robert Hanbury Brown and Margaret Burbidge.

Lovell thought the politicians had made a mistake on the nationality issue: the Act they cited had been rescinded in the nineteenth century. But it was enough at the time for Gold to pull out. The civil servant who flew to the USA to talk to him gave him the impression that the separation of the Directorship from the post of Astronomer Royal was to allow foreign candidates for the Directorship (Schmidt had wondered about that). Gold wanted to know who was in line for the post of Astronomer Royal; the civil servant would not tell him. Although Austrian by birth, Gold was a naturalized British subject, and his election to the Royal Society had been as Fellow, not as a Foreign Member. He found the business 'absurd and somewhat offensive and [I] declined to have any further dealings about it.'[46]

Hoyle had already said he would refuse the post (it might be asked who the five candidates were who were thought less likely to accept than Hoyle); and Hanbury Brown said he would only accept if the job included the role of Astronomer Royal. That left Margaret Burbidge. She and Geoffrey had criticized the Royal Observatories so severely, and she now had the opportunity to run one – a situation in which she might be expected to make some radical changes. She told the civil servant who approached her about the job that she hoped that suitable arrangements could be made 'for her to have the full collaboration and support of her husband in the running of the Observatory and in the development of her plans for the future of astronomy in this country.'[47] After some assurances from Flowers, Burbidge organized a leave of absence from the University of California and took up the post of Director of the Royal Greenwich Observatory in October 1971. She was the first Director of the Royal Greenwich Observatory not to be Astronomer Royal. Just as Schmidt wondered whether his nationality might have been used as a lever for separating the two posts, so many people who were not privy to the discussion within the Science Research Council have since speculated that Burbidge's gender might have been the reason for the separation in her case. But clearly the separation was under way long before Burbidge's name and gender entered the story. The post of Astronomer Royal went to Martin Ryle.

Lovell saw the guiding hand of Brian Flowers in this change, rather than of Hoyle and his Northern Hemisphere Review. 'All this fitted in with the wishes of the chairman of the research council,' he said:

> Whether if it had been another chairman, the Director would have continued to be Astronomer Royal, I don't know. The important point is that the two posts, although they had been similar for 300 years, in principle were entirely separate. This separation was acknowledged in 1970. . . . if you look back on the history of astronomy and what happened in the diversity of researches in astronomy, it was a proper decision to take because radio-astronomy, space and so on became of overwhelming importance particularly in Great Britain, compared with the optical situation. So that was the position in which Martin Ryle was appointed Astronomer Royal.[48]

Only one person since then has been both Director of the Royal Greenwich Observatory and Astronomer Royal, and that was Francis Graham-Smith. But he did not hold the posts simultaneously. By then, he recalled, 'the

Observatory and the office of Astronomer Royal had diverged, there was never any suggestion that they would ever come together again.'[49]

Once Margaret Burbidge was in post at Herstmonceux, she responded to a request from Flowers to give her views on the possible publication of the Northern Hemisphere Review. Burbidge was adamant that the report should be published. It was the result, she told Flowers, of 'much hard work by a group of first-class scientists'; it had undertaken exactly the kind of consultation that was now being proposed in order to solve the on-going problem of the lack of good observational facilities in the northern hemisphere; and there was much speculation in the astronomical community about its contents, given that it had apparently been suppressed. Burbidge felt that a frank discussion among astronomers could not happen unless the report were made public.[50] Flowers did not take her advice, and the report has stayed in the private papers of the participants ever since.

Burbidge's time at Herstmonceux was short and unhappy. She tried, encouraged by the more adventurous of the Observatory's staff, to arrange for the Isaac Newton Telescope to be moved to a better climate, but much opposition was raised by the older staff, who recruited prominent local people including Quintin Hogg, the Lord Hailsham, to their cause (after a brief spell as a commoner, Hailsham had been returned to the peerage by the new Conservative government).[51] Burbidge had expected some support from the Science Research Council, and she felt she did not get it: according to Geoffrey Burbidge, much of what had been promised was not delivered.[52] In August 1972, she told Lovell that 'at the moment I cannot see much point in Geoff and I having left California.'[53] Geoffrey did not feel welcome among the more conservative astronomers there,[54] and he found it hard to live in Herstmonceux Castle, with its rather stuffy traditions and restrictive rules. Their daughter Sarah, now a teenager, was appalled to find she was expected to wear a uniform at her English school.[55] Geoffrey and Sarah went back to the USA, and Margaret followed them when she resigned the post at the end of 1973.

None of the political pressures on Hoyle was enough to keep him from publishing fiction. In *The Molecule Men*, which appeared in 1971, an alien invasion comes, as in *A for Andromeda*, in the form of biological information, which takes material form as animals, plants, and a man called R.A. Adcock who, by taking on various identities, commits crimes to fund his entrée into life on Earth. *The Molecule Men* was dedicated to the Hoyles' friends

'J.B. and Jacquetta': the novelist J.B. Priestley and his wife, who told Barbara they were 'touched and proud to have it dedicated to us.'[56] In the novel, Adcock is arrested but makes his escape during a court appearance by turning into a swarm of giant bees. His eventual triumph, however, is to invade human beings who, despite no change in appearance, are transformed into an alien race: the Molecule Men. The strange case of R.A. Adcock is deciphered by Cambridge chemist Dr Joe West, who nevertheless does not avoid becoming a Molecule Man himself. West attributes the timing of the invasion to terrestrial radio: now that humans are using radio to communicate, and since radio waves can travel very far out into space, alien intelligences now know that we are here. Despite teaming up with a very resourceful journalist for part of his adventure, West mocks the press for being able to fake sensational stories but not being able to recognize them when they actually happen.[57] West also has strong views about Cambridge:

> Cambridge is the most depressing place on wintry days. The east wind slices in across the open fenland turning the portals of learning into cold icy stone. In fact weather conditions are very similar to the feelings that run like hidden rivers under the placid exterior of the city.[58]

So says the fictional Dr Joe West. At the time of writing, the atmosphere for astronomy in the real Cambridge had become decidedly chilly. Dramatic events were in prospect.

In February 1970, the Council of the School of the Physical Sciences reviewed the state of astronomy at the University. Its report addressed two needs: to boost observational astronomy so that it might continue to support the excellent work of the radio-astronomers and the theoretical astronomers; and to find ways of bringing suitably qualified graduate students into observational astronomy. With the financial future of IOTA uncertain, the Council proposed a merger of radio-astronomy, theoretical astronomy and observational astronomy into one unit. Redman, the Director of the Observatory, was due to retire, and his Chair would be filled by the new Director of the merged institution. The appointee should be someone with an interest in optical instrumentation – Hoyle underlined this item in his copy of the report. Also marked are the recommendations for a Coordinating Committee for Astronomy, to oversee radio-astronomy, observation, and theory (including work done in the Department of Applied Mathematics and Theoretical Physics), and the proposal that 'administrative steps should be taken to combine the Institute of Theoretical Astronomy and the Observatories into a single centre with direct Science Research Council representation and a combined budget and establishment.'[59]

Although an eventual merger had been mentioned by the Science Research Council when IOTA was founded, Hoyle found much to disagree with in this report. He did not object to observation and theory being managed together, but he thought the Director should be in charge, and not subject to a coordinating committee. As for the emphasis on instrumentation: this was only the preparation for astronomy, not astronomy itself; and it would surely be far better to appoint an astronomer to the Chair. According to Hoyle, the UK needed observers, not more instrumentation people, of which there were many. Hoyle thought that the professors of observational and theoretical astronomy could take turns at being Director, so that 'it should not be another headless Cambridge organisation'. He recommended that they should wait and see what the Science Research Council had in mind for the funding of IOTA, and in the mean time they should think up a decent plan for bringing together astronomers in Cambridge. Hoyle then applied for leave of absence for the academic year 1970–1971.[60]

By now IOTA staff were very concerned about their future. Peter Strittmatter asked for the proposed merger to be on the agenda at the next staff meeting, and it was a feature of those meetings over the coming months. A survey of reading habits was conducted in IOTA's library to see which journal subscriptions could be cancelled. Hoyle was trying to raise money, but with little luck: in February 1971, the Ford Foundation declined his request for $400,000 to fund the visitors programme.[61] The Conservative Government had appointed Margaret Thatcher as Minister for Education and Science, and in May 1971, the Science Research Council wrote to all grant-holders asking them to think about how they might generate money-making spin-offs from their research. The prospects for theoretical astronomers to rise to this challenge were bleak. Another blow came when Hoyle's ally at the Science Research Council, Jim Hosie, was promoted out of the astronomy division. In June 1971, Hoyle recommended to IOTA's executive board that it agree to the proposed merger .[62] By the end of the year, IOTA's Secretary Peter Vaugon had submitted a request for five years' funding: he asked the Science Research Council and the University to share costs of £1 million.[63]

While the future of IOTA was quietly spiralling out of Hoyle's control, he was informed that he was to receive a knighthood in the New Year's Honours List of 1972, for services to astronomy. As a young man during the war, Hoyle had told a colleague that 'if ever they offer me a knighthood,

I'll turn it down. By accepting it, you put yourself in the same social class as treacle merchants.'[64] But he did accept the award – a sign to some that he had mellowed, or perhaps that he had taken Barbara's feelings into consideration: she would henceforth be 'Lady Barbara Hoyle'. His colleagues in California teased him gently in their congratulatory telegram:

> There was a young man from Yorkshire
> Who appeared to many quite cockshire
> His theories abnormal,
> Are now more conformal,
> And so the Queen dubbed him Fred Hoyle, Sir.[65]

14
The beginning of the end

Resignation, The Inferno, *the Anglo-Australian Telescope*

T he future of IOTA hinged, for Hoyle, on the appointment of the right person to Redman's Chair of Astrophysics, as this person would be overall head of the merged institution. The advertisement for the post specified an observational astronomer with an interest in instrumentation, and Hoyle had a candidate in mind: Wal Sargent. Sargent had made valuable contributions to astronomy in California, and had been an ally for Hoyle on the Northern Hemisphere Review. He had a fine reputation and the required interest in instrumentation, and Hoyle felt that he would be a major force in developing optical astronomy in Britain.[1]

Sargent was aware of Hoyle's plans for him. He had been at IOTA as a summer visitor every year since 1968, and it was clear that the quality of theoretical work in Cambridge was much higher than the quality of the observational work. He had seen the merger coming. In the summer of 1971, Martin Ryle, who was one of the Electors who would appoint Redman's replacement, invited Sargent to dinner at his home. Also at the dinner was physicist Brian Pippard, from the Cavendish Laboratory, who was overseeing the merger plan. Sargent took the occasion to be his job interview, and he left Ryle's house feeling that it had not gone well.[2]

But Sargent was on the shortlist for the post, and the Election was to take place in November. Hoyle went to Caltech in October, and returned in early November to find that there had been some changes to the committee of Electors of which he was a member. Now, alongside Ryle and Richard Woolley, were Pippard and the radio-astronomer Francis Graham-Smith, Ryle's brother-in-law, who had left the Cavendish for Manchester University's radio-telescope at Jodrell Bank in 1964. Graham-Smith's boss, Hoyle's friend Bernard Lovell, found that he had been removed from the committee – a circumstance that emerged only when Lovell found that Graham-Smith was attending a committee meeting to which he had not been invited. Hoyle felt as though a consortium of his enemies had been lined up to vote against him, and indeed his candidate did not win: Redman's chair in Astrophysics was awarded to Donald Lynden-Bell. Lynden-Bell was a former member of the Department of Applied Mathematics and Theoretical Physics, and was now at the Royal Greenwich Observatory at Herstmonceux, where he had done important work with Alan Sandage and Olin Eggen on galaxy formation. Hoyle had mentioned him approvingly during the Northern Hemisphere Review, and, in 1970, had recommended him strongly to Roger Tayler for a visiting chair at Sussex.[3] But Lynden-Bell was neither an observer nor an instrumentalist, but a theoretician.

Shortly after the election, Hoyle was travelling to Australia via La Jolla with Jim Hosie from the Science Research Council, on Anglo-Australian Telescope business, and on the way they discussed the future of IOTA. Hoyle told Hosie that although a theoretician had been appointed to head the merged institution, he would expect to retain control of IOTA. Hosie was surprised: he told Hoyle that 'Cambridge' had told the Science Research Council that Hoyle did not want to be in charge there any longer, and that they had asked for the Council's approval for the appointment of Donald Lynden-Bell as Director, which had been given.

When they arrived in La Jolla, Hoyle went to the Burbidges' house, where his old friends were shocked at how disturbed he was by this news. Margaret Burbidge recalled that she had never seen Hoyle really angry, 'but I did then.'[4] Hoyle saw the appointment as a grave affront:[5] not only did Lynden-Bell not fit the terms of the advertised post, but Hoyle had not been consulted or kept informed during the appointment process. He walked up and down the Burbidges' kitchen all night, and, still jet-lagged, he decided to resign the Plumian Chair and leave Cambridge. According to Geoffrey Burbidge, Hoyle's resignation was inevitable: 'it was going to happen from the time that Institute was first set up, there's no question . . . that was a piece of very nasty politics.'[6]

On 14 February 1972, after securing some funding for IOTA, Hoyle wrote his letter of resignation to the Vice-Chancellor, citing the 'recent appointment to the Chair of Astrophysics' as informing his decision:

> It is unnecessary to spell out the reason for my resignation since the reason will readily be understood by the whole astronomical world. Here I would simply point out that for a substantial block of electors manifestly to have discussed a

name in advance without reference to me, the present Head of IOTA and the senior professor in the proposed new Institute, was a truly monumental discourtesy.[7]

Hoyle copied the letter to Brian Flowers, Lord Todd, Alan Cottrell and Bernard Lovell, and also to Hosie and Lynden-Bell. He enclosed a statement he had written on the same day, to the effect that when IOTA was founded, there had been an informal understanding that its future would be discussed in 1970. This discussion had not taken place because the University had become concerned about the future of the Observatories. Hoyle had cooperated with the merger plan 'because I had no wish to see optical astronomy die in Cambridge. I did so in the expectation that the University would seek to appoint an astronomer with a distinguished record in optical astronomy to the Chair of Astrophysics.'[8] The statement continued:

> It has become clear over recent months that there are forces within the University who have seen the arrangements for the new Institute as an opportunity to downgrade IOTA. [IOTA] emerged in the years 1969–71 as an organisation with an international reputation which some people have found embarrassingly high. The present arrangements will ultimately imply a downgrading from real international status to the level of an ordinary department of astronomy.

Hoyle then wrote to Barbara, using, as usual, her pet name 'Rabbit'. He enclosed copies of his letter of resignation and the accompanying statement, and told her that:

> This has left me no alternative but an immediate showdown. . . . I wrote a letter to the Vice-Chancellor stating unequivocally that I would not stay on except as Head of the Institute. I have slept on this letter for two nights now and have become quite dissatisfied with it. My determination is consonant with the showdown being much more on a national scale. Moreover the disadvantage of such a position is that if Cambridge were to call my bluff, an actual resignation would then have a touch of pettiness in it. Certainly it would lack the explosive quality of the documents I have actually sent out. . . . .
>
> There is also the more personal point; that having taken the crucial step I am now much better able to make enquiries about other possibilities. I ought not to let slip the opportunity of this trip, both in the U.S. and in Australia. Sorry indeed, but there really is no other way to play the cards in this situation.
>
> Love Fred

Hoyle added a postscript asking Barbara to bide her time until he came home: 'let this thing wind around until I return', he wrote, 'I promise you the best roughhouse you have seen.' He blamed 'backstage discussions between Hosie and Flowers and half a dozen people in the University. These people will not be winkled out except by a full-scale explosion.'[9]

Lovell hoped Hoyle might find a way to stay on in Cambridge, but Hoyle was adamant:

> I do not see any sense in continuing to skirmish on a battlefield where I can never hope to win. The Cambridge system is effectively designed to prevent one ever establishing a directed policy – key decisions can be upset by ill-informed and politically motivated committees. To be effective in this system one must for ever be watching one's colleagues, almost like a Robespierre spy system. If one does so, then of course little time is left for any real science.[10]

Lyttleton wrote to Hoyle from Pasadena, saying that he had offered to take on the Directorship in a caretaker capacity until the many uncertainties of the situation had been resolved, but that Lynden-Bell had been appointed anyway: 'this seems to bear out completely the view that LB was told from the outset that he would be the Director and all his actions have been consistent with that.'[11] Some colleagues thought that Lynden-Bell had been used as a means to punish Hoyle; and others saw the appointment as a victory for the 'old-school-tie, minor-public-school set' over the 'grammar school brothers.'[12]

Lynden-Bell replied immediately on receipt of his copy of Hoyle's letter of resignation, expressing surprise at his own appointment, and even more surprise at Hoyle's resignation. He had been looking forward to 'some better understanding of [Hoyle's] unique mind by a few years of close contact.' However, with Margaret Burbidge taking over at Herstmonceux and the prospect of a national centre, 'I can yet see good coming out of a strained situation.' Lynden-Bell took issue with some of Hoyle's criticisms of the Royal Observatories, but promised what support he could from his new position to encourage their revitalization with good students and fresh ideas. He continued:

> For Cambridge your decision is sad indeed and I will find my job all the harder for it. I am sorry that you clearly think that the best man did not win the election but I shall regard it as my duty to prove you wrong in this respect. Meanwhile I feel that you have not acted unreasonably given that you hold those views. I naturally hope that you will find a way to reverse your decision.
>
> Since our personal relations have been pleasant but not at all deep may I propose that in the near future we spend a week or at least a few days walking in the hills together – at present neither of us knows the other and I think this is a pity. I would like you to know that in those far off days at D.A.M.T.P. I was one of those who voted for the good of astrophysics rather than for mediocre

fluid dynamics. Like you I am concerned about the low quality of British Observational Astronomy. Like you I believe that concentrations of theorists working beside each other are essential for good theory, like you I am now finding the turnover of university administration is abominably slow. Just as I have been discontented with the second class standard that pervades some of the thought [at Herstmonceux], . . . so I will be striving to raise the Cambridge Observatories from their present state to one in which members can look any man in the eye and be proud of it.[13]

Lynden-Bell added a postscript congratulating Hoyle on his recent Presidential Address to the Royal Astronomical Society, and raised some technical questions regarding the gravitational constant. Lynden-Bell would later tell colleagues that Hoyle's behaviour towards him at this time had been very unpleasant, but this did not become public at the time.[14] They were still in correspondence that summer, when Lynden-Bell told Hoyle that he 'had nothing to do with this', and pointing out that Hoyle had consistently expressed the intention to retire aged 60 – in three years' time – in any case.[15]

Redman wrote to Hoyle that he was 'very sorry about the misfortunes of recent months and hope that the recent publicity can now be left to die down, since it is not doing astronomers, astronomy or the university any good.'[16] Press coverage had indeed been intense from the moment Hoyle announced his resignation. The *Sunday Times* had swiftly reported that 'Sir Fred Hoyle, Britain's best-known astronomer' had said 'I may be an awkward customer . . . but I can't stand bureaucracy.'[17] The *Daily Telegraph* speculated that Hoyle would take a job in California, and the next day announced that 'Hoyle denies he is taking a job in California'.[18] *The Times* reported strong support for Hoyle in the astronomical community, but failed to find anyone who would be quoted;[19] however, a couple of days later the Vice-Chancellor of Cambridge University, William A. Deer, who had been a member of the General Board's sub-committee when IOTA was set up, wrote to *The Times* explaining that Lynden-Bell was perfectly well equipped to 'develop an observational programme commensurate with the university's existing strength in theoretical and radio-astronomy', and pointing out that the post was advertised as being for someone who would foster instrumental development, and not necessarily for someone who was an instrumentalist.[20] Pippard, another former member of the General Board sub-committee, wrote to the *Guardian* to protest that its 'explanation of Sir Fred Hoyle's resignation as the consequence of Cambridge's extraordinary mediaeval politics needs correction.'[21] Pippard pointed out that not only had Hoyle agreed to the merger, but it had been he who had had persuaded the Electors away from their original intention to appoint an instrumental-ist. Hoyle told Lovell that the newspaper reports and letters 'contain inac-curacies that must have been deliberate, particularly the one from Pippard. This has convinced me that there is little point in trying to ease the situation here.'[22]

The scientific press was dismayed by Hoyle's departure. *Nature* devoted two pages to the story, commenting that 'the unhappy tale of Professor Fred Hoyle's departure from Cambridge . . . will leave a scar on British astronomy and will not conspicuously enhance the reputation of the Uni-versity of Cambridge.'[23] It congratulated Hoyle on his work on the forma-tion of the elements in stars, and his 'provocative and stimulating' work in other aspects of astronomy. It implied that there was still a chance that steady-state theory might turn out to be correct; and congratulated Hoyle both on the establishment of IOTA, and on his committee work that had brought many benefits to the astronomical community. *Nature* continued:

> The temptation to ask who is to blame should be suppressed, but . . . [t]he University of Cambridge is . . . directly implicated. . . . it should surely have been possible for somebody to have recognized that too methodical a grinding of the committee machinery might drive Professor Hoyle away.

A news article in the same edition of *Nature* reported the late changes to the committee of Electors, and noted that these changes had prompted charges that the appointment of Lynden-Bell was 'at best mishandled and at worst, underhand.'[24] The cover story of this edition of *Nature* was about radio sources, and the paper to which it referred was written by Hoyle's rivals in the radio-astronomy group at the Cavendish.[25] One of these, Antony Hewish, later commented that Hoyle's resignation was:

> Absolutely not necessary. It was a bit like a spoilt child, he just couldn't get his way . . . his protégé for the chair at the Institute didn't get the position and Fred was incensed. He had more or less already offered it to [Sargent] and that didn't turn out and naturally he was annoyed. But you can't expect election com-mittees to do what you say, they are there to put the best man in the post. Fred was very influential in getting the Institute going, but he didn't run it in the way that most people run their groups, he just wasn't there. So if things happen when you're not there, to some extent it's your fault isn't it? If you're not around talking to people, saying look, I think so-and-so for the following reasons, that's how it is. It was a pity, he was a very bright man to have around.[26]

Watching the process was IOTA member Sverre Aarseth, who agreed with *Nature* that the bureaucracy had beaten Hoyle:

He just didn't like committees, he wasn't a committee person. People who take it seriously, they prepare before going to committee meetings. He just went in there cold and that was why in the end he just blew his top and said 'enough is enough. They don't pay attention to what I say so why should I be doing this when I can just go off and be a free man.' I don't blame him.[27]

Throughout the spring of 1972, Hoyle was deluged with messages of support. His friend Grace Hubble, Edwin Hubble's widow, told Hoyle that his resignation reminded her of a line from Shakespeare's *Coriolanus*: 'I banish you, there is a world elsewhere'. 'Quite right too', she added, 'when he is confronted by tiresome inferiors.'[28] American summer visitor and climbing companion Don Clayton wrote that, just as Hoyle had not known whether to congratulate or commiserate over Clayton's divorce, so Clayton did not know what to say about Hoyle's resignation. He suggested that they should take a climbing holiday.[29] A postcard for Hoyle arrived at the Royal Astronomical Society from 'wellwishers' at the Radcliffe Observatory in South Africa, which read 'well done Sir Fred, show them which way is up (↑)'. The grisly scene on the card was captioned 'leopard with kill'.[30]

Among the many letters offering professional opportunities was one from Donald Blackwell, of the Oxford University Observatory. He wrote that 'given money there would always be a place for you here', and, after reading the coverage in *The Times*, he wrote again urging Hoyle to withdraw his resignation.[31] Historian Asa Briggs had succeeded John Fulton as Vice-Chancellor at Sussex, and clearly there were no hard feelings: Sussex had recently recruited Martin Rees from IOTA, and Briggs told Hoyle that 'Rees, Tayler and I want you here.'[32] Neville J. Woolf, Director of the University of Minnesota Observatory, wrote, with reference to Hoyle's knighthood, 'I thought you were supposed to receive the flat of the sword on your shoulder, but this seems the sharp end in the back.' He offered a visiting post for a month or so.[33] Peter Strittmatter had accepted a post at Kitt Peak, and he invited Hoyle to visit there. Strittmatter was dismayed at the media coverage of the resignation, and assured Hoyle he would be writing to *The Times* about it.[34] For his birthday in June, the staff of IOTA sent Hoyle a telegram: 'To Sir Fred, our knight in exile, IOTA sends hearty birthday greetings stop please return to leaking ship.'[35]

Hoyle also received letters from some of his many fans. Mrs Warren C. Thompson, whose son was planning to study astronomy, wrote from the Monterey Peninsula to tell Hoyle that 'Californians are hoping you choose us!' She included some postcards and a tourist guide showing the redwood trees and Pacific coastline – 'lures to distract [Hoyle] from other tempting baits.'[36] But despite the many rumours and speculation in the press that

Hoyle was destined for California, not all Californians were hoping Hoyle would choose them: some took the opposite point of view. In March, Fowler had written in confidence to the Division Chair for physics, mathematics and astronomy at Caltech, Robert B. Leighton, and to the Provost, Bob Christy, who was also a physicist, to say that Hoyle had indicated that he would be willing to take up a post in the USA, which 'would be great as far as I am concerned . . . but we all know there are problems.'[37] The problems were many: Hoyle had annoyed a number of colleagues at Caltech by sticking to steady-state theory despite the observational data, some of it produced at Caltech, that indicated otherwise. According to Kellogg Laboratory physicist Charles Barnes, Hoyle was, despite the great success of B^2FH, 'still tarred a little bit with the steady-state universe idea.'[38] Hoyle's opposition to the idea that quasars are at cosmological distances was another irritation. Barnes 'always felt a bit badly that there was this strong current of negative feeling' and suspected that some of it at least was because 'astronomers tend to have a limited view of theorists mainly because the theorists have been wrong so many times in the early days . . . [Hoyle] was well aware that there was a professional hostility on the part of some of the astronomers here.[39] In 1973, Fowler would tell an interviewer:

> always . . . in the background of Hoyle's relationship with the rest of the astro-nomical community is the controversy over the steady-state theory. There are many astronomers who just congenitally believe in some kind of a big bang universe – as I quite frequently say, the old Bible story – and they just cannot accept the concept of the steady state universe, and the denial of a beginning, as it were. So that always and still is something that marks Hoyle as a very special individual in the astronomical community. . . . Fred is the person everyone thinks about in terms of the steady state theory, because he stuck to it.[40]

Hoyle's long-running opposition to the space programme had also alienated him from many in the astronomical community in the USA, and Caltech had close links with NASA: indeed, the Jet Propulsion Laboratory, where Hoyle had made good use of a computer long before such facilities were available in Britain, was a NASA institution. So although he had some close and supportive friends at Caltech, Hoyle had become a marginal figure there – a 'wild man', according to one Caltech colleague,[41] and some staff actively protested against any possible appointment. Caltech's formal response to the situation (in a memo to which Hoyle's new title 'Sir' was added by hand) was to reappoint Hoyle as a visiting associate for one year from July 1972, at an increased salary of '$3000 [£1200] per month while at Caltech.'[42] Perhaps thinking that Hoyle might be expecting something more than a visiting post, Fowler wrote to Leighton that 'I feel a letter to Fred Hoyle is not required (other than to notify him of the salary increase) and that it would be better if you and/or Bob [Christy] would talk to him and express our hope at Caltech that he will continue to spend some time with us.'[43]

In July 1972 it was widely reported in Britain that Hoyle was to be offered a chair at Manchester University, to work alongside Lovell at Jodrell Bank.[44] Lovell had indeed asked Hoyle if he would like a chair at Manchester, and Hoyle had accepted.[45] The appointment was an honorary one, and carried no salary.

But Hoyle was still in charge of the Institute until the end of the year. He continued to press for it to maintain its semi-independent status in relation to the University, and urged the Science Research Council to maintain the Executive Committee on which it could be represented, rather than leaving management to the University. He also argued for the post of Director of the merged institution, which was to be known as the Institute of Astronomy, to rotate among all relevant professors.[46] However, Hoyle had to report to the Executive Committee that there were no salaries available for next phase of the Institute, 1972–1977. He said that this was because the decision about the Institute's future within the University had been made so late that he had been left no time to put together the relevant funding application, but he had asked the Science Research Council to provide salaries for a one year 'holding operation'.[47] Lynden-Bell had however surveyed the staffing situation and by April had informed some IOTA members that they were unlikely to be retained. One such was Chandra Wickramasinghe, who contacted Hoyle after receiving a letter from Lynden-Bell 'followed by a very unpleasant interview.'[48] Wickramasinghe was unhappy about the circumstances of his leaving IOTA,[49] but he was soon to find a new position, and with it a promotion: he accepted a Chair at the University of Cardiff, in Wales, in 1973.

Many colleagues were worried about the future for Stephen Hawking. He had joined IOTA in 1968, and his health had been deteriorating. In June 1972 the Master of Gonville and Caius College, where Hawking was a Fellow, confirmed that Hawking would be supported for as long as he was fit to work, and that he would be retired with 'the usual provision' when that was no longer the case.[50] Hawking moved back to the Department of Applied Mathematics and Theoretical Physics in 1973.

Despite the tensions and uncertainties at IOTA, the summer visitors came as usual in 1972. Among them were regulars such as Willy Fowler and Wal Sargent. Of the 17 visitors, 15 were from the USA, one from the UK and one from Austria. The Executive Committee noted the high proportion of Americans and agreed that the Director should in future encourage applications from British and European Universities.

Aarseth organized a boat trip on the River Cam to mark the end of Hoyle's Directorship on 31 July. Hoyle stayed on for a few more weeks to see off the summer visitors, then he and Barbara hitched up their caravan and drove to Cornwall, where they considered their next move. It had been decided that Hoyle would relinquish the Plumian Chair the end of 1972; but instead of coming back for the new term in the autumn, Hoyle decided to apply for unpaid leave from 1 October to 31 December, which was granted.[51] He could not claim his Cambridge pension until his sixtieth birthday, which was three years away, and the position at Manchester carried no salary, so he had to think about how to make a living. He decided he could accept short-term appointments in the USA, and make money there from writing and lecturing. In the autumn of 1972, on a trip to Washington, Caltech and then Australia, Hoyle wrote to Barbara (on the envelope he used her new title, 'Lady Hoyle'), spelling out his engagements and fees on the trip, and assuring her they would be not be short of money.[52]

Hoyle's protracted absences from the Institute meant that his departure, though saddening for some, was not disruptive. Lynden-Bell settled in for a successful term as Director of the new Institute of Astronomy, a job he repeated some years later. While he was more formal in his management style than Hoyle, and far less enamoured of new technology, the staff still enjoyed the adventurous scientific atmosphere that persisted from Hoyle's time. Martin Rees returned from Sussex to occupy the Plumian chair, and he too took his turns at being Director of the Institute. Nor was there jubilation among Hoyle's adversaries: according to Hewish, life continued unperturbed at the Cavendish Laboratory:

> People didn't bother much over here, to be honest. They got on with their thing. We all were very happy with Donald Lynden-Bell. . . . Mostly people thought around here 'what a silly boy [Hoyle was], to behave like that.' But I don't think it had much impact on the way things happened. People didn't think: 'oh Lord, what now?', or even 'good riddance.' He was around so infrequently that it didn't bother us.[53]

Amid the politics of IOTA's demise, Hoyle had found the time to contribute to another novel, co-written once again with his son Geoffrey. In *The Inferno*, a senior nuclear physicist called Cameron is waiting for the government to decide the funding for his particle accelerator project. The decision rests on similar decisions being made by other European partners. Cameron is growing weary of the wider political context intruding into scientific policy, and starts to plan his retirement: 'he'd be content to leave physics to younger men, especially as the bigger half of the problem nowadays was political and financial rather than scientific.'[54] Just as Hoyle worked with his former classmate Jim Hosie on the Anglo-Australian Telescope, so in the novel Cameron works with his old college friend Mallinson, First Secretary at the Department of Education and Science, to help arbitrate in a project in Australia: the British and Australian astronomers involved cannot agree on the design of their new radio-telescope. Cameron travels to the site of the proposed telescope on Mount Bogung at Wombat Springs, and while he is there something strange appears in the sky. It is a supernova, and it is headed for Earth. When Cameron realizes the dreadful implications of this, and he goes back to his native Scotland – a latitude likely to escape the worst of the supernova's effects – and develops a survival strategy. Under Cameron's leadership, the local people establish some sort of order through periods of intense heat, freezing cold, and then gradual recovery. Cameron finds that a radio-telescope in Scotland has been left running throughout the Inferno, and on the computer tapes he discovers some coherent signals. As the story ends, Cameron walks out of a church service in a ruined cathedral; and unlike the rest of the congregation, he knows that the Inferno was not the wrath of God:

> Cameron had never found a rational explanation for the termination of the great heat or for the descent of total darkness. Now he knew the explanation to be rational but not natural. He knew an intelligence, a creature, had intervened at their direst moment. . . .

This creature was of a different order from man, and in the universe there was 'order on order in infinite progression.'[55]

The story of Cameron's trip to Australia was published in 1973, before the grand opening of the real Anglo-Australian Telescope in the autumn of 1974. While the scientific and technical aspects of that project had progressed smoothly, the political aspect had been rocky indeed. The root of the problem was that no institution had been earmarked to run the telescope. To some astronomers, it seemed to be up for grabs. A telescope needs an observatory, and the question had been raised of whether the existing Observatory at Mount Stromlo, which was part of the Australian National University, might take on the responsibility. Its Director, Olin Eggen, had argued that it would be a waste of money to build a new observatory for the new telescope when such a good one was already available. The Australian National University had a strong claim on the new telescope – many of its staff had worked hard on the project – and they could reasonably expect to have earned a sizeable share of observing time. Not only that, but the agreement under which the telescope was being built could be taken to mean that it was a joint venture between the telescope's Board and the Australian National University.

In 1970 Eggen put a paper before the Joint Policy Committee. It was a funny story about the tribulations of some fictional astronomers who were applying for observing time at the Anglo-Australian Telescope. Eggen was as well-known for his outspokenness and jocular manner as he was for his astronomical achievements – most famously his work with Donald Lynden-Bell and Alan Sandage on galaxy formation. He had always been on good terms with Hoyle, with whom he had much in common: he came from a poor rural background, and had worked his way through college as a cock-tail pianist and waiter; and now he combined his astronomy with driving a bright red Austin-Healey sports car and writing science fiction (under the rather transparent pseudonym Nilo Negge).[56] Eggen was one of the colleagues who had spoken up for Hoyle after the humiliation of Ryle's press conference in 1961.[57] But his manner now seemed to grate on some members of the committee, Hoyle included: with his funny story he had intended to make them think about how decisions about such matters might be made, but instead they were offended by it. Ken Jones, from the Australian Department of Education and Science, calmed the situation a little by presenting the same questions in a more formal style. However, both papers took for granted that the Australian National University would be in charge of the telescope, appointing staff and allocating telescope time. Sir John Crawford, Vice-Chancellor of the Australian National University, set about organizing facilities along the lines suggested in Jones' paper. Crawford was an experienced politician who was used to getting his own way, and he was a fervent advocate of Australian interests. He was sensitive to a persistent colonial mentality in both Australia and Britain: Australia's astronomical community was very strong and its achievements many, especially, most recently, in radio-astronomy; but some Australian astronomers felt that the British saw them as the poor relation,[58] and that on the world stage, Australia was in Britain's shadow. Crawford was not the only Australian to suspect that the new telescope could turn out to be more an imperial out-station than a post-colonial collaboration.[59]

Even once the University's role had been described as that of agent to the Board, Taffy Bowen, who was chairing the Joint Policy Committee, thought that most astronomers, Australian as well as British, would be unhappy with the University having a large share in a facility that had been promised to all. The UK members of the Committee were also nervous: if public funds were to be spent on a telescope that was available to all universities, it could not be placed under the control of just one of them.[60] Since the Joint Policy Committee was an interim arrangement awaiting the proper constitution of the Board, it could not make policy decisions without reference to the Science Research Council, so Hosie reported the situation to the Council's Astronomy, Space and Radio Board in the autumn of 1970. The Board decided to canvas the views of likely users of the telescope, and found that astronomers in the UK were also uneasy: they felt that the Mount Stromlo astronomers would be in a very powerful position, and British astronomers would be reduced to guests at their own telescope. An independent Director should be appointed who would report to the Board. The Science Research Council endorsed this, the only dissenting voice coming from Woolley, himself a former Director of Mount Stromlo and an honorary professor of the Australian National University, who felt that Jones' plan could have been the basis for a positive relationship with the university.

There was unease in Australia too, and even in the Australian National University, about what Crawford and Eggen were trying to do in the University's name, and this was exacerbated when the *Sydney Morning Herald* interviewed Eggen at the end of 1970, and described him as 'the man who will be in charge of the new 11 million dollar, 150-inch telescope.'[61] So when the Anglo-Australian Telescope Board was properly constituted and took over from the Joint Policy Committee early in 1971, its chairman, Bowen, was aware of the need for slow, careful negotiation. The University's case was still being championed by Crawford, whose interpretation of the intergovernmental agreement was that the University was a necessary component of the project, rather than just a friendly neighbour. Eggen went so far as to spell out his own qualifications for the job of Director. Surely, they argued, a separate management for the telescope would involve a complicated and expensive duplication of resources. Hosie continued to press the argument that the British did not want to be treated like visitors at a telescope that was as much theirs as the Australians'. He asked Crawford to put the University's case in writing.

That summer, a paper by Eggen was circulated setting out a new plan. While it reduced the extent to which the University had special claims on the telescope, it still sidelined the Board. Another paper emerged from Crawford's office, but it was almost identical to Eggen's. The Board – with Bowen, Hoyle and Hosie against the University's plan and Eggen, Jones and Woolley supporting it – decided to talk once again with potential users and let them advise on how best to sort out the problem. Hosie talked to the UK astronomers and found that their fundamental objection to the Australian National University plan was that the head of the telescope would be junior to the Director of Mount Stromlo – an Australian National University post, at that time held by Eggen. Hosie formulated a principle of cooperation with the University, with some day-to-day management on site by University staff but with control vested in the telescope's Board.

At the end of September, membership of the Board changed. Margaret Burbidge took Woolley's seat when she replaced him as Director of the Royal Greenwich Observatory. Burbidge's views were much more in line with those of Bowen, Hosie and Hoyle: she found the Australian National University's plan objectionable. She thought that the telescope should be run by a Director who was responsible only to its Board. It should have a small resident staff from Britain and Australia who would have a small share of the telescope time, the rest being distributed equally between British and Australian visitors. Once the telescope was showing what it could do, then closer collaborations with other institutions could be considered.[62]

Crawford continued to press the University's case, now with support from Sandage, who was visiting Mount Stromlo from Caltech. Sandage felt there should either be complete integration of the two institutions, or no contact at all. But he reminded Crawford that scientific success had to be the prime ambition for the project, and without that, any other ambitions would be thwarted. The new Australian Minister for Education and Science, Malcolm Fraser, let it be known to the Board that the Australian government had already invested heavily in astronomy at the Australian National University, and it did not intend to spend more money so that the Board could start an observatory from scratch. In the Australian press, reports continued to appear about the marvellous new telescope being built at the University for Olin Eggen.[63]

The Board had had enough. At Jones' suggestion, the members decided to hold a special meeting on neutral ground, and they chose La Jolla, Burbidge's home in California, as a mid-way point. It was on the journey to La Jolla with Hosie that Hoyle first heard of Lynden-Bell's appointment as Director of IOTA in Cambridge. That did not help the mood of the meeting, where tempers frayed. Burbidge's extensive experience of optical telescopes, and her great dignity and impeccable manners, did much to moderate the atmosphere and advance the discussion, but the meeting nevertheless put a great strain on all the participants. For the only time in its history, the Board acknowledged that it would not reach consensus and opted for a vote, and Hoyle, Hosie, Burbidge and Bowen made a majority of four. The 'La Jolla decision' was that the control, maintenance and operation of the telescope would be the responsibility of a scientific director who would be appointed by, and answerable only to, the telescope's Board.

Crawford was furious. He insisted on seeing the Minister. Fraser's career had been turbulent in the preceding months: John Gorton, the Minister under whom the project had started, was now Prime Minister, and Fraser had resigned because he felt Gorton was interfering in his work. Gorton's authority had collapsed, and he had been replaced as Prime Minister by William McMahon, who reappointed Fraser as Minister for Education and Science. Fraser was tall, Oxford-educated and had a patrician manner and a hectoring speaking style that infuriated his opponents. He summoned the Board to Canberra, where Hosie reiterated the British stance, and Hoyle, drained after the tribulations at Cambridge, could not face another fight and tried to lighten the atmosphere with a few jokes.[64] Before they left Canberra in mid-February, the Board members were summoned once again to Fraser's office. According to Hoyle:

> Evening it was, after dinner, the hour of relaxation you might think. But there was no relaxation for Jim Hosie. I could see small beads of sweat on his upper lip. Nor was the meeting an easy one for Taffy Bowen. . . . The other three of us were angry. Malcolm Fraser . . . was angry with the British members of the board – Jim Hosie, Margaret Burbidge and myself. Margaret was angry with Fraser, while I was just generally mad, as I usually am when suffering from jet lag. Besides which, there was no capacity for worry left in me, since I had used up all my store of worry three or four days earlier, when I had resigned from Cambridge University.[65]

Hosie insisted that the UK had a duty to its astronomers which it did not feel could be properly discharged by placing control of the telescope with the Australian National University. Fraser therefore said he was going to go over Hosie's head and discuss the matter with his opposite number in the UK, Minister for Education and Science Margaret Thatcher.

Sir Hugh Ennor, Secretary at the Australian Department of Education and Science and a former director of medical research at the Australian National University, could see that Crawford's case rested largely on his strong bias in favour of the University. Ennor thought that it was time to do as Hosie had done in the UK, and consult Australian astronomers more widely. So in July 1972, he called a meeting in Canberra. The many astronomers who turned up mostly held the University in high regard, and had open minds on the management question. But Eggen gave a speech in which his intransigence and self-interest were clear to all. As a result, the astronomical community in Australia did not give its support to the plan for the Australian National University to control the new telescope, and instead effectively endorsed the La Jolla decision. Crawford told Ennor he now felt that astronomy at the University was under grave threat; but Ennor was more inclined to see the merits of the Board's decision.

In August, Margaret Thatcher travelled to Canberra for a meeting with Malcolm Fraser. She had been asked to explain the situation in Cabinet, and she had turned to Hoyle for advice:

> She came and said look, I have a problem, or may have a problem, with the Cabinet if they say why do we give so much money to astronomy? And so I said well, we know if we do an astronomy spectacular on the BBC the viewing audience is about 10 million, so I think you just say that 10 million viewers is 10 million votes. She said 'say no more.'[66]

Hoyle admired Thatcher: he thought that 'she, as always, did [the job] better than the others.' Thatcher felt that the La Jolla decision had been properly reached, and that any intervention from her as Minister would be improper given the authority of the Board. The two ministers wrote to the Board and the University urging them to deal with their differences and get on with the job. The tensions resurfaced in December 1972, when a new Labour Government, led by Gough Whitlam, was elected in Australia, and Fraser was out of government (a temporary demise: like his predecessor Gorton, Fraser would shortly return to office as Prime Minister). The day after the election Whitlam received a letter from the University, and in January a deputation, including Crawford, came to meet the new Prime Minister. Whitlam passed the matter to the new Minister for Science, Bill Morrison, and Morrison summoned Bowen and told him that the Government expected the University to run the telescope. Bowen reminded the Minister that he did not have the authority to tell the Board what to do. Bowen had recently accepted a very attractive post in the Australian Embassy in Washington and was looking forward very much to being reunited with his American war-time colleagues, and Morrison told him that if he wanted that job he could not also be Chair of the telescope Board. Bowen was shocked, but he was also tired of the wrangling: he wanted the job in Washington, and so he told Morrison he would resign the chairmanship. Whitlam told Morrison to get someone senior from the Australian National University in Bowen's post. What Bowen knew, and the politicians did not, was that the rules for the appointment of Board Chairs meant that the successor to an Australian chair such as Bowen would have to be from the UK team. Thus the person appointed to the chair was Hoyle, who had been asked by the Science Research Council to extend his period of service as chair of Astronomy, Space and Radio in order to keep him on the Board.[67] Hoyle's relationship with the Council had been soured by events at home: 'Quite how the Science Research Council, which could surely have prevented the situation in Cambridge, now had the nerve to ask for my help was, I thought, like the peace of God – it passeth all understanding. Anyway, I accepted the Chairmanship.'[68]

Australian astronomers continued to press for a better spirit of cooperation with the project, and when Morrison called a meeting of senior astronomers in the spring of 1973, he found overwhelming support for the Board's policy. In June, the Australian Government formally endorsed the La Jolla decision. When Hoyle visited Morrison in August, he felt that 'the politics were over at last.'[69]

By now, the Board was looking rather different. Bowen's seat was taken by J. Paul Wild, a radio-astronomer (and Yorkshireman) who had replaced Bowen as Chair of the Radiophysics Division of CSIRO. Ken Jones had gone to Education when the new government had split the Department of Education and Science, and Sir Hugh Ennor, who remained with Science, took his place on the Board. Hosie had been promoted to another department of the Science Research Council – Hoyle had failed to persuade the Council to allow Hosie to see the project through – and his replacement at Astronomy, Space and Radio, Malcolm 'Mac' Robins, a physicist from Harrie Massey's space research group at UCL, took Hosie's seat on the Board. Crawford's term as Vice-Chancellor of the Australian National University had come to an end, and his replacement, R.M. Williams, was able to establish a much more cooperative relationship with the project. In August, Eggen must also have realized that the political battle was lost, for he resigned from the Board, telling Williams that they had sacrificed university astronomy in favour of government astronomy.[70] Eggen did, however, become a key figure in the University's continuing involvement in the project. But this new Board, under Hoyle's chairmanship, was at last able to move forward with the real business of running a telescope.

Despite the political turmoil, the technical aspects of the project had been progressing well. An extended network of astronomers, engineers and technicians had been energetically pursuing the knowledge and expertise around the world to muster a first-class telescope. The mounting, which was made in Japan, had reached Siding Spring in April 1973, and was

assembled over the summer. The primary mirror, which had been manufactured in London, arrived in December 1973 and was given a police
escort, a civic welcome and a lap of honour around the nearby town of
Coonabarabran. Hoyle's enthusiasm for computers was rewarded at the
Anglo-Australian Telescope: it was the first telescope to be designed with
every operation controlled by computer. This was to become a crucial feature in the success of the telescope, and has been influential in telescope
design ever since.[71] The computer was running by January 1974, and the
first observation – 'first light' – was achieved in April.

The Board had been looking for the first director of the telescope
since the autumn of 1972, when it was concerned that anyone with any
sense would be wary of the job while the bad relations with the University
persisted. Indeed, the first two people who were sounded out – Robert
Hanbury Brown from the University of Sydney and the Canadian
J. Beverley Oke from Caltech – had asked not to be considered for the job.
Although the Board's meetings had been private, its business 'had been
systematically leaked for several years past,' recorded Hoyle. 'The leaks
had been obvious at every one of our meetings with the ANU . . . but since
no-one would admit to being the source of the leaks it was hard to do much
about them.'[72] Hoyle was making frequent trips to the USA at this time –
he was the only Board member who did not have a job, and he needed to
earn some money – and in June 1973 he was able to recruit as Director Joe
Wampler from the Lick Observatory in California. Wampler arrived in
September 1974, and his view was sought on a question that was troubling
the Board: where to put the permanent scientific base for the telescope. As
yet, it had only a temporary base in the grounds of a CSIRO station in the
Sydney suburb of Epping, 500 kilometres away from Siding Spring.
Coonabarabran was one suggestion – it was convenient for the telescope,
and given that UK astronomers would have travelled a long way to get
there, they were likely to prefer to stay nearby. Hoyle thought the attraction
of such an excellent telescope ought to be reason enough to live somewhere
remote; but this telescope was very remote, and there was some doubt over
whether staff and their families would want to live nearby. In the 1970s, it
was becoming more common for astronomers to have husbands or wives
with careers, and professional jobs were scarce in Coonabarabran. A base in
Sydney (the temporary location) or Canberra (which was the constitutional
home of the project), on the other hand, would offer a metropolitan
environment with good academic resources. The Board was split once
again, with Hoyle and the UK members preferring Coonabarabran, and
the Australians preferring a city location. Wampler, however, preferred

Sydney, and relations between him and the Board became strained. Worse, the decision would have to be ratified by the research councils in the UK and Australia before Wampler could have his way. The Board eventually agreed to a trial period in Sydney, and the base at Epping was to be maintained until the position regarding facilities at Siding Spring was clearer. (The base has remained at Epping ever since.) But Wampler had not expected, as Director, to have to battle with higher authorities, and he found it hard to fit into a bureaucracy that had been working for some years without a Director. He was offended by meetings with officials who spent more time on lunch and discussing cricket than they did on telescope business. Three months after taking up the post, he resigned 'with a mixture of sadness and relief.' He explained to Hoyle:

> I have tried over the past months to see my role in the AAT organization. I don't think that the Director of the AAT should be placed in the position of continually having to struggle with the Board to achieve his goals. The Board appoints a Director. A struggle between the Director and the Board indicates either that the Board has made a poor choice of Director or that the Board is not correctly performing its function.[73]

Hoyle took up Wampler's offer to stay on until a successor could be found, which took some time: the Canadian astronomer Donald Morton came from Princeton to fill the post in 1976. So Wampler was still in post for the grand inauguration of the telescope in October 1974, when a violent storm threatened the dignity of the occasion: the dash from their cars left the distinguished guests rather dishevelled. ('I noticed', recalled Hoyle, 'that the clever chaps had equipped themselves with combs. I didn't have a comb.'[74]) The Board members were there, along with John Gorton, Bowen, Hosie and Bill Morrison. While a chamber orchestra played, Board members' wives, including Lady Hoyle, showed the guests to their seats in the telescope dome. Prince Charles performed the opening ceremony, regretting his ignorance of astronomy and congratulating 'all those who have worked so hard to set up the whole operation and keep it running, and who must be praying now that the instrument will work when required.'[75] Prime Minister Whitlam then reminded the assembled astronomers, perhaps with reference to Eggen's parting shot, of the long association between the Australian government and the excellent astronomy the country had produced. Hoyle, as Chair of the Board, took a broader view. He said:

> The beauty and grandeur of a large telescope is well matched to the scientific purposes for which it is to be used. Those who step for the first time on the main floor of the Anglo-Australian Telescope building will, I believe, become

immediately aware of an unusual relationship between aesthetic quality and advanced technology. A large telescope is a good example of the things which our civilisation does well.[76]

Three flags flew at the inauguration: the Prince's standard, the Australian flag, and the Australian National University flag. Protocol demanded that the Prince's fly the highest, and the University's the lowest – but since the flagpoles were all the same height, the flags had to tied off at the appropriate level. Twice during the ceremony someone raised the University flag above the Prince's. Each time it was lowered, and on the second occasion the relevant rope was tied off well out of reach. Thus on his third visit to tamper with the flagpoles, Olin Eggen was to find once again that his plan had been thwarted.[77]

Hoyle's work on the Anglo-Australian Telescope earned him much credit among astronomers. The technical staff were impressed by his interest in and ready grasp of their work. He enthused observational astronomers about the project, listened to their needs, and provided them with a first-rate facility. In committee and with the politicians, he explained his position clearly, had a good grasp of the many facets of this complex project, and managed to be both reasonable and forceful when occasion demanded it. According to Margaret Burbidge, his strong influence on the project was not surprising: even in that distinguished company, Hoyle was invariably 'the best brain in the room.'[78]

Hoyle enjoyed the landscape of Siding Spring Mountain, and after the inauguration he and Barbara took a walk on the mile-long ridge:

> . . . the wattles were everywhere in blossom, and there were wedge-tailed eagles gliding overhead. We sat down and looked towards distant horizons, knowing this was the real end point of our former academic life.[79]

15
On the loose

Moving to the Lake District; how to write science fiction; pulsars; controversy over Nobel Prize; Hoyle's 60th birthday

C ambridge no longer felt like home. The Hoyles considered where they might live instead, and settled on a remote house on Cockley Moor near Ullswater in the Lake District. It was the region where Hoyle had courted Barbara in 1939 and where they had spent many holidays since. The *Cambridge Evening News*, the local paper that had followed Hoyle's career, illustrated its report of the move with a photograph of his now empty armchair by the big window – Hoyle's 'window on the world', said the paper – in the living room at Clarkson Close, and pointed out the worn patch of carpet in front of the chair where Hoyle's feet had left their mark.[1] Barbara wrote to their friends telling them of the new job with Bernard Lovell at Jodrell Bank, and enclosing a copy of the newspaper article.[2]

The new house in the Lake District also provided Hoyle with big windows: it was designed to admit as much light as possible, and to take advantage of the dramatic views across the countryside. It came with several acres of land, and the Hoyles set about making a home there. Visitors recall the excellent job Barbara made of transforming the overgrown gardens, and Hoyle would return from long walks in the hills to play his classical music recordings (the television signal was poor there, and they did not listen to the radio). For Hoyle, the scenery, and the freedom from work pressures, evoked the gentler days of his childhood, pondering nature while truanting on the moors.[3]

The Hoyles' new home was close to Jodrell Bank and the honorary post, but according to Lovell, Hoyle only 'used to come here occasionally.'[4] He was still travelling for much of the time, fulfilling his plan to earn money in the USA. Apart from his continuing commitments on the Anglo-Australian

Telescope and at Caltech, he also spent time at Cornell, where Tommy Gold was now Chair of the Astronomy Department, and Carl Sagan, an astronomer who would soon entrance television viewers in the way Hoyle had his radio audience, was the rising star. Hoyle spoke at the dedication of the Kitt Peak National Observatory in Tucson – the model for his proposed national centre for observational astronomy in Britain – and spent time at Rice University in Texas with his long-time colleague and hill-walking companion Don Clayton, an expert on gamma ray astronomy and stellar evolution.

Hoyle had recently published a book about the life and work of Copernicus,[5] and in 1973 he was awarded the Polish Order of Merit to mark the five-hundredth anniversary of Copernicus' birth. As well as an honorary fellowship of St John's College Cambridge, Hoyle was awarded an honorary doctorate of the University of Bradford, a university near his birthplace, Bingley. The award was presented by the Chancellor of the University, the recently returned Prime Minister Harold Wilson.[6]

Hoyle often travelled to London, where he was Vice-President of the Royal Society and President of the Royal Astronomical Society. He was still working in astrophysics: he published a book with former student and IOTA colleague Jayant Narlikar on their gravitational work, entitled *Action at a Distance in Physics and Cosmology*,[7] and he was awarded the Royal Medal of the Royal Society, for 'distinguished contributions to theoretical physics and cosmology.' The President of the Royal Society commented that Hoyle's 'enormous output of ideas are immediately recognised as challenging to astronomers generally . . . his popularisation of astronomical science can be warmly commended for the descriptive style used and the feeling of enthusiasm about his subject which they succeed in conveying.'[8]

A new novel appeared, co-written with Geoffrey: *Into Deepest Space* picks up the story of *Rockets in Ursa Major* where intergalactic warfare threatens to destroy the Earth. Major Dick Warboys and his fellow adventurers travel to edges of the universe dodging all manner of threats. In this novel, quasars are very distant, and, for once, the universe has a beginning, as Dick reports:

> No terrestrial astronomer will ever be able to observe how things began in the fantastic detail that was eventually revealed to us. We saw masses of gas falling together to generate brilliant garlands of stars, which took on delicate patterns, mostly spirals, but with an apparently endless variety of subtle forms. Then there were violent explosions out of which came thousands of millions of stars. Once we had managed to exclude by a filtering process all the nearer more staid galaxies, we had a truly magnificent picture of the 'beginning' laid out before us. Although there might never be anyone to consult them, we taped a vast number

of photographs and measurements of all kinds, and we documented all we had seen in satisfactory detail.[9]

But this was fiction, and in the textbook that would soon follow, Hoyle made no such concession to convention. In *Astronomy and Cosmology: a Modern Course*, Hoyle, giving his affiliation as Caltech, stated that his aim was to bridge the regrettable long-standing division between astronomy and the rest of science: 'Today we see this isolation to be no advantage. The time has come for astronomy to take its place as a major branch of physics.'[10] In over 700 pages, Hoyle took the reader from an explanation of day and night and the seasons, into astrophysics and then on to cosmology. In this last section, Hoyle described the evolutionary cosmology, with its redshifts and cosmic microwave background radiation, and he explained the problems he saw with the conventional understanding of the meaning of these phenomena. At the end of the chapter, Hoyle asked:

> Can the universe in the large really be as simple and as crude as this? The tendency of astronomers is to answer affirmatively, but many would admit causes of disquiet. In the remainder of the book, we will examine suggestions that have been made for changing the picture. . . .[11]

The suggestions that followed, in Chapters 17 and 18, were the steady-state cosmology, and a discussion of some alternative physics, including a discussion of the meaning of the redshift, that would lead one away from the evolutionary cosmology.

So as other scientists settled into the evolutionary paradigm, clearly, for Hoyle, the war was not yet over, and there was plenty of fight left in him. And although he had lost his position of power, he had also been relieved of responsibilities that had constrained him: in later years he commented that 'a senior scientist has a responsibility for 50 or 100 people, and the social and human problems which flow from that put a very strong brake on things'.[12] In some ways, IOTA had been a millstone, and now he was at liberty. It suited him to be on his own. According to Fowler, speaking in 1973:

> Fred is a very independent worker. He, in addition to the collaboration with the Burbidges and me, was carrying on quite independent work in cosmology at one and the same time. Now, he did finally find a collaborator in that area, and for a number of years he has written papers on cosmological problems with Jayant Narlikar, and so Fred has demonstrated that he can collaborate readily with a number of people. But basically I would say that Fred is a lone wolf, . . . Fred has got be in the forefront. He has to be the leader.[13]

In 1974 a *Daily Mail* journalist visited Hoyle in the Lake District, and reported 'The totally happy man . . . Professor Sir Fred, in tune with his mountains.' They discussed astronomy, Russian literature, and over-population, and the journalist remarked on Hoyle's tendency to run against the mainstream. 'I always was a loner', Hoyle said, 'and the hills are the place for it.'[14]

Hoyle was not alone in the hills for long. At the end of 1974 he was off again, headed for Caltech and Christmas dinner at the home of Wal Sargent, then a brief trip to Washington and some Anglo-Australian Telescope business in La Jolla. He chaired a symposium in San Bernardino on 'Man and the Cosmos', and gave a lecture at Caltech on the emergence of intelligence in the universe. He was in distinguished company as a science fiction author at an event at Caltech in February 1975, where he shared the stage with Ray Bradbury. In his talk, Hoyle complained that much science fiction was so formulaic that anyone could write it:

> There's a simple recipe for writing science fiction stories which I am ready right now to give away for free. Start by copying a terrestrial story; however, transplant it to another planet, more or less like the Earth, give humans peculiar names like Atorjak or AJKL, refer to physical devices like the automobile or electric switch or radio by similarly peculiar names, add a few more devices whose mode of operation you don't understand yourself, and you have a science fiction story. This unfortunate thing is I think a fair description of quite a lot of the science fiction books you find if you take the trouble right now to drive down to LA Airport and look in the book racks.[15]

Hoyle also predicted a major collapse of technological and social systems, which would be caused by over-population. Hoyle's rather gloomy prognoses were countered by Ray Bradbury, who had been well known since the early 1950s for works such as *The Martian Chronicles* and *Fahrenheit 451*, novels that explored the unintended consequences of social and technological change. He was just a few years younger than Hoyle and had grown up in Los Angeles, enjoying the relaxed atmosphere of California. He claimed to be fed up with the doom-mongers and preferred to depress people with optimism – optimism, he claimed, is depressing when it is not fashionable. He mocked the proliferation of new religions in the USA by suggesting that solutions to the population problem might be found in some new churches: he proposed the First Church of Christ the Abortionist, and the Church of Guilt Free Screwing. 'If we could just get that going it would go like wildfire across the world,' Bradbury said. For once, Hoyle was the more conventional voice.

When Hoyle was asked about his heroes, he named Beethoven. He explained that when Beethoven was asked as one of fifty contributors to write a variation on a tune for a collective volume, he wrote thirty-three variations. 'It's this sort of driving energy that I regard as the outstanding quality that has to be put into any activity – it doesn't matter whether it's science or music or anything you like,' said Hoyle. He also quoted Beethoven as saying: 'I don't know whether I've achieved very much, but at least I can say that I never had an idea I didn't exploit.' Hoyle concluded that 'To me this is a tremendous lesson – I try in a small sort of way to live up to this.'

While Hoyle was enjoying the limelight at Caltech, Willy Fowler was working behind the scenes to organize a conference to mark Hoyle's sixtieth birthday in June 1975. The event had been some months in the making. The previous summer, Fowler had asked his colleague Charles Barnes, who was to visit the Institute in Cambridge, to assess the likely reaction in England.[16] Fowler had also consulted the new Plumian Professor Martin Rees about the conference, and Rees had advised against holding the party in Cambridge because it would raise too many tensions. At Caltech, some physicists were keen to host the party, but there was, according to Fowler, just 'too much antagonism in the Astronomy Department.' Geoffrey Burbidge agreed with Fowler that Caltech was not the place. Fowler also considered Manchester, Rice, Cornell and Arizona Universities, and encouraged Don Clayton at Rice University to muster an organizing committee.[17] Clayton was shortly to visit Hoyle in the Lake District during his summer trip to the Institute. He replied positively, offering practical help and assuring Fowler that Rice was a possible venue if others with closer ties were unwilling. 'It is ... regrettable,' wrote Clayton, 'that both Caltech and Cambridge, the centers of Fred's intellectual life, find so many personal animosities as to disable wholesome acclaim.'[18]

By February 1975, however, a neutral venue had been found. The conference would be hosted by the International Centre for Theoretical Physics in Trieste, Italy. This centre had been founded in 1964 by the physicist Abdus Salam, who had been a student of Hoyle's just after the war, to provide opportunities for researchers from developing countries to participate in high-level research. It lay just across the bay from Venice, where there was a meeting place in a monastery on San Giorgio Island. The

Centre was holding a summer school on the island, and many of Hoyle's colleagues had already arranged to attend.

So the invitations went out. Ray Lyttleton, writing from the Institute in Cambridge, regretfully declined his invitation in a letter to Geoffrey Burbidge, but he looked forward to another opportunity to discuss '"cabbages and kings" – the former flourishing strongly here.'[19] Martin Ryle was invited, and replied that he was 'delighted to hear of the proposal to hold a conference in honour of Fred's 60th birthday. That is a very nice idea.' Ryle's colleague Peter Scheuer was attending the summer school, and Ryle volunteered him to contribute to the conference.[20] This was not, however, to be Ryle's only connection with Hoyle's sixtieth birthday party. The old antagonism was about to undermine Fowler's efforts.

Ryle's career had brought him many rewards. He had been knighted in 1966, and in 1972 was appointed Astronomer Royal. Then in 1974 he shared the Nobel Prize for physics with his Cavendish colleague Antony Hewish. Ryle and Hewish were awarded the Prize for 'their pioneering research in radioastrophysics: Ryle for his observations and inventions, . . . and Hewish for his decisive role in the discovery of pulsars.'[21] Pulsars, a new type of radio source that emitted regular pulses of radiation, had been discovered by the Cavendish radio-astronomers in late 1967, and their work had been published amid great excitement early in 1968. In the Presentation Speech at the Nobel ceremony, Ryle was also credited with disproving steady-state theory.[22] The news of the award reached Hoyle in the run-up to the inauguration of the Anglo-Australian Telescope.

Hewish went to Stockholm in December 1974 to collect the Prize:

Martin Ryle didn't go. He was of a rather nervous disposition. . . . he loathed any sort of public ceremony. He hated social occasions, and being any sort of display figure. . . . the idea of being part of that Nobel ceremony was just so distasteful, and his health wasn't good at that time. . . . when Ratcliffe was running the group he said he was a bit reluctant to put Martin Ryle forward for the Nobel Prize because while totally deserving he felt that the award might so disturb Martin Ryle as to upset his life. . . . when it came to the crunch, I think Martin Ryle felt that he just couldn't go through with it. . . . it was just felt that it would be just too damaging to his mental state to go to Stockholm – you have to give lectures and meet all the other people. And so I was detailed to do it.[23]

So Hewish's Nobel lecture covered both his and Ryle's work in an hour: the development of the radio-telescopes, the surveys of sources, and the discovery of pulsars.

Although the different specialisms in astronomy in Cambridge worked in close physical proximity, they did not always discuss their work with each other. Ryle's group was particularly insular, to the point where its secretiveness had become notorious. According to John Faulkner, three of the four astronomy groups did talk together about their work, often at their collective lunch on Fridays; but the fourth group, Ryle's, 'would absorb but never emit.' Faulkner thought that this was a response to Hoyle having told Bernard Mills in Sydney about the problematic source count data in 1955:

> Martin Ryle was actually paranoid because of the problems that the radio-astronomy group in Cambridge had had with the Australians finding out that Cambridge had a real problem ... in interpreting the 2C survey. Ryle gave instructions that no member of the radio group should give the slightest information or talk with anyone outside the group; and his personal graduate students were not allowed to even talk to anyone in the rest of the radio group about what they were doing. It was a very bizarre situation.[24]

At one Friday lunch in February 1968, however, a member of Ryle's group had told the members of the other groups that there would be a seminar the following week that would be unmissable. He could not say what it was to be about, but they should watch for the announcement. Faulkner recalls:

> So on Monday or Tuesday there appeared a notice at the Institute saying 'discovery of a new type of pulsating radio source'. It turned out that Hewish was going to give a seminar on the Thursday about something that had not yet been published. This was a first in Cavendish radio history. . . . they already knew that the discovery of pulsars would actually appear in Friday's *Nature* and therefore Ryle had relaxed his rule that nothing should be talked about prior to it appearing in hard print. He was actually allowing a seminar to be given on Thursday about something that wasn't going to appear in *Nature* until Friday.[25]

One of Hewish's graduate students from the pulsar team, Jocelyn Bell, noticed Hoyle at the seminar:

> Every astronomer in Cambridge, so it seemed, came to that seminar, and their interest and excitement gave me a first appreciation of the revolution we had started. Professor Hoyle was there and I remember his comments at the end. He started by saying that this was the first he had heard of these stars, and therefore he had not thought about it a lot, but that he thought these must be supernova remnants rather than white dwarfs. Considering the hydrodynamics and neutrino opacity calculations he must have done in his head, that is a remarkable observation![26]

Shortly after that seminar, Hoyle went to London for a policy committee meeting and sat next to Bernard Lovell. Hoyle lamented the absence of the positions of the newly discovered pulsars, both in the Cavendish seminar and in the *Nature* paper. Without these positions, it was extremely difficult for anyone else to study pulsars, as they would first have to scan the sky to find them. However, having done all he could with the Cambridge telescope, Ryle had given the positions to Lovell so that he could work on pulsars at Jodrell Bank. Lovell also told Hoyle that Ryle had given the positions to Alan Sandage at Caltech, who was interested in studying pulsars optically. Sandage had been told, however, that he was not to pass on the data to anyone. According to Faulkner, Hoyle asked Lovell if he had been asked to keep the positions secret. Lovell said he had not. So Hoyle asked Lovell for the positions, and Lovell agreed to send them. Then Hoyle called Margaret Burbidge about the pulsars, and she regretted that the positions were being kept secret. Hoyle told her not only that he had the positions, and could give them to her, but also that Sandage believed he was the only observer who had them. Hoyle suggested to Burbidge that she call Sandage, and ask him if he had the positions. Sandage would then deny knowing the positions, at which point Burbidge could offer them to him. Burbidge called Sandage, and the conversation followed the lines that Hoyle had predicted. Since Sandage did not want to admit to knowing the positions of the pulsars, he allowed Burbidge to dictate them to him over the telephone, and then expressed his gratitude before hanging up.[27]

Gold had also had trouble finding out the positions of the pulsars. He was using Cornell University's radio-telescope, at the Arecibo Observatory in Puerto Rico. He would soon explain pulsars as rotating neutron stars; but for now he just needed to know where they were. He visited Ryle in Cambridge in the hope of finding the locations, but was sent away empty-handed. Gold was staying in college rooms, and that evening he answered the telephone to a voice that suggested he pick up a pencil and paper. The locations of the pulsars were dictated to him, by, Gold assumed, one of Ryle's group who had decided to break the silence.[28]

It was in the context of this culture of secrecy that, in March 1975, Hoyle made a provocative remark. He had left Caltech for a short trip to McGill University in Montreal to give some lectures. At a press conference after one of these lectures Hoyle was asked about pulsars, and he had commented on the involvement in their discovery of the graduate student, Jocelyn Bell. Bell, who was now using her married name, Burnell, had been the first to spot the regular blips in the radio signals that are characteristic of pulsing sources. In the wake of Ryle and Hewish's Nobel Prize, Hoyle suggested that Hewish, who was Bell's supervisor, had 'filched' her discovery.[29] Hoyle later recalled the Cavendish seminar just before the publication in *Nature* of the pulsar research:

> There was nothing in the proceedings to tell me that a graduate student, Jocelyn Bell ... had been the discoverer, nor was there any such attribution in the way the discovery of pulsars was presented to the world thereafter. Several years later it happened to fall to me to explain, in somewhat brusque terms, what the truth had really been ... the truth was not welcomed in respectable quarters. ...[30]

Hoyle's 'somewhat brusque terms' at the Montreal press conference generated newspaper coverage around the world. Hewish described the charges as 'an astonishing fabrication'; Hoyle countered that no one else in British science would have said anything because people were 'scared stiff' by the Cavendish's reputation.[31] The local newspapers had started the ball rolling: according to the *Montreal Gazette*, Hoyle said:

> I was a fellow professor with Hewish at Cambridge at the time, but they never told me anything. There were about eight of them and they kept the whole thing under wraps for six months. ... No one in the group was allowed to speak to anyone else outside. It was just like a wartime situation, where you know the date and time of the invasion but are not allowed to speak. They were busy pinching it from the girl, that's what it amounted to.[32]

Back in Cambridge, Antony Hewish was:

> ... absolutely amazed. It took me totally by surprise, because he had not a clue about what really went on in the work, he just wasn't around. The fact that he came out with that outburst ... was just ridiculous.[33]

Bell was also taken aback: she was quoted in the *Montreal Star*, saying she did not feel 'squeezed out' of the Nobel Prize.[34] As the *Cambridge Evening News* put it:

> The knights of old were chivalrous men who rode to the aid of damsels in distress and rescued them from the clutches of monsters. There can be very few instances of the damsel telling her eager would-be rescuer to clear off and mind his own business, but we had one such example this weekend.[35]

In the USA, Caltech was pestered by journalists looking for Hoyle. A memo on file there in the hand of Fowler's secretary Evaline Gibbs records the situation thus:

> At a press conference in Montreal arranged by McGill University, Fred was asked what his opinion is of Hewish and the pulsars. He replied that Burnell

made the discovery and was not being recognized. The report in the Montreal paper did not make clear that Fred was *asked* the question – the report makes it seem like Fred had made a gratuitous statement. Fred's reply is 'contact other astronomers and see what they think and what is generally felt by other astronomers.' Fred says there's a very large group of astronomers who believe she played the major role in the discovery. Mrs Dotto of the *Toronto Star* has gotten several confirmations – in agreement with Fred – one of these is Prof Gold at Cornell (he is not afraid to speak out – most scientists are). Fred said that Mrs Burnell should have stood up for her rights in the very beginning, and there is a very powerful organization in the UK and she dare not stand up for her rights.[36]

Another memo quoted the *New York Times'* coverage, emphasizing Hewish's indignation and Burnell's support of him. Burnell had told the *Times* that Hoyle had 'drastically exaggerated the situation and overstated to the point of being factually incorrect'.[37] In a note to Hoyle, Gibbs told him that the *New York Times'* science reporter was trying to find him, and that 'your reputation was at stake and he urgently requests you to give him your comment. Willy and I decided not to give him any information as to your whereabouts, except that you would be away from Pasadena until about the middle of April.'[38] Gibbs noted that Hoyle told her: 'About the *NY Times* (or anyone else) – tell McElheny (if he calls again) that it's their job to dig out the truth – and that their reputation is at stake not his (Fred).'[39] Another note in Gibbs' hand records that Hoyle had hired a lawyer in Montreal:[40] some of his colleagues have speculated that, even before this episode, the tension between Hoyle and Ryle's camp had meant some resort to the law. Hewish had indeed thought about suing Hoyle; after all, 'it was libel':

> I toyed with the idea of actually suing him for libel, but I don't want the lawyers running all the way to the bank, and it doesn't do anybody any good in the end, that sort of thing. But what he said was so off-beam, . . . it was just plain wrong. I think he was trying to justify offhand remarks he made which were taken up by the press and made a big story of, but he just wanted to be nasty to the Cambridge group and it came out in my direction. . . . Whether he thought I was going to sue him I don't know.[41]

The journalists contacted Gold. He told the *Montreal Star* that Bell was shy, devoted to Hewish, and would never complain. Hewish should have spoken out for her. Gold wrote to Burnell, inviting her to speak at the next Texas Symposium, and telling her that there was strong applause for the idea when he suggested it at the last Symposium.[42] Twenty years on, by which time Burnell was firmly established in the leading ranks of British astronomers, the episode still bothered Gold, in the context of his relationship with the staff of the Cavendish:

> Tony Hewish got the Nobel prize for the discovery of the pulsars in which Jocelyn Bell had a very major part. . . . I had arranged to have her invited by the Texas Symposium and she gave a speech on the discovery of the pulsars, in which she explained exactly what her part was and what Hewish had said at different times but never with any slightest feeling of bitterness towards Hewish, which of course people might well have suspected her of harbouring, because Hewish did not mention her name in the Nobel speech. Nor in the speech he gave at my request at the National Academy of Sciences here. But nevertheless I am still on good terms with Hewish. . . . with Ryle long gone now, all that is forgotten. But I still think Hewish could have been more outgoing towards Jocelyn Bell.[43]

Hewish's Nobel speech, as held in the Nobel archive, does mention Bell, and describes her role in the research in some detail.

Brian Pippard, who by 1975 was Director of the Cavendish Laboratory, also joined the fray. He told the *Daily Telegraph* that Hoyle's accusations were 'absolute nonsense' and 'almost actionable.' Hewish asked why Hoyle had not challenged him back in 1968, at the Cavendish seminar, and added 'I just don't agree with anything said by Hoyle.'[44] Hewish and Hoyle, who by this time was visiting Don Clayton at Rice University in Texas, corresponded through the letters column in *The Times*, where Hewish gave a step-by-step account of the discovery of pulsars.[45] Hoyle and Clayton revised draft after draft of Hoyle's reply in a rented beach house on a week-end trip to the coast. The published version credited Bell with the persistence to pursue a phenomenon others would have dismissed as an anomaly: Hoyle noted her 'willingness to contemplate as a serious possibility a phenomenon that all past experience suggested was impossible.'[46] But he insisted that: 'my criticism of the Nobel award was directed against the awards committee itself, not against Professor Hewish . . . the committee did not bother to understand what happened in this case.'[47] Even the reticent Ryle spoke out, also in a letter to *The Times*. He countered Hoyle's charge about the lack of openness in his research by encouraging more scientists to follow their work through thoroughly before publishing it: 'The present overloading of scientific journals by the publication of preliminary and half-thought-out ideas would be relieved if this view were more widespread today.'[48]

In one of his letters to *The Times*, Hoyle said that he had made his remarks in Montreal in a 'conversational style' and had expected the reporter to check with him before printing anything controversial – 'normal American practice', according to Hoyle. Hewish was not impressed:

It could be [that] he hadn't expected to be quoted verbatim. That seems rather naïve, if the press talk to you and you say something, to not expect [to be quoted]. He said . . . that he had expected to get editorial review of what was going to be published, which I thought was an amazing remark to make, bearing in mind what the press is like.[49]

Hewish thought that it was appropriate for the Nobel Committee to have mentioned steady-state theory in the citation for the Prize, even though astrophysicists had by then found other reasons to abandon steady-state theory, and there were plenty of other achievements for which the radio-astronomers deserved credit:

[The Committee] are trying to stress the importance of the work in context. It really was the first real evidence that this main theory [steady-state theory] wasn't tenable. That's fair enough. That was Martin Ryle's major contribution. . . . It was a very critical, crucial question to answer in those days. It was [Hoyle's] pet theory, it was such a beautiful theory, but it turned out to be wrong. I'm not a sociologist, I'm not a psychologist, but when a very bright guy like that doesn't get a prize and two upstart observationalists get it, you're narked.[50]

Reflecting some years after the event, Hewish described a 'nasty streak' in Hoyle's personality:

. . . he was nasty to Donald Lynden Bell when he got the chair at the Institute. When all that happened [about the Nobel Prize] I got a nice little note from Donald saying 'I've been through all this before', of course that wasn't public it was private. But Fred was just like that, just a nasty streak, a most unpleasant streak. I'd got on with Fred reasonably well up till then, I hadn't been the butt end of his vindictiveness, but after that I began to understand more of what Martin Ryle could have had in private. . . . When it became public, [Hoyle] went down in my estimation as a human being.[51]

The radio-astronomers had kept pulsars a secret when they first found them, but for what they considered to be a good and very serious reason. In the early days after the discovery, the end of November 1967, the group called the regular pulsing signal LGM, which stood for 'Little Green Men.' They thought they might have discovered aliens. Hewish explained:

You're getting . . . totally artificial-looking signals from a body that's smaller than the Earth or comparable, and at a distance of a few hundred light years. What else could it be? [Signals] like that are broadcast from the BBC or something, you don't get them from stars.[52]

The press had invaded the Cavendish when the radio-astronomers had detected Sputnik in 1957, and that had been a mere artificial satellite that everyone knew was up there. So the radio-astronomers were dreading what would happen if the press got wind of the Little Green Men, especially now that, unlike in the Sputnik days, most people had a television. That, said Hewish, 'was one of the reasons why I kept my mouth shut.'[53] They felt that 'the thing had to be sorted out before one mentioned it outside.' Hewish was making measurements to see whether the source of the signal was in a planetary orbit around a star, which he felt it would have to be if it were supporting intelligent life. But these measurements would take several weeks. In the mean time, there were enough experienced scientists in their immediate circle who took seriously the possibility of the pulses being an intelligent signal for them to feel that they could have something momentous and highly disruptive on their hands. Hewish took the idea very seriously 'for quite a while':

> . . . because I couldn't see any other explanation. I discussed it with Martin Ryle: 'look, what the heck do we do with this if it turns out that that's the most likely explanation? You can't just publish it in *Nature*, that there is intelligence elsewhere,' and so we were wondering what the heck to do.[54]

Ryle's social conscience surfaced once again, according to Hewish:

> Martin Ryle was slightly crazy, but he said if it is [intelligent], the best thing to do is to burn the records. Because if it is announced that we are picking up intelligent signals from such and such a place, what's going to happen next is that there are going to be radio signals from Earth launched in that direction, and that's just what they want because they are looking for somewhere to go. He said 'Look what happened to the Red Indians. Primitive civilisations being taken over by more advanced ones doesn't do them any good normally. Who's to know they are benevolent?' With his sociological thinking he said things like that, but he couldn't really have meant it. But he was serious when he said that.[55]

Eventually Hewish and Ryle decided that they would have to go to London and talk to the 'top scientific brass' at the Royal Society: 'first of all convince them that what we are doing is OK, and then . . . get their advice on how to handle it.'[56] But the crisis passed: Hewish's measurements eventually showed that the source of the pulses was not in a planetary orbit, and as other pulsars were found, it seemed unlikely that the universe was suddenly awash with technologically advanced aliens. So they decided that the signal was not intelligent. In their paper for *Nature* the team wrote that they had discounted aliens as a possible source of the pulses, and even the absence of aliens was enough for the newspapers. According to Burnell:

> . . . the press descended, and when they discovered a woman was involved they descended even faster. I had my photograph taken standing on a bank, sitting on

a bank, standing on a bank examining bogus records, sitting on a bank examining bogus records: one of them even had me running down the bank waving my arms in the air – Look happy dear, you've just made a Discovery! . . . Meanwhile the journalists were asking relevant questions like was I taller than or not quite as tall as Princess Margaret . . . and how many boyfriends did I have at a time?[57]

Hewish felt that the stories of secrecy about pulsars had been exaggerated, and that the radio-astronomers' motives had been misunderstood:

> I was acutely aware of the problems we were running into of being invaded by the press if this [alien idea] got out, and it was a well kept secret actually. The Americans got totally the wrong end of the stick over this, they thought we'd kept it secret for six months when actually it was three weeks. They thought we discovered pulsars in August. . . . They got totally the wrong end of the stick and they wanted to get it wrong. They were narked because they had instruments that could have done it too but they didn't. They wanted to be a bit nasty just like Fred, and funny, strange stories started circulating. But that didn't bother me because we achieved our objective and got a jolly good paper published straight away with everything in it. I think to do that a month or so after the first pulse comes in is not bad going actually. So I would do it all over exactly the same. But if you bring the press in you've lost it.[58]

Bell did accept Gold's invitation to the next Texas Symposium, which was held in 1977. In an after-dinner speech there, she said:

> It has been suggested that I should have had a part in the Nobel Prize awarded to Tony Hewish for the discovery of pulsars. There are several comments that I would like to make on this: first, demarcation disputes between supervisor and student are always difficult, probably impossible to resolve. Secondly, it is the supervisor who has the final responsibility for the success or failure of the project. We hear of cases where a supervisor blames his student for a failure, but we know that it is largely the fault of the supervisor. It seems only fair to me that he should benefit from the successes, too. Thirdly, I believe it would demean Nobel Prizes if they were awarded to research students, except in very exceptional cases, and I do not believe this is one of them. Finally, I am not myself upset about it – after all, I am in good company, am I not![59]

While the controversy about pulsars did raise some general questions about the distribution of credit for scientific work, most scientists thought that Hoyle had behaved badly. Fowler's carefully assembled audience for the sixtieth birthday celebration in Venice started draining away. With only two months to go, letters from people withdrawing from the meeting started to arrive on Fowler's desk. Among them was a note from Jesse Greenstein who, despite having been on the organizing committee, now declined his invitation because 'with my present feelings on Hoyle's lack of tact concerning the last Nobel Prize award to Ryle and Hewish I would not go as far as the Kellogg Laboratory [i.e. across campus] to attend a symposium honoring him.'[60] Another Caltech member of the organizing committee, Hoyle's friend and ally Wal Sargent, did the same: he said he had hoped that the meeting would be 'a time of reconciliation and not just an occasion for Fred and his old cronies to sit and congratulate one another.' He felt the meeting should be cancelled.[61] Don Clayton took Sargent to task. He pointed out that:

> Fred's current opinions, be they misguided, as you believe, or courageous and correct, as they may turn out to be, are irrelevant to the unique role he has played in the growth of astrophysical thought. . . . Some years hence you may not be filled with pride at your public dissociation if its causes haven't the integrity you profess them to have.[62]

Clayton reassured Fowler that he did not expect many more defections, and they should press ahead with the programme.[63] He also wrote to Martin Rees, who would in effect be representing Cambridge at the conference, acknowledging that 'Fred is persona non grata within the Cavendish', and encouraging Rees to look to 'the future of our science and to the roles Fred has played in its development. I sincerely hope the Nobel-Prize flap will go unmentioned. . . .'[64]

Despite all the bad feeling, the party went ahead in Venice, with the title 'Frontiers of Astronomy – 1975' after Hoyle's 1955 textbook. The talks were mostly about nucleosynthesis, and the many fruitful research programmes that had opened up in that area since B²FH in 1957. Lady Hoyle and their daughter Elizabeth were there to hear Martin Rees announce that the original IOTA building, now part of the Institute of Astronomy at Cambridge, was to be named the Hoyle Building. There were absent friends as well as absent enemies: Hoyle received a number of birthday telegrams, and from world leaders as well as colleagues and fans.[65] Fowler had asked the Prime Minister, Harold Wilson, to write, but the reply from his office was that while the Prime Minister knew 'Sir Frederick Hoyle' very well and had great respect for his work, no birthday telegram was possible until Hoyle reached the age of 100. But telegrams did arrive at the meeting from Pierre Trudeau, Prime Minister of Canada; Giovanni Leone, President of Italy; Indira Gandhi, Prime Minister of India; from the University of Minnesota, thanking him for lectures in 1973 and advice since; from Dale

Corson, the president of Cornell University; Bob Christy and the Kellogg Radiation Lab at Caltech ('wish we could all be there to help make a big bang to celebrate your birthday'); St John's College, the Royal Society, the Royal Observatory at Edinburgh, the Sydney Radiophysics Group, the Institute for Advanced Study at Princeton; colleagues at Sussex, Kyoto and Imperial College London; from the Science Research Council and its US equivalent the National Science Foundation, and from *Scientific American* and *National Geographic*. Hoyle also received a telegram from his old friend Bernard Miles, of the Mermaid Theatre, who had produced *Rockets in Ursa Major* in 1962. The *New York Times'* science editor Walter Sullivan wrote:

> We have all heard the jingle on Hoyle and on Dingle
> And we know full well about Jocelyn Bell
> Fred's endurance as a walker matches his as a talker
> His famous Black Cloud wrapped the sun in a shroud
> His opera with Smit was rather a hit
> But who would carp? Not Fowler or Arp
> It was Fred's intention to stretch our perception of the universe –
>   all that is great or small
> In the vernacular his success was spectacular
> But we, unlike Burbidge, Salam and Schramm, Bondi, Shapiro
>   and Narlikar
> Must send warmest greetings from afar.[66]

A few weeks later BBC Radio 3 broadcast a discussion to mark the occasion. Presenter John Maddox – in 1975, he was both former and future editor of *Nature* – characterized the conference thus:

> First, it wasn't an old man's birthday, even if half the people there were once Hoyle's students – younger by a generation; in one way or another, everybody acknowledged that Hoyle was the most imaginative person among them. It wasn't so much a farewell to a grand old man, he isn't that in any case, as a chance to guess what he'd be up to next. Second, it wasn't a stuffy academic conference but the opposite. People asked questions openly and didn't bite off their tongues for fear that they might ask foolish questions. Third, it was a comradely conference. Hoyle has lots of friends but also lots of enemies, who by their absence made the occasion seem a little conspiratorial. A group of like-minded people, gathered in an Italian monastery to try to puzzle out what the future of astronomy will hold.[67]

Hoyle argued that the universe we perceive is but a tiny part of a bigger structure, and we must look to understand the bigger structure if we are to

separate 'the physical laws which are the true laws . . . from the environmental effects of the universe.' Maddox described continuous creation as
'in a strict technical sense, a failure. It doesn't work like that and it's fair to
say that Hoyle wasn't the first to admit it.' But:

> . . . to assert that matter is continuously created is of course anybody's privilege.
> What Hoyle and [Bondi and Gold] did was strictly within the rules of orthodox
> science. They worked with all the equations of General Relativity, they made the
> mathematics come out, and in the process they accomplished feats of algebra
> that were in themselves remarkable. So it seems to me there's a case for saying
> that even Hoyle's failures have been successes. They've changed the way people
> think.

Maddox concluded:

> Sir Fred Hoyle's programme for the next 30 years in cosmology won't appeal to
> everybody. Some will say 'Why go looking for trouble? Supposing that what we
> think are the laws of physics are just a kind of accidental consequence of the
> piece of the universe in which we happen to be?' Philosophically, Hoyle is a
> Pythagorean, always on the lookout for more perfect symmetries. But he's also a
> Machian, convinced that microscopic and macroscopic phenomena are united.
> But like the old steady-state theory, the mere idea that what we now think of
> as the accessible part of the universe is just a piece of the larger whole has set
> people thinking on new lines. Even if Hoyle is wrong, his new theory will be
> worthwhile, which is why he's so many people's hero. And he may, of course,
> be right.

16
Apocalyptic visions

Academic isolation; more on interstellar dust; life from space;
diseases from space; championing nuclear power; The
Incandescent Ones; *dust grains are bacteria; the last*
Munro; 'there must *be a God'; the next Ice Age*

1976 was another year of travels. The Hoyles abandoned Britain's driest
summer for decades for a road trip around the USA, from California to
Minnesota, Connecticut, Kentucky, and on to California, in a car stuffed
with maps and guidebooks and organized for them by Caltech colleagues.
By November, Hoyle was lecturing in Hawaii. When they returned to their
remote lakeland home, Hoyle contributed to a BBC television programme
called 'According to Hoyle,' and in an accompanying article in the *Radio
Times* his former student John Taylor of King's College London hinted that
the isolation of Hoyle's retirement was beginning to tell on him: having
'apparently dropped out of British science', he no longer enjoyed the checks
and balances provided by a community of critical colleagues. Taylor asked
why a scientist of Hoyle's eminence and accomplishments in science and
popularization would have forsaken 'Establishment Science':

> Academic scientists regard the too unfettered use of imagination in science as
> dangerous, especially if it is paraded before them on their television sets or in
> their newspapers by a colleague. Nor do they take lightly to its use in organisa-
> tional matters. There is surely some justification in such an attitude, since
> caution is necessary in the guidance of complex affairs or the justification of
> complex scientific ideas. I still sympathise with Sir Fred, whose main fault lay
> in the richness of his imagination, a talent more to be supported than attacked.
> But it is the balance between imagination and discipline that makes a scientist
> great or not. If Sir Fred can achieve this balance best when removed from the
> academic toils then he has surely taken the right step.[1]

The dispersal of Hoyle's close colleagues from IOTA meant that collaborations could now continue in other places. In 1973 Chandra Wickramasinghe had been appointed Professor of Applied Mathematics and Astronomy at Cardiff University, and he had taken a provocative stance in his inaugural lecture: he argued that just as the Copernican revolution had displaced Earth as the physical centre of the universe, so we might have to experience another revolution in which we come to understand that Earth is not necessarily the centre of the biological universe.[2] Hoyle was awarded an honorary chair at Cardiff in 1975, and thereafter work with Wickramasinghe took up an increasing proportion of his time.

The focus of their attention, interstellar dust, had become a fashionable topic within the astronomical community, and many researchers were interested in it. The dust had also gained wider currency during the 1960s, due to the idea that it was the original, universal matter from which everything, and everyone, is made. This common origin of all things fitted in well with the growing calls for equality and tolerance of the alternative movements that had sprung up in the 1960s, and stardust, once the stuff of magic and romance, now peppered radical politics and protest songs.[3]

Because of their very early origin in the universe, in some academic studies the dust grains are called 'presolar grains', because they were formed before the solar system – indeed, they were the material of which it was built. So they offer insights into what the early galaxy was made of, and possibly also into the evolution of the planets. Thus for some astrophysicists, the study of these grains was a natural extension of the ideas of B^2FH – it was about what happened next to the elements. Don Clayton was one such: he had been working on nucleosynthesis, and spent time at Cardiff in the mid-1970s working with Wickramasinghe on the formation of dust in supernovae.[4]

While Hoyle had been away during 1973 and 1974, Wickramasinghe had decided that the discrepancies between spectroscopic observations and the reference spectra would disappear if the grains were organic, and he had found a good match with the polymer polyoxymethylene, POM. When Wickramasinghe discussed this with Hoyle, Hoyle made what he later described as 'an unguarded remark':

I said, 'But Chandra, if the interstellar material is organic, if that is true, then, there is so much of it that this will be a better precursor material for biology than to do it on the Earth in Urey–Miller fashion' [that is, inorganic, non-living chemicals coming together to form living things]. That was the unguarded remark. That set him off and then he must have looked through hundreds and hundreds of spectra to fit the infrared data among organics.[5]

So if there were organic materials in interstellar space, perhaps it was there, and not on Earth, that life began.

Wickramasinghe's search through the reference spectra revealed some good matches between the spectra from space and those of another range of organic compounds, the anhydrous polysaccharides, one of which is cellulose – a major component of plants.[6] Hoyle and Wickramasinghe then felt justified in pursuing a path towards living things, although this was not a necessary step. In the history of organic chemistry, the description 'organic' had been chosen because such substances were understood to be the result of biological processes, and it had come as a shock to chemists when they had found that they could make organic compounds from non-living ingredients in the laboratory, and in the absence of any biological action. So while organic compounds are made in living processes, there are other ways, independent of biology, to account for them. But Hoyle and Wickramsinghe chose to consider the former route: in their history of this work, they wrote 'From organic polymers in space to pre-biology in space cannot be considered an unduly revolutionary step. Indeed it would seem the most logical follow-on to consider.'[7] They then outlined the orthodox 'primordial soup' hypothesis of the terrestrial origin of life, in which the necessary elements and compounds come together by chance in ponds and seas to form the building blocks of life, but added: 'astronomical data now suggested to us that these building blocks, and perhaps more complex materials still, were assembled in the interstellar clouds rather than in a terrestrial organic soup.' Precursors of amino acids, which are the basis of proteins, are, they argued, found in clouds in interstellar space. Stars are formed when the dust clouds collapse, and during the collapse bigger molecules (such as POM or proteins) form and coalesce on to the surface of dust grains.

This idea echoes Hoyle's description of another cloud: the Black Cloud from his 1957 novel. There, fictional astronomer Chris Kingsley explains to his colleagues that the Cloud is made of dust grains coated with organic material. Kingsley's grains are made of ice, but in the favourable conditions inside the cloud – much more favourable than conditions on Earth – 'complicated molecules get together when they happen to stick to the surface of these lumps.' When a sceptical colleague points out the huge amount of energy spent by living creatures on Earth on turning the raw materials of their food into the complex molecules of life, Kingsley points out that this need for 'molecule building radiation' may be exactly why the Black Cloud has come into the solar system to soak up sunlight. It will leave the solar system, Kingsley predicts, before it collapses into a star – and leave it does, vindicating Kingsley in the process.[8]

In the real universe, Hoyle and Wickramasinghe now suggested, grains with their organic coating collapse together as a star forms, producing clumps of grains. Once the star has reached a certain stage of its growth it starts to produce gas which blasts the grain clumps out into space. These grain clumps, 'within which complex organic molecules are elaborated, might then develop into prebiological "protocells." . . . Within each "protocell" all the essential ingredients of life may be present.'[9] These protocells spread through space, become incorporated in comets, bombard the Earth and 'could . . . have led to the start of all life on our planet.'

This was not a new idea. Theories of Earth being seeded from space, known as panspermia, were proposed in the late nineteenth century, and were supported by scientists as distinguished as Lord Kelvin. In the early years of the twentieth century, panspermia came to be closely associated with the Swedish chemist Svante Arrhenius, who, like Kelvin, was attracted to the idea as it allowed him to reject Darwinian explanations of life on Earth. Arrhenius thought that the distribution of organic matter through space meant that all organic beings throughout the cosmos would be made of the same stuff, and have a universal character. But as the century progressed and it became clear that space was awash with intense radiation, it became more difficult to see how living things or organic matter could survive the long journeys between hospitable sites across the universe – as Jeans explained in 1930. By the 1960s, Carl Sagan, for example, had decided that this was a good enough reason to conclude that if there was life elsewhere, it was either going to be very close to Earth, in which case it might be chemically and biologically similar to life on Earth, or it was going to be isolated and have arisen independently, in which case it might be very different indeed.[10]

Hoyle and Wickramasinghe were undeterred by the fact that the tide was against them. Their next step was to argue that organic polymers might be found instead of ice in the solid nucleus of comets.[11] Since prehistory, comets have been seen as the harbingers of doom in many cultures around the world, and Hoyle and Wickramasinghe wondered if the reason for this was the biological material that comets brought to Earth. What if this material could affect our genes, or was poisonous or infectious? They studied the history of comets to see whether their flights near Earth coincided with outbreaks of illness. They asked colleagues, among them

Bernard Lovell who is an expert on comets and meteors, for data. Lovell helped Hoyle and Wickramasinghe to look for a correlation between meteor showers associated with a comet and outbreaks of the viral illness 'red flu', but Lovell could not see one:

> [Hoyle] got Wickramasinghe to phone me about the appearance of the great Leonid meteor show in November, about the dates of it, because they were wanting to prove that the red flu came from the impact of meteors associated with this comet which produced the great Leonid meteor shower. I once wrote a book on meteor astronomy and had all our records for many years, and I gave Wickramasinghe these records which were absolutely antagonistic to this theory that the impact of red flu could have anything to do with the appearance of the Leonid shower. Well, it didn't suit their theory so that was that. . . . [Hoyle] was absolutely determined to preach the gospel that [red flu] arrived from outer space. It was a pity really.[12]

Lovell felt that Hoyle's biological interests rather got in the way of any fruitful collaborations at Jodrell Bank, where his honorary post went largely unoccupied.[13]

Wickramasinghe continued the work during Hoyle's absences in the USA. In 1976 he undertook research and published it under his and Hoyle's names while Hoyle was busy elsewhere. In October, he wrote to Hoyle at Caltech, telling him:

> I discovered an interesting effect which could be due to organic matter in grain clumps. Since you were involved in both the present suggestion and the original grain idea, I took the liberty of including you amongst the conspirers in this theory! I was stimulated to look for UV signatures [the matching spectra] mainly by your letters.'[14]

Two weeks later Hoyle got another letter from Wickramasinghe informing him that the paper had been published in *Nature*, and that it 'had wide coverage in the popular media – stressing the origin of life aspect.' In the same letter Wickramasinghe confessed to having submitted another paper in Hoyle's name without getting his approval first.[15]

In 1977 Wickramasinghe persuaded Bernard Dixon, the editor of *New Scientist* magazine, to publish a feature article under the title 'Does epidemic disease come from space?'[16] An accompanying cautionary editorial by Dixon acknowledged that the article would attract criticism because it was so far at variance with established theories of epidemiology. Dixon also noted that Hoyle and Wickramasinghe, while they were distinguished scientists, were not biologists, and were therefore 'far from their field, making staggeringly heterodox extrapolations.' But the criticisms the article would attract would be 'responses of orthodoxy, the reasonable criticisms that have always characterised periods before the setting of new Kuhnian paradigms.'[17] Hoyle and Wickramasinghe's article explains how comets bring diseases to Earth from outer space, and concludes with a warning that 'A continual microbiological vigil of the stratosphere may well have to be necessary to eliminate the havoc which may well ensue from extraterrestrial invasions in the future.'[18]

Dixon had expressed the hope that Hoyle and Wickramasinghe's article would be 'fully appraised and criticised by the scientific community,' and appraisal and criticism duly appeared in the letters column of *New Scientist* over the next few weeks.[19] One criticism was a biological objection: why would extraterrestrial biological material be similar enough to life on Earth to have any effect here? Hoyle and Wickramasinghe responded that of course extraterrestrial biology was similar to ours: life on Earth is a descendant of extraterrestrial biology, and has been modified by influxes from space throughout its history.[20] The idea of material from space being incorporated into terrestrial genomes, thus altering the course of evolution, was to outrage evolutionary biologists, and lead to some of the most vitriolic exchanges of Hoyle's career.

Visions of pollution, disaster and apocalypse were common in popular science and in intellectual culture in the 1970s. The seed had been sown by the new environmental and anti-nuclear movements of the 1960s, and it came into full flower in the early 1970s in books such as Gordon Rattray Taylor's *The Doomsday Book*, which explored the threat of imminent environmental catastrophe from over-population and the abuse of resources; E.F. Schumacher's *Small is Beautiful*, which sought an alternative economics to avert the collapse of the industrial world under the weight of its own inefficiency and inhumanity; and the Club of Rome's report *The Limits to Growth*, which spelt out the urgent need to constrain the burden the human race was placing on the environment. Many people wanted to put the world to rights, and Hoyle was one of them.

Energy was the focus of Hoyle's campaign: he was an enthusiastic supporter of nuclear power. The issue was a fraught one. In the wake of the oil-producing nations in the Middle East stopping exports to western countries in protest at their support of Israel in the Yom Kippur War, and the crippling miners' strike in Britain that plunged the nation into darkness, nuclear power was seen by some as providing welcome independence from both the Arab nations and the trade unions. Environmentalists saw both coal and oil as dirty fuels, but nuclear power was no solution to the problems either of energy or of pollution. An inquiry into the fire in 1957 at the Windscale nuclear power station was opened in June 1977, and over three months it exposed many of the problems of the management of technology, and brought opposing points of view into open conflict. In this tense atmosphere, Hoyle published a book called *Energy or Extinction? The Case for Nuclear Energy*, in which he argued that nuclear power was essential to the survival of the human race (though he thought better reactors would be needed). He also suggested that environmentalists were being paid by the KGB to oppose nuclear power generation.[21] He told the *Birmingham Post*:

> If you are Russian, you start your vociferous friends in the West baying against nuclear energy. You instruct your friends to operate through a mild, pleasant 'save the whales' movement which you observe to be growing popular throughout the Western democracies. And all this they do, right to the last letter of your Kremlin-inspired instructions.[22]

Another book, *Commonsense in Nuclear Energy*, appeared in 1980, along with a flurry of media interest in Hoyle and his astrophysical theories.[23]

Hoyle had joined the nuclear controversy out of 'deep patriotism', he told the *Yorkshire Post*:

> . . . the West will eventually have to turn to nuclear energy. It is a question of whether we remain the tail-end Charlie that Britain has become since the beginning of the century or whether we go back to having some weight in the world.[24]

At this time Hermann Bondi was Chief Adviser to the Secretary of State for Energy, Anthony Wedgewood Benn; and for several years, Hoyle and Bondi appeared to be running parallel media campaigns in support of nuclear energy.[25] Opposing them once again was the Astronomer Royal, Sir Martin Ryle, who was fervently opposed to nuclear power. Their salvos on this topic recaptured some of the intensity of their earlier cosmological exchanges, but this was to be the last fight: Ryle's health was deteriorating, and he died of cancer in 1984, at the age of 66.

Problems of the Cold War and energy supply were the theme of a novel published in 1977, once again co-written with Geoffrey and with an editorial credit to Lady Hoyle. *The Incandescent Ones* is set three centuries into the future, where Cold War squabbles about energy have been calmed by the Outlanders, humanoid extraterrestrials who have set up a base on Earth and who control the supply of energy from their own source – deuterium in the atmosphere of Jupiter – to the nations of the world. As the central character, Peter, explains:

> I knew how humans, in their overwhelming greed for energy, had run them-selves out of accessible fissile material [nuclear fuel] by wantonly burning it before a satisfactory breeder technology had been developed. This was more than two hundred years ago now, in the early twenty-first century. I knew how, as decay spread everywhere over human institutions, the Outlanders had come to earth, supplying power beams to both major political blocs, the so-called West and East. And I knew how, by exploiting the threat to cut off one or the other of the political blocs, they had enforced an unprecedented peace upon the world.[26]

Peter becomes a secret agent for the Outlanders: one of their special batter-ies has gone astray, and a return to the old Cold War order threatens as the superpowers compete to get hold of it. Peter is charged with returning the battery to its rightful owners. These turn out to be not the Outlanders, who are mere robots serving the real masters: the Incandescent Ones, a race described by the scientist in the story as follows:

> 'Imagine the evolutionary difference between an insect and a human. Imagine a similar difference between the human species and incandescent ones. Imagine a vast difference of perception of the world. Then you will have the beginning of the right idea.'[27]

Peter's journey to return the battery ends at Jupiter, where he 'knew at last that [he] had come home' – he had found out *en route* that he was himself an Outlander.[28] His guide on this journey is a man called Sam, who sports a pigtail, a pork-pie hat and a strong accent that Peter at first finds baffling. When Peter asks too many questions, Sam replies: 'Never thee mind. Sithe, just coom along. And be nippy abaht it. . . . Outlanders or Earthlanders, they're all t'same to me. Ah coom fro' Yorkshire.'[29]

Hoyle published another contribution to the doomsday literature in 1977: *Ten Faces of the Universe*, a book of essays giving his views of various approaches to understanding the universe, including a critical look at the expanding universe.[30] In the chapter 'The Biological Universe' Hoyle took a very physical perspective – he wrote that 'Animal ≡ Chemical Replication + Electrical System' – and he concluded that the chemical starting point for life on Earth could have been reached in many other places in the universe. This led him to a consideration of contact with other life forms, for which he proposed the use of radio-telescopes. In the final chapter, 'Everyman's Universe', he speculated whether other intelligent species have found a way to avoid the fate that awaits us here on Earth: self-destruction.

Hoyle concluded his book by reiterating a theme from *Decade of Decision*, published in 1953, and the novel *Seven Steps to the Sun*, from 1970: over-population, which will lead to the collapse of technology and society. One reviewer suggested that Hoyle's motive for urging the human race to save itself was to allow enough time for scientists to sort out the thorny problems in cosmology.[31]

The life-from-space story emerged in the newspapers in April 1978, when it was reported that Hoyle and Wickramasinghe's funding application to the Science Research Council had been refused – a common experience for scientists and one they do not usually publicize. The articles contain clues that Wickramasinghe had contacted journalists; and Hoyle was still good copy. The *Guardian* drew comparisons with Darwin's struggle for the acceptance of his work, and quoted Wickramasinghe, who condemned Hoyle's former allies at the Science Research Council as 'fourth-rate minds.'[32] *The Times*, which also quoted Wickramasinghe, reported that Hoyle and Wickramasinghe had complained about the grant rejection to the Prime Minister, James Callaghan: they claimed that they were not being supported because their work was being seen as a challenge to orthodox Darwinism.[33] Callaghan had other things on his mind: he was struggling to maintain the minority Labour Government that would soon fall to the Conservatives, under their new leader, former Science Minister Margaret Thatcher. In May, however, the Medical Research Council asked for a meeting with Hoyle and Wickramasinghe; the record of that meeting remains closed.[34]

Hoyle and Wickramasinghe were later to write of this period that:

> [E]ven the most modest grant applications [were] thrown back in our faces by S.E.R.C. [the research council], an organisation which in a time span of no more than a decade and a half managed to go from a beginning of rich promise to one of the outstanding Gilbert-and-Sullivan operettas of the twentieth century. . . . S.E.R.C. [is] a fun organisation which a Government supposedly given to thrift endows with nearly £300 millions annually, used it seems to pay an orchestra that exhausts itself with every crescendo.[35]

In August 1978, Hoyle and Wickramasinghe presented at a conference the results of a study they had undertaken of the spread of illness in boarding schools in Wales, which, they claimed, revealed patterns that could only be explained by their theory of diseases from space. The illness had appeared simultaneously in separate, isolated places, and could not possibly have been spread by contact with infected people. The *Daily Telegraph*'s science correspondent Adrian Berry reported the conference under the headline 'Bugs from space spread flu, says Fred Hoyle,'[36] and said that 'Doctors reacted with interest.'[37] The *Telegraph*'s coverage of this issue generated some letters to the Editor: Hoyle responded with some technical arguments, and said:

> We did not arrive at this position capriciously, but from what we regard as an overwhelming body of fact. Every radically new idea provokes incredulity from the existing establishment; and sometimes the establishment is right and sometimes it isn't.[38]

The *Guardian* published a long interview with Hoyle in September 1978, in which he outlined the rationale behind his research. In the first instance, he said, he was just one of many astronomers who took an interest in the question of life on other planets. The orthodox hypothesis of life originating on Earth was not testable: however if life – organic material, bacteria, viruses – were still arriving on comets in the same way as it arrived in the first place, then his theory could be tested (though as the editor of *New Scientist* had pointed out, the theory could be 'verified, though not, alas, falsified'[39]). Epidemics, both contemporary and historical, were the test material. Links between comets, atmospheric disturbances and epidemics in history were apparent, and contemporary epidemiological data fitted the theory, Hoyle argued. The interview concluded with a suggestion from Hoyle that there is in the cosmos a super-intellect which is so far superior to humans that we cannot possibly imagine what it might be like – a suggestion reminiscent of the description of the fictional Incandescent Ones, and the cosmic forces that populate Hoyle's other novels. When asked if this idea might discredit him among scientists, Hoyle said he was sure that it would, but that he was sure he was right. 'He was sure?' asked the *Guardian*. Hoyle replied: 'Certainly, yes.'[40]

The *Daily Mail*'s aerospace correspondent took up the story: 'The Cosmic Connection' was subtitled: 'Behind Hoyle's latest hunch, a staggering find in outer space'; three subheadings were 'shock', 'surprise' and 'soup.' 'Shock' was Hoyle's reaction to his discovery that comets bring diseases to Earth. He told his interviewer:

> I suddenly felt, 'This is terrible – this is going right to the heart of mediaeval superstition.' But this work was no sudden whim. I had reached the position that life must have come from comets in the most logical way. I was satisfied that it could not have come from anywhere else.[41]

The 'surprise' was the discovery of cotton in the Orion Nebula – which demonstrated the existence of cellulose in outer space. The 'soup' was the primordial soup, which Hoyle dismissed in favour of comets as the source of life on Earth. Comets brought not only life but also disease, and showed that the Universe abounds with life.

Such descriptions of the theory in public over-reached by far the scientific claims Hoyle and Wickramasinghe were making in their academic papers. But they were convinced that 'both the silicate and ice grain theories had lost enough ground for their main proponents to be driven into a silence of despair. . . . In sharp contrast the organic theory had already shown its merits in a decisive way.'[42] They then began to search for evidence of organic, or what they call 'prebiological', molecules in carbonaceous chondrites, which are a form of meteoritic stone, and noted evidence that suggested the presence of the ring-shaped group of organic molecules, and of some nitrogen-based compounds, such as are found in proteins and in genetic material. They also discussed evidence that interstellar grains might be found in carbonaceous chondrites, and decided that grain clumps 'may be identified with' tiny inclusions in the stones.[43]

Hoyle and Wickramasinghe published a book on the subject in 1978. *Lifecloud* covered subjects ranging from cell biology to intergalactic communication technology. It included spectra, diagrams, molecular formulae and numerical data, and despite its dramatic and colourful cover, the book read like a textbook. It explained how grains were organic, and that life fell to Earth; and discussed possible communication with other intelligences in the Universe. But other aspects of the theory were missing: for example, although by this time Hoyle and Wickramasinghe had speculated about epidemics from space in the pages of *New Scientist*, there was no mention of disease or bacteria in *Lifecloud*. Hoyle and Wickramasinghe rejected the criticism that their work was opposed to Darwinian evolutionary theory: after charting the reception of Darwin's idea, from initial hostility to eventual acceptance to its present position as 'the cornerstone of modern biology',[44] they said that there was no conflict: their ideas accounted for the period in the development of living things before organisms are sophisticated enough for Darwinian evolution to take over.

By 1979 Hoyle and Wickramasinghe had addressed the question of how the grains were uniform throughout the universe in a wide range of physico-chemical environments, and yet their 'prebiological' model required very specific conditions for the complex molecules to form and endure. Their solution was to propose that biological replication accounted for the uniformity of interstellar grains. This, however, presented a new problem: where did the first biological entity come from? According to Hoyle and Wickramasinghe: 'living systems are too complicated for assembly by random processes in a terrestrial primordial soup.'[45] Instead, carbon, nitrogen and oxygen could have been processed in space:

> Episodes of rapid biological conversion would be associated with the formation of new stars where conditions in the outer regions of a solar-type nebula permit liquid water and organic nutrients to be present within comet-type objects for millions of years. . . . We estimate that a large fraction of all the [carbon, nitrogen and oxygen] in our galaxy would thus have been biologically processed.

The result of this biological processing, over millions of years, would be microorganisms, some of which would be expelled into interstellar space, and taken up by comets. Our planetary system is surrounded by a halo of comets, whose orbits are sometimes perturbed to the extent that they pass the Earth and deposit material there, and 'we know . . . from spectroscopic evidence, that comets are very similar in their chemical composition to biological material.' Since life is only generated by other life, 'could not life have originated on Earth by contamination from space-borne microorganisms?'

Hoyle and Wickramasinghe then proposed that interstellar dust grains are bacteria. This idea came about when Wickramasinghe sent Hoyle some books from the Cardiff University library, and Hoyle, in front of the fire at home in the Lake District, noticed, in one of the books, a drawing of a dried-out bacterium. Bacteria have strong cell walls, and when they dry out they do not collapse, but rather they become hollow, as holes open up inside them as their contents shrinks. Hoyle wondered whether thinking of grains as hollow might remove more of the discrepancies with the spectroscopic data. Wickramasinghe checked, and he reported to Hoyle that he had found a perfect fit.[46] This was biologically reasonable also, since, they wrote:

> . . . bacteria possess many properties that do not have explanations in terms of Darwinian fitness to the terrestrial environment, as for instance their ability to survive heavy X-ray damage, their ability to withstand crushing pressures . . ., as well as to survive at essentially zero temperature and pressure, properties which are relevant cosmically but not terrestrially.

Hoyle later commented:

> It's my nature – I recognise that it must be an accident in my upbringing and the turn of the century when I was at the University – I just go from observation. I don't say, 'it's absurd that there should be bacteria in space.' I don't say that. It fits the observation, so it's the best theory we have. I don't care if it's absurd. So

I didn't hesitate to publish it. That of course was the beginning of the disaster ... they [our critics] know! They are born to know that the particles in space are not bacteria. God has told them.[47]

The bacteria idea did indeed seem beyond the pale to other scientists, and Hoyle and Wickramasinghe found themselves running up against much, albeit passive, opposition.[48] Funding was even harder to find, and the 'refereed journals, to which a wise scientist touches his or her forelock,' stopped considering their work. The early stages of the project had fared well, they felt, but attitudes had changed when they had focused on the organic polymers, 'because of their biological association apparently.' Hoyle and Wickramasinghe described the opposition to their work as 'anti-Copernican', and dismissed spectroscopic counter-evidence from other scientists;[49] and, when they found more evidence of their own,[50] they acknowledged it, they reported, 'more in awe than in triumph.'

As usual, Hoyle found respite in the hills. He had begun climbing 'Munros' in the late 1960s – Scottish mountains higher than 3,000 feet, of which there are 280. It is a tradition among dedicated walkers to climb all 280 Munros. A group of Hoyle's walking friends had planned a celebratory trip up his last Munro, Blaven, on Skye, for the early summer of 1972, with a champagne picnic. But events in Cambridge had made any kind of celebration impossible then, and the trip was postponed. In the event, Hoyle climbed his last Munro alone, in 1980, and characteristically he recorded the menu of his mountain-top picnic: no champagne, but a box of the 'Mr Kipling' brand of apple pies, and a vacuum flask of tea.[51]

Despite the hostility to their life-from-space theory, and the lack of funds, Hoyle and Wickramasinghe pursued their cause. They reiterated the main elements of their theory, often in a tone of surprise at their own astounding claims despite the long history of some of them. Ideas they had long since presented in a public lecture or a popular book were revealed as though for the first time in academic papers.[52] And as their critics grew angrier, the tension, and the profile of the issue, rose ever higher.

In May 1980, Hoyle and Wickramasinghe published a series of pamphlets through the university press at Cardiff. *Influenza – a Genetic Virus with the External Trigger* generated some press reaction: under the headline 'Invaders from space "trigger evolution",' Adrian Berry of the *Daily Telegraph* outlined the claim that evolution is continually driven by the arrival on Earth of genetic material from outer space.[53] Then came *The Relation of Biology to Astronomy*, and *The Origin of Life*: the first of these told the story of the life-from-space theory from its spectroscopic beginnings in the 1960s to the conclusion that cellulose is abundant in space. Hoyle then argued for the bacteria theory:

> I am now going to make a hypothesis: *Interstellar grains are bacteria.* This hypothesis may be said to be peculiar, or drastic, but it can hardly be said to be complex, because I have used only four words to state it, and I think you will all have understood what I meant by those four words. Moreover, having stated the hypothesis, little or no room is left for subsequent manoeuvre, because the properties of grains now become determined by the known laboratory properties of bacteria.[54]

He continued: 'for instance, we have no freedom any more to choose the sizes of interstellar grains,' and he presented a size distribution which showed the grains to be mostly between 0.5 and 1 microns in diameter. Since bacteria reproduce rapidly, the universe would be full of them if they were all active; therefore some of them must be dormant. This means they would have a low water content, and as the water was withdrawn a cavity would develop inside each bacterium. These cavities meant that the grains matched the spectroscopic data.

Hoyle argued that it was often claimed that the chances of life arising on Earth were exceedingly small. An analogy he used to describe this at a symposium for Willy Fowler's seventieth birthday in 1981 was printed in a *Nature* news report, which immediately gave it a wide audience as well as immortality: in typically vivid style, and in the context of an argument for why the universe needed the expanded timescales offered by steady-state theory, Hoyle described this small chance of life arising on Earth as similar to the possibility that 'a tornado sweeping through a junk-yard might assemble a Boeing 747 from the materials therein.'[55] Biologists said, therefore, that life was an incredible fluke. Hoyle, on the other hand, suggested that the chance being exceedingly small of life arising on Earth was an indication that it did not arise on Earth – the assumption that it did was a pre-Copernican notion.

Then Hoyle asked:

> How much of astronomy do we really understand? . . . very little, in spite of the sustained efforts of a generation or more of astronomers. The reason I suspect for our persistent long failure . . . is simply that we haven't got hold of the right end of the stick. The heart of the matter has been missing.

This 'heart of the matter' is that the cosmos is fundamentally biological. He continued:

> You may wish to follow this line of thought with the question: Is the biological control over astronomy to be an intelligent control or is it to be a product of blind evolutionary processes signifying nothing? My personal speculation would be that the control is intelligent.[56]

Given the degree of intelligence that has been generated on mere Earth, surely the whole universe 'should be able to manage an intelligence to which the manipulation of astronomical processes would be reasonably straightforward.'

Hoyle and Wickramasinghe considered 'the question of the origin of life on our planet in a somewhat wider context than is usually thought to be necessary.'[57] They suggested that despite experiments that have produced 'biological' molecules inorganically, the conditions on Earth, the time available and the complexity of living organisms mean that life simply could not have started inorganically. Hoyle criticized the argument that since the probability of life is so low, it could only happen once in the entire universe; and since it has happened here it could not have happened anywhere else. He said this was not logical, and that the geological record showed that 'the Earth was showered with living cells from the very dawn of its creation.' The interiors of comets were well suited to the initiation of life, and, statistically, 'living' comets were more likely to be outside than inside our solar system.

So living entities, 'in the form of bacterial cells, viruses and viroids, were first housed in comets.' These comets then crashed into the Earth, 'seeding' it. Even now, the debris of comets is deposited on the Earth at a rate of 100 metric tons per day. Much of the biological material is dead or gets burnt up in the atmosphere, but some is active or viable microbes. If these microbes were pathogens, their arrival on Earth would be manifest in the patterns of diseases. Hoyle and Wickramasinghe offered supporting cases of diseases that had disappeared for centuries and then reappeared; and of diseases that had occurred simultaneously in populations that have had no earthly contact. Their study of Welsh boarding schools had shown that the incidence of influenza could not be explained by horizontal transmission: they claimed that 'incidence at ground level was intimately connected with meteorological factors.'[58] The influenza agent was being periodically resupplied, from space, at the top of the Earth's atmosphere. In regions where there is mixing of the upper and lower atmospheres – such as wet and windy Britain in winter – the flu agent is brought down to ground level, making flu common in winter.

They argued that this influx has had a positive role in evolution: 'Viruses add on to our genomes over geological time and viroids unlock hidden potential.' Diseases caused by viroids, though bad for the individual, are good for the species, 'and appear moreover to be absolutely necessary for its evolution.' Hoyle and Wickramasinghe concluded, in an echo of Wickramasinghe's inaugural lecture seven years earlier, as follows:

> The ideas we have described in this essay both singly and taken together, represent radical departures from currently accepted beliefs in science. Scientists as a rule are reluctant to accept such change, but, as we have indicated, the changes of dogma are dictated, almost inevitably, by the facts that have emerged.
>
> Throughout our long history as a thinking species we have always been loath to accept a theory of the world in which we ourselves did not occupy a central and most important place. The view that the Earth was the centre of the solar system was held for centuries, and was abandoned only with great anguish after Copernicus. We have now come to accept the position that the solar system itself is not the centre of the galaxy, and that our galaxy itself is not the centre of the Universe.
>
> Likewise, this same Copernican-style revolution must be applied to life. Life, with its extraordinary subtlety of design and its exceedingly intricate complexity, could only have evolved on a scale that transcends the scale of our planet, the size of the solar system, even perhaps the size of the entire galaxy.[59]

These ideas were not hidden away in pamphlets for long: another non-fiction book, *Evolution from Space*, appeared in 1981. It was more strident than *Lifecloud* – one chapter heading was 'The evolutionary record leaks like a sieve' – but the conceptual content of *Evolution from Space* was the same, except that it also dealt with the cosmic nature of terrestrial evolution, and concluded with a chapter entitled 'Convergence to God.' It identified the cosmic intelligence in an equation: 'God $\equiv$ universe.'[60] The *Daily Express* reported 'There *must* be a God':[61] according to Wickramasinghe, 'We were hoping as scientists that there would be a way round our conclusion, but there isn't. Logic is still hopelessly against that.'

Reviews of *Evolution from Space* ranged from the bemused and dismissive in the *Sunday Telegraph* to the charge that Hoyle and Wickramasinghe were 'so arrogantly wrong' in *New Scientist*.[62] The book made explicit reference to steady-state theory, which is 'in much better fettle than its detractors over the last fifteen years would have one believe,' and which offered the expanded timescales needed for biological processes.[63] Thus when biologists countered with arguments that, to use Hoyle's metaphor, the tornado did not have to make the 747 in the junk-yard in one fell swoop, Hoyle's reply was that it could do, if the junk-yard were not on the scale of Earth, but of the entire cosmos.

Evolution from Space was marketed for the general reader, and an academic version that emphasized the epidemiological aspects of the theory was published the same year: *Space Travellers: the Bringers of Life* did not mention God, but it did point out the advantages of steady-state theory when considering the origin and development of life.[64]

Clouds of dust were not just affecting biology on Earth: as in the paper Hoyle had written with Ray Lyttleton in 1939, which considered how interstellar dust could prevent sunlight from reaching the Earth, and in *The Black Cloud*, where the Earth is plunged into perpetual, freezing night, they could potentially play havoc with the weather. In 1981, while out walking in the Lake District, Hoyle considered the possibility that tiny ice crystals forming on dust in the upper atmosphere could scatter sunlight back into space, precipitating an Ice Age on Earth. Were the atmospheric dust to be generated suddenly, for example from a volcanic eruption or a meteorite impact, the Ice Age could start suddenly and catastrophically. This Ice Age, wrote Hoyle, 'is not so much a possibility as a certainty.'[65] His book on the subject, *Ice*, would, according to *The Times*, 'cause a stir extending far beyond the academic world of climatology.'[66] Hoyle's solution to the problem was, according to the *Sunday Express*, 'worthy of one of his own science fiction novels':[67] the idea was to use the ocean as a 'night storage heater,' by pumping cold water up from the deep ocean and letting it absorb heat from the Sun. The scheme would take 3,000 years to store enough heat to fend off the Ice Age, but since meteorite impacts happen only once every 10,000 years, it might still provide a solution in time. Hoyle himself contributed two extensive feature articles – extracts from the book – to the *Sunday Times* magazine, each one illustrated by photographs of Hoyle dressed for the coldest of weather and surrounded by the snowy peaks of his favourite walking country.[68]

Scientists' responses were, according to *New Scientist*, 'glacial'. It collected comments from a range of experts in geology and climatology who expressed considerable reservations about the theory.[69] Astrophysicist and writer John Gribbin, reviewing *Ice* for *New Scientist*, noted his colleagues' amusement that after 'life from space' had come the 'ice age from space,' and added with regret that much of Hoyle's theory had been long since

developed by experts in the relevant fields – research of which Gribbin
thought Hoyle seemed unaware:

> *Ice* is simply the latest vehicle for Hoyle's continuing obsession with the notion
> that everything comes from space. He may well be right, but this time his
> concentration on his own genius seems to have blinkered him to the fact that
> almost everything in this book is already well known, and although he may have
> arrived at these ideas independently, for once he is strikingly unoriginal. . . . The
> saddest thing . . . is that it undermines my faith in the rest of Hoyle's recent
> work. . . . Sir Fred . . . has displayed a great ignorance of the subject.[70]

Gribbin had noted a habit of Hoyle's that had been annoying colleagues in
physics and astronomy for some time. If he could arrive at an idea, or derive
a formula, or undertake some other operation, from first principles, then he
did not feel the need to refer to work already done by others. Much of the
rhetorical power of his popularizations lies in this resort to fundamentals.
But it isolated his work from the main body of scientific knowledge, and, as
a serious breach of academic etiquette, it cost him allies in the scientific
community. This habit was exacerbated by this time by a practical problem
that dogged Hoyle after he left Cambridge: as Geoffrey Burbidge put it,
'one of Fred's mistakes was to live in places where there were no accessible
libraries, unfortunately.'[71] Hoyle simply did not have to hand the latest work
from the fields in which he was now writing. He did have some discussions
about *Ice* with a geologist, Clark Friend, who had experience of working on
glaciers in Greenland. His connection with Hoyle was a personal one: he
knew Hoyle's daughter Elizabeth and her husband Nick Butler. Elizabeth
was now was living in Oxford and studying geology, and the book is dedicated to her. Friend recalls that Hoyle quizzed him about some aspects of
the book: 'Some of it I thought was interesting, but other lines of enquiry
I thought a little flimsy. . . . My impressions of him were that he was very
provocative and sometimes argued from a devil's advocate stance saying
"prove me wrong".'[72]

   Indeed, while Hoyle's proposed solution to the climatic disaster may have
been imaginative, the basic idea of *Ice* was a common one at the time, after
a decade of ideas about environmental catastrophe, and Hoyle was not
completely out of touch. He considered as an open case the idea published
in 1980 by geologists Luis and Walter Alvarez, who proposed that a cata-
strophic change in climate, caused by the dust raised in a meteorite impact,
was responsible for the sudden extinction of the dinosaurs.[73] The idea had
sparked a long-running controversy.[74] And after Hoyle, the idea took on
a topicality and urgency that Hoyle's work lacked, when scientists from a

variety of fields published work on the effect of the dust that would be added to the atmosphere by a nuclear conflict. In the USA in 1983, under Present Reagan's hawkish regime, chemists John Birks and Paul Crutzen considered the smoke from the extensive fires that would burn after a nuclear exchange, and Carl Sagan, whose knowledge of atmospheric conditions on other planets had prompted his own thinking about climate change on Earth, considered the dust raised by nuclear blasts. The scenario envisioned in this work was named by Sagan's co-author Richard Turco as the 'nuclear winter.'[75] Sagan took the idea to the policy community, and it remained influential – and was a significant boost for the peace movement – for almost a decade. In Hoyle's hands, though, the idea was hard to tell from science fiction. Out there on the hills, his thoughts had taken on an air of fantasy.

17
Evolution on trial

*The Battle of Little Rock; the cosmic intelligence; controversy
over another Nobel Prize;* Comet Halley *and Comet
Halley; feathers fly over 'fake' fossil; autobiographies;
astrobiology*

While some of Hoyle and Wickramasinghe's earlier work had
claimed kinship with Darwin, who, like them, had struggled
against the disapproval of his peers, by 1980 Darwinian evolution
had become their target. In December 1981, in the wake of the publication
of *Evolution from Space*, when they took part in a legal battle in Little Rock,
Arkansas, where a law had been passed requiring that Creationism – the
idea that God created the universe and its creatures as described in the Bible
– be given the same attention in schools as Darwinian theory. The law was
being challenged by the American Civil Liberties Union on the grounds
that it infringed the academic freedom of teachers. Hoyle did not attend the
trial, but wrote about it for *The Times*, where his picture was printed alongside one of Darwin.[1] Hoyle explained that his position was not necessarily
one of support for Creationism nor even for God, but since Darwinian
evolution was a theory, not a proven fact, it had no greater claim to the truth
than any other theory of the development of life on Earth, including that of
the Creationists. Hoyle also argued that the hardworking taxpayers of
Arkansas had a right to expect school courses 'that do not make a mockery
of the beliefs many of them hold.' He suggested that the American Constitution, which states that religion should not be taught in schools, would be
undermined, 'in an inverted way', if Darwinism, and Darwinism alone,
were taught, for it is anti-religion. Creationism should therefore be taught to
provide the balance implied in the Constitution, which surely should be
taken to read that no single religion should be taught in schools. Hoyle
concluded his article by suggesting that Darwinian evolutionary theory was
in any case flawed – he wrote that his 'own recent work has caused me to

doubt not that evolution takes place, but that it takes place according to the usual theory of natural selection operating on randomly generated mutations.' What those seeking to champion Darwinism over Creationism were doing was 'to impose on the people of Arkansas [a theory that] may be scientifically wrong.'

Wickramasinghe went to Arkansas as a guest of the state, to testify on behalf of the Creationists. Willy Fowler wrote to him urging him not to go, but the letter did not reach Wickramasinghe before he left for the USA. Fowler told Wickramasinghe that he was writing 'as a good friend', and reminded him that he had already told Hoyle that he did not agree with the science:

> I would bet 50–1 that he (and you) were wrong. . . . Arkansas is not the proper forum . . . for settling or even discussing the issue. I have equal disdain for the Creationists and for the [American Civil Liberties Union]. . . . [T]o become involved with either group is incredibly poor judgement. . . . It is with great relief for me that Carl Sagan has withdrawn from the proceedings . . . It is my very strong advice that, no matter what commitments you have made to date, you should withdraw from the Arkansas proceedings.[2]

Fowler reminded Wickramasinghe that he had helped him and Hoyle to find proper scientific forums for discussion of their work, but said that he felt that the Arkansas connection would be a step too far. Fowler's letter bears a postscript: 'I am enough of a coward that I do not want my name mentioned in any of this.'

The trial strayed, during its ten days, from the issue of rights and freedoms, and instead put the competing theories in the dock. Wickramasinghe was one of the last witnesses to be called, and he used the occasion to expound the theory of life from space, and made frequent references to Hoyle. Under questioning by the state's lawyers, he read answers from notes, and one of the lawyers for the American Civil Liberties Union lodged an objection: he wanted spontaneous answers. Wickramasinghe countered that he was an experienced lecturer and that he never lectured without notes. The judge allowed him to continue, and Wickramasinghe read from his notes, uninterrupted by questions, for some considerable time.[3] He told the court that as a child he had been impressed with Darwinian theory:

> I can hardly remember my first encounter with Darwin's Theory of Evolution, but it was surely taught in a class room context at a very early age, long before I was in any position to assess the facts. . . . It impressed me as a great theory, seductive and compelling, even though it ran counter to my own cultural inheritance of Buddhist beliefs that the Universe is eternal and that the patterns

of life within it have a permanent quality. Neo-Darwinist ideas became firmly imprinted in my mind and they became part and parcel of my scientific inheritance.[4]

However, he recounted his work with Hoyle on life from space from the early 1960s, and argued from that basis that Darwinian theory – the 'conventional wisdom' – was unsatisfactory:

This conventional wisdom, as it is called, is similar to the proposition that the first page of Genesis copied billion upon billions of time[s] would eventually accumulate enough copying errors and hence enough variety to produce not merely the entire Bible but all the holding[s] of all the major libraries of the world. The two statements are equally ridiculous. The processes of mutation and natural selection can only produce very minor effects in life as a kind of fine tuning of the whole evolutionary process. There is above all an absolute need for a continual addition of information for life, an addition that extends in time throughout the entire period for the geological record. Frequent and massive gaps in the fossil record and the absence of transitional forms at the most crucial stages in the development of life show clearly that Darwinism is woefully inadequate to explain the facts.[5]

Wickramasinghe's testimony explicitly related life-from-space theory to the steady-state cosmology. He said:

The facts as we now see them point to one of two distinct conclusions: an act of deliberate creation, or an indelible permanence of the patterns of life in a Universe that is eternal and boundless. For those who accept modern cosmological views as gospel truth, the latter alternative might be thought unlikely, and so one might be driven inescapably to accept life as being an act of deliberate creation. Creation would then be brought into the realm of empirical science. The notion of a creator placed outside the Universe poses logical difficulties, and is not one to which I can easily subscribe. My own philosophical preference is for an essentially eternal, boundless Universe, wherein a creator of life somehow emerges in a natural way. My colleague, Sir Fred Hoyle, has also expressed a similar preference.[6]

Wickramasinghe's lecture did not spare him cross-examination. When asked whether any rational scientist would support the Creationists' view that the Earth is much younger than science suggests, Wickramasinghe replied that they would not, effectively undermining the case he was there to support. He became agitated under interrogation, and told the American Civil Liberties Union's lawyer that his questioning was 'low level.'[7]

There was much press coverage of the trial on both sides of the Atlantic. The *Arkansas Democrat* reported that 'Wickramasinghe said he had not published in established scientific journals because the editors of those journals are "collectively misled" and his theory goes "fundamentally against the grain of current opinion".[8] In the UK, Wickramasinghe's colleagues at Cardiff University wrote to the newspapers to dissociate themselves from his testimony.[9] Wickramasinghe found Fowler's letter when he got back to Cardiff. He replied, saying that while the letter might have stopped him going, he had not found the occasion as sinister as Fowler had supposed. He pointed out that he had not thrown his support blindly behind the Creationists: 'My support in the trial was only for those creationist ideas that I thought were consistent with my own views resulting from my work with Fred.'[10]

The judge published his decision a year later, in January 1983. He said that Wickramasinghe's testimony had so discredited the Creationists' case that 'The Court is at a loss to understand why Dr. Wickramasinghe was called in behalf of the defendants.'[11] The judge decided that American Civil Liberties Union's complaint against the new law be upheld.

In January 1982, Hoyle gave a lecture at the Royal Institution in London that added a new twist to life-from-space theory – while at the same time taking it one step closer to the fictional scenario of *The Black Cloud*. Appropriately, the lecture was sponsored by *Omni*, the popular magazine that mixed science and science fiction. In his lecture, Hoyle said that the building blocks of life that circulate throughout the cosmos are the remnants of an intelligence that had been destroyed by an environmental catastrophe. This intelligence is now being reconstructed, from those remnants, on Earth. According to *The Times*, 'Sir Fred . . . would have us believe he is in earnest.'[12]

The story was taken up in a long feature in the *Sunday Times* a few days later.[13] Hoyle told the reporter that he realized that adding a cosmic intelligence to his life-from-space theory would do nothing to improve his position with his critics. However, the odds against life arising spontaneously on Earth were too high, and so Hoyle, with 'a real sense of shock . . . invoked an unknown form of cosmic intelligence, threatened . . . with cosmic catastrophe, thinking up a new form of life capable of surviving it – the carbon-based life we have on Earth – and seeding the universe with spores.' The article concluded by acknowledging that while Hoyle's ideas were unorthodox they did not detract from his important contributions to astrophysics, and that 'maverick ideas' were useful 'in goading people out of ossified patterns of thought.'

By the end of 1982, Hoyle and Wickramasinghe had decided that 'this scattering of arguments has by now made it awkward for others to see how all the pieces fit together, to a point where we have felt it necessary to put together the whole thing between the same covers.'[14] The result, *Proofs that Life is Cosmic*, was published in Sri Lanka, and included bacteria, comets, diseases, evolution and the origin of life, but did not mention God or the cosmic intelligence. Shortly afterward, Hoyle and Wickramasinghe announced that they had evidence derived from data gathered by the satellite Explorer for the presence of massive quantities of protein in deep space. While *The Times* described the work as 'remarkable,' the *Sunday Times* reported that 'the two astronomers find little or no support from fellow scientists.'[15]

In May 1983, Hoyle gave a lecture at Balliol College Oxford in which he outlined his objections to Darwinian evolution. Jeremy Cherfas, biology consultant to *New Scientist*, reviewed the lecture, and after correcting what he saw as Hoyle's biological and historical inaccuracies, he concluded in strong terms:

> I have done [Hoyle] the singular honour of taking him seriously, but only because I am fed up to the teeth with malodorous mouthings such as he regularly foists on an unsuspecting public. Sir Fred is undoubtedly distinguished, having made major contributions to astronomy. His intellect in that field has been recognised by his peers, but these attributes alone do not entitle him to pontificate on other matters. Neither a knighthood, nor conviction, is a substitute for thought or scholarship. . . . If I have been disrespectful to Sir Fred, it is at least partly because he has grossly misrepresented someone [Darwin] who is dead; but more because he shows the living, who listen to him in expectation of wisdom, such disrespect.[16]

Hoyle's book *The Intelligent Universe* was published in October 1983, and the *Mail on Sunday* gave readers a preview. The article was subtitled 'a startling new scientific theory of the universe', and headlined 'In the beginning there *was* a creator'. It explained that Hoyle's thinking had led him inevitably to the belief that a supreme intelligence within the universe is guiding and controlling everything that happens. According to Hoyle: 'Some other scientists are worried by the way everything fits just right. They don't admit it, but they are very close to the idea of some intelligence at work.' Hoyle argued that this idea was something scientists ought to investigate.[17] *The Intelligent Universe* covered the same ground as *Evolution from*

Space: the steady-state theory was prominent here, as was God – indeed, some years later the book was much quoted in a special edition of the Jehovah's Witness publication *Awake!*, which presented the cosmological case for the Creator.[18]

Evolution from space and a cosmic intelligence made good copy for journalists. However, another story permeated this coverage: in October 1983, the Nobel Prize for physics was announced, and it was to be awarded for advances in our understanding of the evolution of stars. Half of the Prize went to Subrahmanyan Chandrasekhar, of the University of Chicago, for his work on the physical processes that determine the structure and evolution of stars. The other half of the Prize was awarded to Willy Fowler, for 'theoretical and experimental studies of the nuclear reactions of importance in the formation of the chemical elements in the universe.'[19] The Royal Swedish Academy of Sciences noted that: 'Many scientists have studied these problems, but Chandrasekhar and Fowler are the most prominent.' Fowler's work, it said:

> . . . deals with the nuclear reactions which take place in the stars during their evolution. In addition to generating the energy which is radiated, they are of importance because they lead to the formation of the chemical elements from the original matter, which chiefly consists of the lightest element, hydrogen. Fowler has done extensive work on the experimental study of nuclear reactions of astrophysical interest, as well as carried out theoretical calculations. Together with a number of co-workers, he developed, during the 1950s, a complete theory of the formation of the chemical elements in the universe. This theory is still the basis of our knowledge in this field, and the most recent progress in nuclear physics and space research has further confirmed its correctness.[20]

Many in the scientific community had felt for a while that stellar nucleosynthesis was an achievement worthy of a Nobel Prize. But in the wake of the Swedish Academy's announcement, a *Nature* headline summed up a widespread dissatisfaction with the award when it asked: 'where was Hoyle?' The *Observer* newspaper, commenting on Hoyle's recent thinking on a cosmic deity, suggested that this 'unexpected conversion may bring him some comfort during a vexing time – for the diminutive, mercurial physicist has just been soundly snubbed by the world of science.'[21] It noted that 'even the normally reticent journal *Nature*' was 'incensed'. *Nature*'s editorial by John Maddox connected what it saw as an oversight by the Nobel committee with Hoyle's theory of life from space:

It is . . . possible that the committees, or even the Royal Swedish academy, were impressed that Hoyle differs from Fowler (and most other prize-winners) in that, always subversive as the best scientists are, he has recently been over-subversive, advocating with too much passion and too little evidence the unnecessary theory that living things on the surface of the Earth originally arrived from intergalactic space. By general consent of this journal's referees, that is where the weight of opinion lies. Yet it would be a shabby circumstance if such considerations or fears of the embarrassment there would be if an unrefereed account of this theory were emblazoned across the world at some prizegiving ceremony had deflected the committees from the logic of what might have been their intentions.[22]

The editorial concluded that whatever Hoyle's wilder ideas, his important contributions to astrophysics should not be overlooked.[23] *New Scientist* published an article subtitled 'a defence for Sir Fred Hoyle,' which pondered the paradox of a likeable, inventive scientist who manages nevertheless to repeatedly raise the hackles of the scientific establishment, and suggested that Hoyle's 'breadth of vision' made other scientists uneasy.[24]

Tommy Gold thought Hoyle deserved a Nobel Prize for stellar nucleosynthesis, and he recalled a conversation with Fowler in which Fowler had told him: ' "you know I would still be a small timer in low-energy nuclear physics here at Caltech if Fred Hoyle hadn't come along and prompted me to experiment on the carbon-12 nucleus." But it was not Fred Hoyle but Willy Fowler who got the Nobel Prize.'[25] Gold offered three possible reasons for the oversight: first, Hoyle's independence, which deprived him of institutional backing; secondly, lack of understanding by the Nobel committee of the significance of the work described in the monumental paper B²FH; and thirdly, Hoyle's unpopular theories:

. . . if you go through the list of Nobel Prize winners, the entourage in the place where they work is all-important. . . . of course at a place like Caltech you get a lot of support. . . . Hoyle at that time, many years later after he did the work, was not attached to any university – he was a freelance person and didn't have anybody rooting for him. And they didn't recognise [B²FH] early on . . . that's understandable. . . . But it was after all one of the major advances in astronomy, to tell us where the elements come from. What could be more important than that? . . . But Hoyle had probably made himself unpopular in many other fields . . . too.[26]

Other colleagues have since speculated that Hoyle destroyed his chances of a Nobel Prize when he protested about the exclusion of Jocelyn Bell from the award to Ryle and Hewish. Hoyle thought that too: towards the end of his life, he told a journalist:

Quite simply, the Nobel Committee doesn't like having its decisions discussed. [In 1974] the prize for physics was given to Martin Ryle and, jointly, just to Antony Hewish for the discovery of pulsars. I was clearly opposed to that last choice. Hewish was director of the research but all the work was done by his student Jocelyn Bell, and she should at least have been cited too.[27]

Very soon after the announcement of Fowler's Prize, Hewish sent Fowler a congratulatory telegram on behalf of the Cambridge radio-astronomers.[28]

Fowler wrote to Hoyle early in November:

Dear Fred

After the initial elation and excitement I have had a heavy heart for two weeks. It is impossible to understand why the prize was not given to you or shared between us. I realize that nothing I can write will help but this personal note to you helps relieve my own feeling at least.

However I can not turn my back on my colleagues in Kellogg especially Barnes, Kavanagh and Whaling. The citation stresses experimental research as well as theoretical and they did it. I have told everyone I consider it to be an award to Kellogg and it is most important for the future of the laboratory.

This is just between you and me but I was pleased to give *New Scientist* permission to publish my tribute to you in the Family Journal.

Sincerely
Willy[29]

Hoyle's reply was written immediately and posted soon after, but did not reach Pasadena for almost a month:

Dear Willy

Thank you for your letter. Everybody at Caltech must be immensely pleased with your Nobel award, on which my congratulations.

It has been clear I think to many people that the subject of nucleosynthesis was likely to attract such an award sooner or later. In so far as I have thought about the matter myself, it had seemed to me that the two of us might share the Prize, with my special claim resting on the pioneer work of the period 1945–53, and yours on having continued through to the present date in what has become a major branch of astronomy. But it didn't turn out that way. Otherwise it would have been nice to be together once again.

Best wishes
Fred[30]

The tribute that Fowler mentioned was published in *New Scientist* within a few weeks of the Nobel announcement, and not long after Cherfas's scathing review of Hoyle's Oxford lecture had been published in the same magazine. Fowler had written the piece earlier that year, before he knew of the award, at the invitation of two of Hoyle's granddaughters for an edition of the 'Family Journal' they had produced to mark Hoyle's sixty-eighth birthday in June. *New Scientist* added a preface that noted 'the great contribution that Sir Fred Hoyle has made' to our understanding of stellar nucleosynthesis. Fowler surveyed Hoyle's achievements in astrophysics and mentioned Hoyle and Wickramasinghe's 'revolutionary and provocative new ideas in cosmic chemistry and biology'. He concluded that Hoyle:

> . . . is truly a Renaissance Man, just like the great intellectuals of the 16th century who first glimpsed the full range of science and art and human culture and who brought men and women out of the Middle Ages and into the modern world. But underneath it all Fred Hoyle is a Yorkshireman – down-to-earth, combative, intolerant of nonsense, generous to a fault with friends, contemptuous of detractors. It was a great day for me when I got to know Fred Hoyle and became his friend and collaborator.[31]

While *Nature* and *New Scientist* could extend their sympathy to Hoyle but not to life from space, the Royal Astronomical Society did find room for the theory. In November 1983, the Society called a meeting under the title 'Are Interstellar Grains Bacteria?' According to Hoyle, data from the UK Infra-Red Telescope which was presented at the meeting by those opposed to his theory had not yet been published, and so he had been unable to build it into his argument. Despite their apparent advantage at the meeting, he was dismissive of the opposition's case:

> It was . . . helpful to our position that opposition speakers frittered away the equal time they were given, not in presenting an alternative theory of grains that matched the observational data as well as the bacterial theory, but in attempting to score polemical points that could at best be of ephemeral advantage.[32]

A *Nature* news report of the meeting regretted the lack of time for free discussion: 'many direct questions were left in the air for lack of time. An hour of unplanned debate might have done everybody involved a great deal of good.'[33] This attempt, then, by the scientific establishment to engage with life-from-space theory seems to have been half-hearted at best.

In the summer of 1984, Hoyle gave a long interview to the BBC's *Listener* magazine in which he explained his commitment to the theory:

> It might be asked: what would be required to make me change my views that there is life in space? I find it difficult to conceive of what could be discovered that would negate this point of view. I think that once one has conditioned one's

mind to be open to think about the matter and to consider all the details, it is so overwhelming that I can't, myself, feel that there could be anything wrong.

I am reminded in this connection of a remark Einstein made, which I will slightly change. He said 'the universe may be subtle, but it's not malicious.' What he really meant was that when the evidence points towards a conclusion, and it points fairly and heavily towards that conclusion, the universe doesn't trick you, and it doesn't turn out that you're wrong. And I feel this very much in regard to this present position. Of course, one can press this point a little further and just conceive that something might come along that suggests one is wrong. Would this be important, would it matter? Well, of course it would. It would be like one's house falling down. It would be a disproof of Einstein's statement, in my view the world would be malicious, God would be malicious, if that would happen.[34]

Hoyle differentiated between disagreeing with colleagues and disagreeing with the universe:

One of the things that you have to have, is a sense of obstinacy inside, because if you don't, you're not going to go against the crowd, and if you don't go against the crowd you're not going to have any real successes. But the question then is: can it interfere with one's judgment? Well, let me make it absolutely clear that obstinacy is only of value in so far as it allows you to discount the opinions of other humans. The sense of obstinacy that goes against the facts is deadly. It is terrible, and you must not have that at all.

You see, the real thing in science is not the talking among humans, the real confrontation is between the human brain and the universe. And once a person becomes controlled by their relation to other humans, then they lose that essential connection, and that's really what I mean by obstinacy.

For myself the satisfaction comes when out of the brain I've managed to deduce that such-and-such should happen. Then, what I do is actually go away and take a look; and if I find it is so, I have a feeling that I've scored a point over the universe.[35]

Now aged 70, Hoyle was showing little sign of retiring, and the visit of Comet Halley in 1986 offered him an opportunity for both fact and fiction books about comets. Sometimes the two were difficult to distinguish, and reviewers remarked on the overlap. The fiction was a novel, *Comet Halley*, and according to its dust-jacket:

Embodying many of the ideas contained in his best-selling view of creation and evolution, *The Intelligent Universe*, Sir Fred Hoyle's splendidly entertaining new novel is not only an engrossing story with a rich variety of characters, it is also a delightful satire on the ethics of the nuclear age.

According to the *Guardian*, the novel 'exudes a positively adolescent desire to follow trains of thought off mental precipices so dire you'd rather not look at the facts from which they've been constructed.'[36] The novel is awash with criticisms of Cambridge politics and Cambridge plumbing; of government and of the scientific establishment. An early tirade concerns funding for science:

> Would the world ever again be transformed by the ideas of one man seated at his desk, or by ingenious but simple bits of equipment . . .? Or did the future belong entirely to mammoth projects and to mammoth organisations like CERN, and like the multi-national corporations which now dominated the field of electronics? . . . As the need grew for larger and more expensive equipment, it had seemed sensible for Governments to concentrate resources into national centres . . . [and] as the system had evolved down the years, the central organisations had arrogated more and more power to themselves, eventually leaving the universities pretty much in the condition of a school of beached whales. Nor was this the worst of it. . . . The world had evolved implacably from small to big, with the little fellow going inevitably to the wall.[37]

In *Comet Halley*, the world is thrown into chaos when it appears that the comet is heading straight for Earth. The astronomers at the centre of the story all have troublesome relationships with the scientific establishment: an American scientist experiences science as an outsider because she is young, foreign and a woman; an academically successful young physics professor is in turn ridiculed, feted, loved and loathed; and a young researcher who registers intelligent signals coming from the comet is first silenced, then has his funding withdrawn, and is finally murdered.

The doomed researcher, Mike Howarth, has detected signals from the approaching comet, but fails to convince the research council, 'CERC', that his work is worth pursuing. The Council has decided that Howarth's conclusion – that comets are alive – is nonsense.[38] Eventually the charismatic young professor Isaac Newton takes up Howarth's cause, but when it becomes apparent that the human race will be wiped out by the impact of the comet, everyone settles down to ponder their demise. Then at the eleventh hour Halley changes course, and the Earth is saved. So Howarth must have been right all along; and as normality returns, Newton muses on the inability of scientists to face up to the real world:

> In politics the world is the way people believe it to be. In science, the world is also the way people believe it to be, at any rate in the short term, at least as far as

grants for research and publications acceptable in the literature are concerned. There's actually very little perception of the existence of a real truth in the world, a truth irrespective of human opinions and emotions.[39]

Like *The Black Cloud*, *Comet Halley* set great store by the ability of lay people to see sense. Newton remarks on 'the unerring ability of ordinary folk to arrive at the truth of a situation,' and at a press conference he tells a journalist:[40]

> What you appear to stand for is an anti-intellectual attitude of mind, which you then proceed to confuse with the attitude of ordinary people. Ordinary people are not anti-intellectual when things are explained to them in a proper way, which is surely the job of the media? . . . [contacting the comet is] something which ordinary people would give more than tuppence for . . . because ordinary people have supported unusual things from time immemorial, whether at Stonehenge, or in building the temples of ancient Greece, or in modern scientific laboratories. Your trouble, as I have already explained, is that you persistently confuse ordinary people with anti-intellectual people – of which there are nowadays a fair number, I'll grant you, but not a majority, I'm happy to say.[41]

The real Comet Halley brought with it new data about comets, and opportunities to collect epidemiological evidence for diseases from space. In the spring of 1986 an array of scientists lined up for and against the theory: an epidemiologist wrote a booklet with them, and another dismissed it.[42] Astronomers in Australia – among them, Wickramasinghe's younger brother Dayal, who had studied under Peter Strittmatter at IOTA – were collecting data from the Giotto probe that had intercepted Halley, and they suggested that carbon, nitrogen and oxygen were common elements in the comet; while others countered that this hardly amounted to evidence of bacteria in space.[43] By the summer of 1986, Hoyle and Wickramasinghe were once again complaining of blocked funding applications and hostile critics.[44]

At this time Hoyle and Wickramasinghe suggested that AIDS had arrived from space – in a shower of organic material in central Africa in the mid-1970s.[45] It was originally passed to humans from rainwater via cuts in people's feet – hence it was common in parts of the world where people tend to walk about without shoes. This idea raised much critical reaction from journalists, and Hoyle and Wickramasinghe protested that surely AIDS was dreadful enough 'to make us look seriously at the evidence that pathogenic microorganisms are falling from the skies.'[46] The *Evening Standard* expressed its scepticism by comparing the theory to other suggestions, such as that HIV was manufactured by the Americans to subdue the Third World, and that AIDS was caused by the fluoridation of water. 'We still don't understand all the factors involved,' reported the *Standard*, 'but it seems pretty safe to say that the fancy of AIDS from the sky is pie in the sky.'[47] Hoyle and Wickramasinghe were to make similar claims about other new diseases, including the 'mad cow disease' BSE, which hit English and Welsh cattle herds so hard because, unlike in other countries, the cows spend the winter outside, making them vulnerable to 'small particles of bacterial and viral sizes [which] descend through the Earth's stratosphere mostly during the winter months.' Hoyle and Wickramsinghe concluded a letter to *The Times* by suggesting, in regard to the politicians' management of the issue, that 'We live nowadays in a blame culture, but in our view there was no culprit, not unless blame be equated with ignorance.'[48]

In September 1984, Hoyle had received a letter from an Israeli physicist, Lee Spetner, to whom he had been recommended by his war-time colleague Cyril Domb. Spetner had become interested in evolutionary theory in the early 1960s when he spent a year in the Department of Biophysics at Johns Hopkins University in the USA. He was sceptical about Darwinian theory, particularly with regard to the element of probability in the mutation process. In 1978, while studying the evidence for evolution, Spetner had examined a fossil specimen of *Archaeopteryx*, the celebrated 'missing link' between dinosaurs and birds, in the Natural History Museum, in London, and had discussed his reservations about it with its curator, the palaeontologist Alan Charig. In 1980 Spetner told a scientific conference in Jerusalem that *Archaeopteryx* was a fake. Spetner had tried to arrange tests on the specimen at the Natural History Museum, but had been refused access to the fossil because the tests would have damaged it. Now he was coming to London, and hoped that Hoyle could help him.[49] Hoyle replied quickly, offering his support: Wickramasinghe would meet Spetner in London. Hoyle warned Spetner that 'the intellectual investment in the UK in the Darwinian theory is so enormous . . . for which reason it will be best to keep the position low-key until as much information as possible has been obtained.'[50] Hoyle was unable to arrange tests on an actual specimen, so he undertook an analysis of photographs of the fossil. Between March and June 1985 he published a series of papers in the *British Journal of Photography* which claimed that the Natural History Museum's *Archaeopteryx* was a fake: crucial details had been carved into a false surface layer on parts of the fossil.[51] The news spread around the world.

Spetner was unprepared for the press attention, and thought it would ruin their chances with the Natural History Museum. In April 1985, he apologized to Charig: 'This was certainly not the way I intended the matter to proceed', he wrote; 'I had no idea that the newspapers would get onto it and I apologise for the way the publicity got out.'[52] Hoyle suggested to Spetner that they should offer the Museum a further chance to participate in tests, as then, whether the fossil was genuine or fake, the Museum would come out looking honourable. In May 1985, Charig told Wickramasinghe that the Natural History Museum's *Archaeopteryx* was not available for tests. In response, Spetner asked Hoyle if he would approach the Museum in person.[53] When Hoyle visited Charig at the Museum shortly after-wards, Hoyle blamed the *British Journal of Photography* for the publicity: it had issued a press release that had not been discussed with either him or Wickramasinghe.[54] But Charig was still not prepared to release the fossil for testing, and after the meeting Hoyle told Spetner that the only way to put pressure on the Museum to release its specimen for testing was to build up public support for their cause.[55]

The following spring Charig and his colleagues published their response to Hoyle's charges in the prestigious journal *Science*.[56] But Hoyle and Wickramasinghe had written a book on the subject which was already in press, and was set to raise the temperature again. *Archaeopteryx, the Primordial Bird: a Case of Fossil Forgery* argued that the fossil had been manufactured in the nineteenth century by a Bavarian fossil collector to cash in on the public interest in Darwin's *Origin of Species*.[57] Reviewing the book for *Nature*, zoologist and curator Tom Kemp wrote:

> The authors go about the task of proving their theory with all the comprehen-sive enthusiasm of fictional detectives, although of the Inspector Clouseau rather than Sherlock Holmes variety. For, whether through perversity or stupidity I do not know, they exhibit a staggering ignorance of the nature of fossils and fossilisation processes. . . . The motive [for writing this book] seems to relate to the authors' own theory of evolution . . . it seems that major evo-lutionary change occurs as a result of 'genetic storms,' whence showers of virus particles from outer space invade Earth and become incorporated into the genomes of living organisms. . . . As to the effect of the book, I am afraid that it has a superficial appeal and will undoubtedly gain currency among Creationists. . . . At best this book is mischievous; at worst it might well be described as evil, because it betrays that morality of objectivity that science cherishes. Perhaps *Archaeopteryx* is a fake. Certainly the possibility should be

investigated. But it should be done by people who actually understand fossils . . . not by a couple of people who exhibit nothing other than a Gargantuan conceit that they are clever enough to solve other people's problems for them, when they do not even begin to recognise the nature and complexity of those problems.[58]

The Natural History Museum felt that it had been accused of condoning forgery, and the Head of its Palaeontology Laboratory, Peter Whybrow, protested in the letters column of *New Scientist*.[59] In the mild disguise offered by his home address, Whybrow also wrote another letter, this time a spoof arguing that *Archaeopteryx*'s feathers had evolved because they protected the beast from showers of genetic material falling from space.[60]

Spetner by this time had stepped away from Hoyle and Wickramasinghe, and for some time he corresponded with Charig about palaeontology. Just after Hoyle and Wickramasinghe's book was published, Charig wrote to Spetner:

> Last year this controversy was conducted in a reasonably friendly and scientific manner. This year H. and W. have thought fit to change its whole tone, introducing hostility and bitterness and making outrageous accusations of improper behaviour, not only against me but against almost the whole of the palaeontological profession, geologists and zoologists too. . . . Indeed, I understand that some of the comments in their book are technically libellous, but we have no intention of taking action against them . . . it is the saddest thing of all that these two men, rightly renowned for their scientific achievements in their own field, should now be doing such irreparable damage to their own reputations. I am glad that you have not thought fit to associate yourself with their activities; I do not agree with your ideas, but at least I respect you for holding them sincerely and for believing that we are not guilty of cheating or lying or any other form of dishonesty.[61]

By 1987 the Natural History Museum had mustered, with input from Hoyle, an exhibition entitled *The Feathers Fly: Is Archaeopteryx a Genuine Fossil or a Fake?*[62] Hoyle and Lady Barbara accepted an invitation to the Museum to see the exhibition in development. When it opened, the exhibition challenged visitors to 'decide for yourself if *Archaeopteryx* is a fake,' but it also offered Creationism as the only alternative to Darwinism; and the Museum's staff were not coy about their agenda: Charig told a press conference that Hoyle had used 'faulty observation, incorrect data and untrue statements, arrived at through sheer carelessness, lack of knowledge and false logic.'[63] The exhibition brought public attention to the issue once again, and the *Independent* newspaper expressed concern that while 'Sir Fred and his colleagues are not creationists, . . . their allegations could be used the next time that born-again Christians challenge the teaching of Darwinism in US schools.'[64] Hoyle was unmoved: in 1991 he told the *Independent on Sunday* that *Archaeopteryx* is 'an obvious fraud. "Either they haven't looked at the fossil, or they are liars. I prefer to believe they haven't looked at the fossil." '[65] Hoyle's colleagues were baffled by the episode: John Maddox would later write in *Nature* that 'Even his best friends were at a loss to know why he behaved in such a way. . . .'[66]

Strong reactions appear not to have deterred Hoyle and Wickramasinghe: *Cosmic Life-Force*, another non-fiction treatment of the life-from-space theory, was published in 1988. It spent less time on the gritty details than its predecessors, and concluded with a chapter on 'The concept of a Creator.' One reviewer noted that 'We were about due a reappearance of [this] hypothesis . . . [the] argument's already eccentric trajectory looks set to spin out of worldly orbit . . . we shall probably never see its like again.'[67] He was wrong on the last point: *Our Place in the Cosmos* retold the story in 1993.[68] It contained little that Hoyle and Wickramasinghe had not published before on life from space, but it was resoundingly condemned. DNA chemist Robert Shapiro of New York University, and author, in 1987, of *Origins: A Skeptic's Guide to the Creation of Life on Earth* in which he rejected Hoyle's work, reviewed the book for *Nature*. Shapiro's view was that 'This book cannot be taken seriously as a work of science,' but he thought it an interesting document of a great scientist's decline.[69] In *Science and Public Affairs*, the house journal of the Royal Society, Oxford chemist Peter Atkins wrote:

> . . . this book is not worth the paper it is written on . . . it is a travesty of science . . . a personal diatribe against established science . . . one can only hope that it will . . . expose the authors' hubris and incompetence to the point that publishers will cease pestering us with their nonsense.[70]

Hoyle had come up against an influential critic at a difficult time. Atkins was one of a group of senior scientists, based mostly in Oxford and including a number of distinguished biologists, who were, like the sceptical movement in the USA, actively protesting against – instead of simply ignoring – a variety of unorthodox scientific ideas and alternative theories and beliefs, particularly those that might be confused with science or that traded off science's reputation. Their targets included science fiction, parasciences such as parapsychology, alternative therapies such as homeopathy and many common elements of popular culture such as horoscopes, UFO sight-ings and New Age therapies. This group became, like the Royal Society which had published Atkins' review, prominent activists under a banner of 'public understanding of science', their rationale being that the public should only be exposed to 'good' science.

According to the assistant editor of *Science and Public Affairs*, the Royal Society chose to publish Atkins' strident review rather than simply ignoring the book because Hoyle was supposed to be 'one of us' – he was a Fellow of the Royal Society – and could not be allowed to bring the scientific com-munity into disrepute in this way. The book's publisher, Dent, was also to take note that this was not the sort of book it should publish if it wanted the cooperation of the scientific community.[71] Hoyle felt very bruised by the review, coming as it did via the Royal Society – he was after all, 'one of them' – and his wife took up his cause: Lady Barbara voiced her objections to a member of *Science & Public Affairs*' editorial board.[72]

But while Hoyle's ideas were beyond the pale, he was not. He celebrated his seventieth birthday in 1985, and the newspapers paid tribute to his long and colourful career.[73] He published an autobiography in 1986: *The Small World of Fred Hoyle*. It was well received, although the *Observer* commented that 'Since Fred Hoyle has always leant heavily on his person as a blunt, plain speakin' Yorkshireman, it comes as no surprise that his memoirs have something of the aura of a Hovis advertisement' – a reference to a long-running television commercial that showed a young boy in a mistily quaint northern town coming home to a sustaining meal of the Hovis brand of bread.[74] The life-story offered in *Small World* concentrates on Hoyle's Yorkshire childhood and his undergraduate years in Cambridge, and con-cludes with his whirlwind courtship of Barbara at the outbreak of World War II.[75] A second, fuller, autobiography, *Home is Where the Wind Blows*, was published in 1994.[76]

The subtitle of this second autobiography, *Chapters from a Cosmologist's Life*, acknowledges that it is not a complete record. The book was the result of a long process of distillation in which a passionate and often bitter rant against the iniquities of politics and the scientific establishment was trans-formed into a charming, sometimes mischievous story. Early drafts had been rancourous, and reopened old wounds.[77] But the end result was genial enough, if vocal in its omissions. Colleagues and fans hoping for some glimpses behind the scenes of the more colourful episodes of Hoyle's life were to be disappointed: it is a book that reveals few secrets, and does not name names. Some of those involved with the founding and demise of IOTA and Hoyle's departure from Cambridge do not recognize his version of these events.[78] One story told with conspicuous brevity is that of the theory of life from space, to which Hoyle devoted only two paragraphs:

> Over the decade from 1975 to 1985, I became interested, together with my colleague and former student Chandra Wickramasinghe, in the big problem of the origin of life. We came to think that life is a cosmic phenomenon and not the outcome of a number of highly improbable events that took place locally here on the Earth. It will seem strange to the nonscientist that this clearly interesting question should have provoked a high degree of opposition and resentment, but very certainly it did so. Curiously, it is perfectly all right to say that life exists elsewhere in the Universe, provided it is considered to have arisen separately in each place of its existence, even though to have arisen separately would require a repetition of the same highly improbable events as on the Earth. What must not be done is to regard life in many places as manifestations of the same cosmic process. This, it seems, deeply offends against the scientific culture.
>
> Although – as yet, at any rate – we have not received any plaudits for a decade's quite hard work on this question, at least on one score we were success-ful. It was a consequence of our views that much of the material on the outside of our planetary system would have to be of an organic character, and also much of the solid material in interstellar space, a prediction that is now acknow-ledged fairly generally to have been correct.

These two paragraphs suffice to summarize what even by Hoyle's account was a decade's work; perhaps the criticism had hit home. Colleagues have suggested that in later years the impetus for the work came solely from Wickramasinghe, but that journalists and reviewers foregrounded Hoyle because of his familiar face and wide appeal.[79] Wickramasinghe had made the cosmic extent of life the topic of his inaugural lecture at Cardiff in 1973, and has continued to publish on the subject after Hoyle's death. He is now Director of the Centre for Astrobiology at the University of Cardiff, and involved in experiments with high-altitude balloons that are designed to collect samples from the upper atmosphere. The aim is to detect material there that could only have come from space. He is collaborating in this work with scientists in the UK and in India, including Jayant Narlikar. Wickramasinghe's ambitions for the project echo the 'stardust' ideas of the 1960s: the acceptance of the theory of cosmic life would, he argues,

> . . . herald a new worldview that I hope will unite the many nations and races of people that inhabit our planet. To realize that our life is only a minute compon-ent of an all pervading cosmic living system, and the knowledge that our genetic brothers and sisters still lurk amidst the stars will, I hope, change the way we look at our trivial differences and ourselves.[80]

As Hoyle had noted in his autobiography, other researchers have now also found organic chemicals in space, and centres for research in astrobiology are becoming more common – NASA first discussed astrobiology in 1995.[81] So Hoyle and Wickramasinghe's work may yet come to be seen as valuable, if unsteady, early steps in an important new field. But while Hoyle believed that once he had set off along a line of enquiry he should not abandon it just because the data was sending him in uncomfortable directions, he had become less committed to the work. In an interview in 1993 he concluded the discussion of life-from-space theory with the words: 'I wish I'd never got into that.'[82]

18
A new cosmology

Steady-state theory turns 40; the redshift problem; the quasi-steady-state theory; Hoyle's legacy

A fter 15 years in the rugged landscape and capricious weather of the Lake District, the Hoyles were finding life less comfortable in their isolated home. So in 1988 they moved to Bournemouth, a seaside town on the south coast of England much favoured by tourists and retirees for its sandy beaches and gentle climate. They had found a spacious flat on top of the cliffs, with broad views over the sea; and Hoyle settled into an armchair by the living room window. Despite his great enthusiasm for computers as calculating machines, he never took to the PC: he preferred the pad of lined paper that had been his chalkboard and medium throughout his career. He did, however, install a fax machine, and in Bournemouth, this was to prove his intellectual lifeline. In this final phase of his life, Hoyle returned to old friends and a new cosmology.

Much had happened in the 20 years since Hoyle left Cambridge: his retirement had coincided with the establishment of big-bang theory as the organizing principle for research in cosmology. Although the discovery of the cosmic microwave background radiation in 1965 is often cited as the defining moment in big-bang's victory, cosmologists tend to refer to the early 1970s as the time when their work finally settled around the evolutionary paradigm. Indeed, in December 1972, during Hoyle's last days as a member of Cambridge University, *Nature* published a 'Steady State Obituary?', and concluded that the question mark was not justified: the evidence against the theory was 'entirely convincing.'[1] So the fighting was over, and now the work was in the details. As Hoyle left the Plumian Chair and what he called 'the cosmological scene', he felt he was 'leaving it to the younger generation to overgraze the infertile pasture of big-bang cosmology.'[2]

To Hoyle, this complacency was madness. The 1960s had produced a torrent of startling data and new phenomena – quasars being merely typical of the surprises and mysteries that decade had produced – and this was surely not the time to be hiding in the theories of the past. Rather than squeezing these new observations into a theory that strained to accommodate them, argued Hoyle, should not the fundamentals be reviewed? In 1972, repeating a theme from his Presidential Address to the Royal Astronomical Society that same year,[3] he told an audience of the American Astronomical Society in Seattle of the 'developing crisis in astronomy': it was time to rethink fundamental physics.

> For each of us there is a decision to be made. Do we cross a bridge into wholly unfamiliar territory or do we try to remain safely within well-known concepts? This depends on how each of us sees the data. For me personally, the exact state of the data at any given moment is less important than the trend of the data. There seems to me no doubt that the trend is toward forcing us, whether we like it or not, across this exceedingly important bridge. Either the bridge must be crossed or one must judge the data of the past five years to be extremely freakish.[4]

Through the 1970s, even as Hoyle busied himself with ideas a long way from fundamental physics, the crisis he saw in astronomy was still on his mind. In his 1975 textbook, *Astronomy and Cosmology: A Modern Course*, he continued his campaign for a reassessment of steady-state theory.[5] Parts of the book were reprinted verbatim in 1980 in *The Physics–Astronomy Frontier*, which was co-written with Jayant Narlikar, who was now at the Tata Institute of Fundamental Research in India. In *The Physics–Astronomy Frontier*, perhaps in acknowledgement of the prevailing climate and with an eye on potential sales, evolutionary cosmology was given its adopted name of big-bang theory, and the steady-state cosmology was relegated to Appendix D.[6]

Steady-state theory did, however, take centre stage in a short book Hoyle published that same year via University College Cardiff Press. Called *Steady-State Cosmology Revisited*, it elaborated upon the oscillating cosmos Hoyle had first proposed in the late 1960s. In this model, the age of the universe calculated from its expansion does not describe the full extent of the history of the cosmos, but only of one cycle of the oscillation of the universe. Hoyle also explored the weaknesses of the original steady-state models, and found an analogy for his new idea:

My 1948 form of the SS theory was rather like a gramophone record stuck in the groove, playing endlessly the same phrase, while the Bondi–Gold form of the theory was like a record that plays only one note, a sort of test record one uses to check a stereo system.

One does much better to listen to a recording of Beethoven's 5th. If one analyses the music rotation-by-rotation of the turntable, nothing much usually happens in a single cycle (although occasionally there is a sudden outburst when something drastically different really does happen). Over the course of a whole movement, involving very many rotations, the information content of the music changes – more so if we substitute a late quartet for the 5th. Yet from beginning to end through the movement there is a coherence to the structure of the music, analogous I would say to the working through of the consequences of the values of the physical coupling constants which specify a particular subspace of the whole universe. Changing the subspace would be like changing the record from Beethoven to Mozart.

Possibly reinterpretation of SS theory along these lines will receive a more sympathetic reception than did the theory of 1948. In retrospect I can see that not everybody welcomed the idea of listening to a record in a stuck groove. Nor does it take forever to test a stereo system.[7]

Big-bang cosmology flourished during the 1980s, with IOTA graduates prominent among the many scientists contributing to an increasingly elaborate body of theory. The gaps in the theory were being filled, often by new phenomena. One such was the black hole, an idea to which Stephen Hawking and his University of London colleague Roger Penrose made a crucial contribution in the late 1960s. Black holes, which are formed when a dense object such as an elderly star collapses in on itself, and from which nothing can escape, provided a means for understanding the big bang itself: whereas when a black hole forms, matter and energy collapse into an object of ultimate density, the big bang was the opposite – energy and matter burst out of a tiny dense core. Hawking and Penrose found that the two processes were indeed mirror images of each other, and could work equally well in either direction – a conclusion that gave weight both to the idea of the black hole and to the theory of the big bang.

More support for big-bang theory came with the revival of the idea of dark matter. This strange stuff had been proposed in the 1930s to account for a problem in the gravitational environment across the universe: clusters of matter, such as gas clouds or groups of galaxies, appeared to behave as though the gravitational field they inhabited was stronger than could be accounted for by the matter one could observe in the vicinity. This extra gravitational force must have come from somewhere, and so it was proposed that it originated from some invisible matter (hence the name 'dark'). The idea lay neglected for more than thirty years until big-bang theorists came to struggle with the same problem again: how does a cluster of matter in expanding space stay together when its gravitational field ought to be insufficient? They adopted the idea of dark matter, and it became an important feature of the big-bang cosmos – indeed, by some estimates, 95 per cent of all matter in the universe is dark, and the stuff we encounter every day is relatively insignificant.

Another important contribution was provided by Massachusetts Institute of Technology physicist Alan Guth, in 1981. The initial expansion of the classical big-bang universe had been beset by problems: among them were the question of how the universe could initially burst forth and expand in seeming defiance of its own gravitational field which should instead pull it in on itself; and why the character of the universe is so constant across such great distances. Guth proposed, in an idea known as inflation theory, that the universe began as a very small, very dense bit of matter that achieved a uniform character by virtue of all parts of it being in close contact. This dense bit of matter then entered an unusual state in which its gravitational field was reversed. So instead of gravity holding the matter together, this matter pushed outward, expanding with great force, opening up a vast space and spreading its uniformity across it in the process. The negative gravitational energy was transformed into matter, and it became more negative as more matter was created. Once it had expanded, fluctuations in the energy density across this universe provided the points at which large structures such as galaxies could form. In this peculiar and special condition of inflation, the sudden expansion of the universe seemed much more reasonable: instead of all matter being contained in a tiny, ready-made core, like Lemâitre's primeval atom, here the expansion is what creates the stuff of the universe as it goes along, and mass–energy is conserved. Hoyle thought this work the only exception to his claim that 'cosmology has achieved nothing of real importance in a quarter of a century,'[8] and he was not the only scientist to notice that Guth's mathematics was very similar to the work he had done with Narlikar on the C-field in the early 1960s: replace their C by Guth's term Θ, and the theories are very similar indeed. Hoyle later reflected that the lesson from Guth's success is that:

... it is a good maneuver to begin by identifying new weaknesses of the existing paradigm and then to show how the proposed new idea removes them. If you simply point out the weaknesses and ask for the revision of the standard paradigm nobody will take you seriously, because doing so would imply a threat to existing beliefs.[9]

Work on black holes, dark matter and inflation theory fuelled a thriving industry around cosmology, which in turn generated more ideas. The frontier between astronomy and particle physics – a frontier crossed by Hoyle when he had walked into Fowler's office in 1952 – was long gone, and big questions in cosmology were being answered in giant machines such as cyclotrons and supercolliders, the 'atom smashers' that scientists used to try to mimic the extreme conditions of the early universe. More jobs were created, more scientists considered these problems, and the science popularizers of the 1980s all talked about big-bang cosmology. Hawking's *Brief History of Time*, published in 1988, became the biggest-selling science book in history. When, as part of a poll conducted in 1989, the British public was asked to say 'true' or 'false' to the statement: 'the universe began with a huge explosion', the right answer, according to the pollsters, was 'true.'[10]

Not everyone was comfortable with the situation. Was big-bang theory leading scientists to great discoveries of otherwise unimaginable phenomena such as inflation with its reverse gravity, and the vast quantity of dark matter that clogs up the universe but yet remains elusive? Or were these oddities convenient fictions to shore up a theory on shaky ground? After all the investment and careers made on big bang's back, would any amount of counter-evidence suffice to pull scientists away from it now? At Cornell University, Tommy Gold was among the sceptics:

> Everyone is so convinced now that this huge edifice has been built up. So much was built up on the presumption that the big bang must be right. So then this must have happened, and this must have happened, and the other must have happened, and they are all pure inventions just to make it possible for the big bang to be right. A huge superstructure is built on this without having done anything to examine the foundations. This is what I paraphrase as my own version of the 'Confucius say'. Confucius say: 'Never judge strength of foundation by size of building.'[11]

Hoyle too thought big-bang theory's foundations were weak, which was good reason to remind people of its rival, steady-state theory. That theory celebrated its fortieth birthday in 1988, and Hoyle was reunited with the third member of the Cambridge Circus, Hermann Bondi. Bondi had long since given up on steady-state theory, but he was happy to reminisce at a meeting called in Cambridge for the anniversary. He had left King's College London in 1983 to become Master of Churchill College Cambridge, and Hoyle and Lady Barbara stayed at his home for three weeks around the meeting.[12] (Reflecting Hoyle's other interests at this time, Barbara used the back of Bondi's letter of invitation to note down some data on the infrared absorption spectrum of DNA.) Bondi's talk at the Cambridge meeting was confined to the early post-war years, but Hoyle brought the story up to date, with an assessment of the evidence against steady-state theory. He reminded his audience of the two original papers, and dissociated himself from Bondi and Gold's version: Hoyle described it as a corset that had unnecessarily constrained the idea, making it too vulnerable to criticisms.[13] Hoyle was prepared to sacrifice the 'strong predictive quality' of the Bondi–Gold theory for the 'greater resources' of his own version.[14]

According to Hoyle, the cosmic microwave background radiation could now be accounted for by the thermalizing action of iron whiskers in space. Hoyle and Wickramasinghe had considered many candidates for this thermalizer, including solid hydrogen,[15] but now they were committed to iron, and would present the idea in support of steady-state theory at a conference later that year.[16] The counts of distant radio sources, as presented by Ryle in his challenges to steady-state theory, could be discarded, since it was now known that radio sources were mostly either galaxies or quasars, and for Hoyle there were clearly problems in deciding whether they were distant or not.

Hoyle's list of charges against the big-bang universe also included it being an inappropriate environment for the development of life. He had told the symposium for Fowler's seventieth birthday in 1981, where he had first used his 'tornado in a junk-yard' analogy for the unlikelihood of life arising on Earth, that the cosmos would have to be several times its supposed age to explain the advanced state of life. Now he concluded his reflections at the Cambridge meeting with a challenge to the 'evolutionary' aspect of big-bang theory:

> To deny evolution in the sense in which big bang cosmologists have considered evolution is no loss at all. . . . The proper philosophical point of view, I believe, for thinking about evolution cosmologically involves issues that are superastronomical, as one inevitably gets as soon as one attempts to understand the origin of biological order. Faced with problems of a superastronomical order of complexity, biologists have resorted to fairy tales. . . . To have any hope of solving the problem of biological origins in a rational way a Universe with an essentially unlimited canvas is required. . . . It is to provide just such an unlimited canvas that steady state theory is required, or so it seems to me.[17]

Hoyle had come to regret using the word 'steady': it was taken too much to mean that nothing ever happens, whereas what he wanted instead was a universe that kept to a pattern, because 'only by having such a Universe as an unwavering backdrop could life evolve, because an unwavering physical environment is an essential requirement for such a superastronomical development to take place.'[18]

Hoyle argued that the creation of matter, instead of happening smoothly throughout space-time as he had proposed in 1948, happens in 'little bangs' that give rise to:

> ... whole clusters of galaxies, ... with the physical processes occurring at the genesis of each such unit being closely the same as those which are in modern cosmology taken to apply to the whole universe. ... Each new object makes room for itself among previously existing units, forcing the previously existing units to move apart from each other, and so providing a physical raison d'être for the expansion of the universe.[19]

Unlike big-bang theory, 'in which nothing is ultimately explained, ... the steady-state theory has the immense task of explaining everything in terms of ongoing processes':

> The theory requires that all observed physical properties must be capable of regenerating themselves, otherwise a property would already have died out over the indefinitely large number of previous generations of the Universe. ... Particularly, each generation of clusters of galaxies must be capable of regenerating itself, as in biology each generation of plants scatter seeds some fraction of which germinate and reproduce the next generation of plants. ...[20]

In their paper on iron whiskers, Hoyle and Wickramasinghe explained that this process of regeneration means that galaxies or associations of galaxies can have any age, up to about twice the age calculated for the big-bang universe. They proposed that our galaxy is one of the older ones. In this scheme, the universe has no age as such, but the fact that Earth is very old, in a very old galaxy, is demonstrated by the advanced state of evolution of nearby stars. In his thinking about cosmology, Hoyle had often used anthropic arguments: that is, he believed that the universe must have certain properties (among them, his predicted state of the carbon-12 nucleus) because without these properties, there would be no people around to think about such things. In this spirit, he and Wickramsinghe asked:

> ... why do we live in an exceptionally old system? If we suppose that upwards of 10^{10} years are needed to evolve an animal of our intelligence we have no option but to be in an exceptionally old system. The requirement would be an anthropic necessity.[21]

✳*✳

Through the 1980s, Hoyle was in contact with his friend Halton 'Chip' Arp, whose work on the association of supposedly far-off quasars and nearby galaxies had become an important part of Hoyle's challenge to big-bang theory. Over the thirty years since he had first discussed with Hoyle and Geoffrey Burbidge the idea that some quasars might be local, Arp's career had been dogged by resistance to this work. A boost for the idea that quasars are local had come in 1971 when some quasars were found to be receding, according to their redshifts, at speeds faster than the speed of light – a physical impossibility that suggested that redshifts were not a reliable indicator of speed and therefore not of distance either. But conventional explanations soon followed, and the idea of distant quasars remained dominant. Arp had however found that he was able to photograph a physical connection – a luminous bridge – between quasars and nearby galaxies. This connection surely made it impossible for the galaxy to be nearby and yet the quasar to be very far away. Again, colleagues were sceptical, demanding more examples; but at the same time Arp was finding it harder to get observing time at the Caltech telescopes in order to satisfy this demand, even though other observers, among them Margaret Burbidge, were finding similar phenomena.

Like Burbidge and Hoyle, Arp had come to the conclusion that quasars were produced in galaxies, and were then ejected out into space. Hence, the broad distribution of quasars. Some, however, were ejected only weakly, and stayed near their galaxies, hence the associated objects. The quasars that showed the luminous bridge leading back to their parent galaxy were being photographed at the moment of birth.[22] Often, the galaxy and the quasar had very different redshifts. With such a close physical connection between the two objects, the assumption that their different redshifts meant that they were at different distances away was, to Arp, simply nonsense.

Arp pressed on, finding, through the 1970s, more and more evidence to support his case. He spoke in a debate on the problem organized by the American Association for the Advancement of Science in 1972, but did not carry the day. While many astronomers saw that event as putting an end to the matter, a few of his opponents kept looking, and they continued to fail to find the same evidence in their observations of the same objects.[23] Towards the end of 1981, the committee allocating observing time on Caltech's Mount Palomar telescope advised Arp that he should reorientate his research programme if he wanted to continue observing. The following year his allocation was much reduced, and by 1983 Arp was removed from the observing schedule. This made front-page news in the *Los Angeles Times*, which reported that 'the dispute is a classic example of a dissident voice being stifled by a powerful majority.'[24] At the time, a committee of the National Research Council was reviewing the provision for astronomical research, and on quasars it concluded:

> ... the long-standing controversy over the nature of red shifts came close to resolution [over the past ten years]. ... It seems likely that quasars as a class really are very distant and hence have extraordinarily high luminosities. ... Despite the enormous energies involved in some of the outbursts observed in quasars and active galaxies, it has been possible to devise dynamically self-consistent models of these outbursts. There is thus at present no strong theoretical reason to doubt the cosmological nature of the observed red shifts or to believe that 'new physics' is required to understand these objects.[25]

An observational astronomer is lost without telescope time. Arp left Caltech and pursued his work at the Max Planck Institute for Astronomy and Astrophysics in Munich, Germany.

Arp's exclusion, and Hoyle's support for him, marked a deterioration in Hoyle's relationship with Caltech; it was swiftly followed by Fowler's warning to Wickramasinghe that participation in the Arkansas creationist trial would be a step too far, and by Hoyle's exclusion from Fowler's Nobel Prize. Burbidge knew that 'Fred wasn't very keen on Caltech in many ways, he thought it was a pressure cooker. Fred liked to sit and write and think and talk to his friends, people he respects. He didn't really like all the ballyhoo at Caltech.'[26] Hoyle's last formal appointment there had ended in 1982, and his subsequent trips to the USA were more often to see Don Clayton in Texas.

In 1987, a conference to celebrate Arp's sixtieth birthday brought together supporters and opponents of his work, and he published a book, *Quasars, Redshifts and Controversies*, which covered the history of the debate and the latest data. One reviewer was Hoyle, and he was in a pugnacious mood. He argued that the book demanded humility in the reader – humility towards the Universe. 'Either one must know enough about physics to be aware of the limitations of one's own knowledge, or one must have the innocence of heart to believe the evidence of one's own eyes.'[27] Hoyle noted that even now, with more than 3,000 quasars to study, they still did not, like the first 40, fit neatly into an expanding universe. However, he said, 'the mob' at scientific conferences would hear Arp's claims and then 'relapse into a kind of collective amnesia'; nevertheless:

> It is my opinion that this book presages a revolution of scientific thought. There have been major revolutions in the past and in every case the younger

generation always says to itself: why didn't they see it before? Answers are usually bland with no shortage of face-saving excuses. Chip Arp has made certain, however, that there can be no face-saving excuses on this occasion. Once the earthquake comes several hitherto famous institutions are going to disappear in the rubble.

Hoyle took an obvious swipe at Caltech in his review:

This is not a book for the bustling products of famous, know-it-all graduate schools, where students have been required to learn too much too quickly, where they have been obliged to accept their instructor's views on trust, and where they have been given to understand on entering research that their best hope of advancement is to support conformity in every controversial issue of the day.[28]

Hoyle's analysis of Arp's demise at Caltech was that 'he was too far out, too close to the uncertain frontiers of knowledge, to be accommodated into the dependency culture science has become.' Arp had written in his book that:

If the goals of research are subordinated to any other consideration, this can lead to the rapid decline of excellence in even the most prestigious institutions. One danger today . . . is that with science tied increasingly closely to expensive equipment, which in turn is tied to prominent institutions, . . . science may progress less rapidly than its potential. . . . In fact, it might even become so that, as in the arts, the truly most creative and important achievements in science will not take place within the universities or institutions at all.[29]

In 1988, Arp found evidence that the luminous bridge between a quasar and a galaxy was made of hydrogen. The redshift of the bridge itself was the same as the redshift of the galaxy, meaning that the hydrogen was coming from the galaxy.[30] Arp sent Hoyle 'the latest summary of the news which the prelates are trying to avoid;'[31] he appeared on the BBC's *The Sky at Night* programme, and *New Scientist* carried news of his work. A paper by a team of American astronomers about a galaxy and the adjacent quasar was published in *Nature* in 1989,[32] and included a picture of the bridge; but Arp felt that the picture had been poorly reproduced, so that the evidence for the bridge was not clear. Arp wrote to *Nature* to protest that the picture should have been on the cover of the journal, and then it would have been big enough for readers to decide for themselves whether the quasar was connected to the galaxy. 'After all, if it is,' he wrote, 'there goes at minimum 25 years of extragalactic astronomy down the drain.'[33] Arp told Hoyle that:

. . . they probably did it to try to draw attention away from the result by running on the cover the faddish and completely empty distraction of 'colliding galaxies'. Perhaps this is the moment for each of us to let *Nature* know they are not hoodwinking all their readers. . . . I find myself spending a lot of time arguing for continuous creation these days. But it will only make slow progress until this fantasy structure built on redshifts is blown away and we can make a fresh start.[34]

Hoyle studied the picture, and decided that:

> . . . the diagram that they published there is unreadable – all that is blurred. It had been deliberately blurred by *Nature*, so that the reader who moves fast will not notice this. We noticed it of course, and if the information is still in the drawing, you can get it out again – you can use a light pen to get all the contours out and you can expand it in the computer and so on. . . . But if you actually look at *Nature* you will see that it has been deliberately totally degraded so the reader can't see what is there.[35]

In 2004, *Nature*'s editor, Philip Campbell, considered this interpretation of *Nature*'s treatment of the picture, and dismissed it as 'totally false.'[36]

While Hoyle's name was often seen in *Nature* during the late 1980s, both on research papers about the spectroscopic work with Wickramasinghe and in correspondence and reviews, his cosmology remained too far from the mainstream for publication in such a journal. Science had lost its spirit of adventure, according to Hoyle, and surprising discoveries like Arp's were being missed all the time, or not followed up, because there were too many pressures on scientists to conform:

> [It] is not a healthy situation. . . . it's entirely due to the way [science] is organised. We find lots of young people who when they make discoveries like [Arp's] they would like to follow it up but they are quietly told that their career will end if they do. That the studentships or temporary jobs or whatever will not be renewed. So you can't get any young person who's in the inventive ages to follow up strange new things today.[37]

Hoyle felt that journals had great power to implement and preserve the scientific community's conservatism:

> Science today is locked into paradigms. Every avenue is blocked by beliefs that are wrong. And if you try to get anything published by a journal today, you will run up against a paradigm, and the editors will turn it down.[38]

But Hoyle was now in close touch via the fax machine at home in Bournemouth with Jayant Narlikar at the Inter-University Centre for Astronomy and Astrophysics in Pune, India, and Geoffrey Burbidge in La Jolla in the USA. According to Burbidge, the three friends' research strategy followed the principle Hoyle had been spelling out since the early 1970s: observations should be allowed to raise questions about theory, rather than theory being allowed to constrain observation. New observations might mean that old and cherished theories have to be discarded, or new theories found. For example, there are groups of galaxies that appear to be coming apart, and convention, as Burbidge explains, is:

> . . . to conclude that they are not coming apart, they are bound together, and in order to do that you have to put in large amounts of dark matter to bind them together . . . the kind of matter that one had never heard of before. The alternative is to say they really are coming apart. . . . Fred and I considered they might really be coming apart. It's that kind of idea.[39]

The point of the exercise, according to Burbidge, is to expand our understanding of the physical world, rather than to pick over the details of what is already known. And while the particle physicists battle for ever more expensive atom-smashers on Earth, a ready-made laboratory is there in the sky:

> A large part of physics came from studying astronomy – the obvious case is Newton, and Newton's laws. But we are now at a point in time where most people are taking the position that the laws of physics are completely known . . . they won't allow the possibility that you may learn something new about physics from astronomy. The only new physics which people will allow is if it's early in the big bang – they will not allow it to go on now. One of the rods that has been used to beat us is that 'Geoff and Fred believe in the new physics.'[40]

As Hoyle had put it, as far back as 1976, an astronomer who wanted to make progress had two options: either he could sit around idly 'waiting for the physicist to scale some new peak of achievement', or he could 'be willing to risk some thoughts of his own.'[41] Hoyle, Burbidge and Narlikar were in the latter camp.

Burbidge was powerful ally in the battle for publication – according to Hoyle, he was:

> . . . a tremendous fighter: he's hardly ever had a paper not printed because he just has a tremendous voice – a Pavarotti sort of voice and build, and these big heavy fellows seem to have these huge voices which will fill an auditorium without the slightest effort, and he has the capacity to terrorise editors of journals by telephoning every day if necessary . . . secretaries of societies and so on, they ultimately decide that to print the paper would be the simplest solution! To stop him persecuting them. Because he's so logical, devastatingly logical at the same time. And so I am fortunately in that position now of not having to

worry about that – only having to worry that the arguments are correct, and not having to persuade people.[42]

The first fruits of this alliance appeared in 1990, in a paper by Arp, Burbidge, Hoyle, Narlikar and Wickramasinghe, for which John Maddox, *Nature*'s editor, instigated a new category of paper, called 'Hypothesis.' This category was intended for papers 'that fail to win the full-throated approval of the referees to whom they have been sent, but which are nevertheless judged to be of sufficient importance to command the interest and attention of readers – sometimes, as in the case of at least two of the referees of the first document in this series, by the sceptics concerned.'[43] *Nature* described the authors as a talented group of distinguished scientists, and noted the time and effort they had spent on responding to referees' criticisms. The paper discussed 'evidence to show that the generally accepted view of the Big Bang model for the origin of the Universe is unsatisfactory,' and proposed instead a steady state in which there are many creation events over time, just as Hoyle had described at the fortieth anniversary meeting in 1988.[44]

In the summer of 1992, Hoyle visited the Institute of Astronomy in Cambridge, where a life-sized statue of him was unveiled in the garden. The bronze Hoyle, in trademark open-necked shirt and sweater, steps purposefully towards his building (the real Hoyle accepted the honour in jacket and tie). Although he had visited the Institute since resigning, the unveiling was at last an occasion on which, according to *Nature*, 'many hatchets were elegantly buried.'[45] The atmosphere was friendlier now, and the *Nature* 'Hypothesis' was just a start in the career of the new cosmology: in 1993, Hoyle, Burbidge and Narlikar published a paper in the *Astrophysical Journal*, the prestigious American research journal.[46] Entitled 'A quasi-steady-state cosmology', it described a universe that grew through cycles of expansion and contraction. Hoyle explained to a journalist that it was 'not yet a complete theory, but more of a new vision which is intended particularly to shift this inappropriate fixation with the big bang.'[47] This universe had no starting point, but cycles succeeded one another as do the generations. 'In our model,' said Hoyle:

> . . . the universe is in perpetual oscillation, a contraction phase followed by an expansion phase. Associated with each contraction phase is creation of matter, in a sort of mini-big bang. The creation of matter is like a perturbation, such that in the contraction-expansion cycle of about 30 billion years, it is superposed on a general expansion over at least many hundreds of billions of years. In that sense, we have built a quasi-steady theory. Not every moment is the same – you can tell if the universe is expanding or contracting – but still, over a long enough period, the properties of the universe are repeated.[48]

Unlike the continuous creation of 1948, creation in this universe happens in brief, distinct episodes, in the intense gravitational fields associated with the contraction phases. Unlike in big-bang theory, where the creation of matter was not explicable in physics, in quasi-steady-state theory it was controlled by 'very strict mathematical laws' with the introduction of the C-field from Hoyle and Narlikar's earlier work.

A new cosmology from Hoyle was news: the UK press reported that he was back in the fray.[49] Hoyle felt that the paper had been received 'pretty well.'[50] To a journalist who asked about possible observational tests of the new theory, Hoyle showed that he had learned the lessons of proposing a theory that predicts too much:

> One of the main tests would be the direct detection, which is unhappily still out of reach, of gravitational waves associated with the mini-big-bangs. . . . Unlike big-bang theory where the creation of matter is a unique, invisible and unstoppable event, we predict the appearance of certain new particles . . . whose behaviour is still poorly understood. Perhaps that's a good thing: a theory that is too predictive has a high risk of being prematurely and unjustly invalidated, since the exceptions that prove the rule are always the ones discovered first![51]

Other scientists struggled to understand the new theory. At Caltech, physicist Charles Barnes, Hoyle's colleague from the carbon-12 experiments, read it:

> . . . as the evidence gets stronger and stronger for the big bang then the theories that Fred concocts with his collaborators have to get more and more complicated. . . . I have had enormous trouble understanding [the paper] – it's not very long but I had great trouble understanding it.[52]

Some colleagues commented that the new cosmology simply replaced big-bang's one big miracle by lots of little miracles.[53] But Hoyle was invigorated by the idea, and was pleased to be in close collaboration with Burbidge and Narlikar once again; picking up the threads of their long collaborations, their work flourished. In 1994, Hoyle's life's work brought him a share in the Balzan Prize, a substantial cash award and medal that celebrates achievements in science and the humanities. Another share of the prize went to Martin Schwarzschild of Princeton University, with whom Hoyle had worked in the early 1950s on the evolution of the stars.

Another echo of Hoyle's time at Princeton was captured in a little book called *The Origin of the Universe and the Origin of Religion*, the text of a lecture Hoyle gave in New York in 1993.[54] He proposed a thesis reminiscent of the one described in 1950 by Immanuel Velikovsky in his controversial book

Worlds in Collision, where earthly disasters, such as those described in the Old Testament and other scriptures, were caused by astronomical phenomena. At Princeton in 1953, a sceptical Hoyle had advised Velikovsky, who was attending astronomy seminars there, that in astronomy, mathematics matters more than scriptures. Now though, astronomers Victor Clube and Bill Napier of Edinburgh University had suggested that Earth had periodically been showered by fragments of a giant comet.[55] So Hoyle had scientific grounds for believing that such astronomical events, which might well have been accompanied by lights in the sky or other unusual sights in the heavens, could have been the trigger for events on Earth that disrupted and reshaped civilizations and led to beliefs in a higher power. Hoyle ended his lecture on an upbeat note, however: he remarked that our understanding of this situation had come just about in time for us to prepare to fend off the next visit, around 2100, from Clube and Napier's giant comet.

Hoyle celebrated his eightieth birthday in 1995. He had recently been reminded of his own mortality: on a trip to the Lake District he had gone out walking and had failed to return, and he was found collapsed on the bank of a stream, critically ill from hypothermia. For a while it seemed that he would not live. Friends wondered whether such an experienced walker could have slipped, or whether Hoyle might have been brought down in more sinister circumstances.[56] But he was back on form for his birthday, which brought fond attention from journalists: *The Times*, referring to his new cosmology, reported 'a second dawn for the universal rebel.' Hoyle told *Scientific American* that he deserved a pat on the back for living so long. 'When I was young,' he said, 'the old regarded me as an outrageous young fellow, and now that I'm old the young regard me as an outrageous old fellow.'[57] Outrageous or not, the value of his work was becoming clearer, and 1997 brought another prize: this time, the Crafoord, which again carried a substantial sum of money, and was awarded, 40 years on, for stellar nucleosynthesis. Hoyle shared the honour with Ed Salpeter, who had pioneered the triple-alpha process in the early 1950s. The Crafoord Prize is awarded by the Royal Swedish Academy of Sciences in areas of science not covered by its Nobel Prize; but that qualification does not seem to apply in this case, as the citation was similar to Fowler's Nobel citation: the award was for Hoyle and Salpeter's 'pioneering contributions involving the study of nuclear reactions in stars and stars' development.' Colleagues saw this as some redress for Hoyle having been excluded from Fowler's Nobel, but Fowler did not live to see it: he had died in 1995, at the age of 83.

By now, faxes were flying between Bournemouth, La Jolla and Pune: new work was in progress. Hoyle, Burbidge and Narlikar were writing the inside story of post-war cosmology – a story in which they had played a major part. They wanted to explain both the history and the ideas, and show just not what was done, but how it was done, and why. They wanted to show where astrophysics had taken the wrong path, and how to set it right again. The writing was done mostly through 1996 and 1997, with the authors sometimes visiting each other, but mostly corresponding by fax. At the end of a day's work in Bournemouth, Hoyle would fax Burbidge's secretary Betty Travell in La Jolla, who would find Hoyle's handwritten script when she started her day's work. Travell would type the manuscript and fax it back to Bournemouth in time for Hoyle to start work the next day. As always, Lady Hoyle provided editorial support, on this occasion with the help of Geoffrey Hoyle's daughter Nicola, a mathematics student. Margaret Burbidge provided a critical eye and expert proof-reading in La Jolla.

The result of this collaboration appeared in 1999: *A Different Approach to Cosmology: from a Static Universe through the Big Bang Towards Reality*. As with his recent autobiography, so with this book: Hoyle found it disturbing to revisit the old controversies. Burbidge worked hard to moderate his friend's temper:

> I think in many ways he was quite bitter about it, and he used to relate it all to religion and mediaeval monks and all that. I toned down some of the things that were written in that book. . . . We decided we would talk about the way it really was, and by and large we did.[58]

Hoyle was convinced that Cambridge University Press – which is constituted as a Department of his old university – would refuse to publish the book, but publish it Cambridge did.[59] The book assessed the development of cosmology since the war, and offered some alternatives to the conventional picture, including the quasi-steady-state cosmology. Its concluding remark was 'If nothing else, we hope that we have made both theorists and observers aware that observations remain primary in this field.'[60]

The book was, according to Burbidge,

> . . . the result of what we've all been doing. It's very unpopular, still extremely unpopular. Nobody wants anything but a big bang. [The book has] been received quite well and it's sold well, better than anyone expected. Most of the reviews have been very reasonable, but there are some reviews where the

reviewer, if he is a cosmologist, he can't believe in it. . . . there have been some rotten reviews because people don't want the universe to be that way. . . .[61]

Some reviewers were hostile to the book, which is not always subtle: at one point it illustrates big-bang community with a photograph of a flock of geese on a country lane, each goose tottering after the one in front. The book also mocked the recent much-publicized claims made about the background radiation as a result of the COBE mission, which for some scientists had found 'the Hoyle Grail of Cosmology' and which for these authors was 'irrelevant to the point of absurdity.'[62] *New Scientist*'s review commented that the authors were just as guilty of the sins for which they condemned their opponents, and that 'a lingering bitterness pervades this book.'[63] Other reviewers were more positive: Maddox described it as

> . . . a scholarly rather than a polemical work, in that it is a well-documented guide to extragalactic evidence against the Big Bang. My own conviction is that Hoyle's scepticism [is] well-founded. But it is too soon to tell how the Big Bang will be replaced by some other cosmology. That is heterodoxy.[64]

According to University of Sussex astrophysicist Leon Mestel, a colleague of Hoyle's from the early 1960s at the Department of Applied Mathematics and Theoretical Physics:

> To most other workers in cosmology, it may appear as a labour of love. I would myself say that at a time when – to be sure, as a consequence of accumulating observational evidence – standard big-bang theory has acquired the explicit or tacit support of most of the astronomical community, it has been good for the health of the cosmological enterprise that an intellect as powerful as Fred's should ensure that possible alternatives do not go by default.[65]

Hoyle knew that times had changed in science: he missed the 'looser structure' of science in the pre-war years – fewer conferences, much cheaper research, and the intellectual space to take risks – which he thought was more stimulating and productive than more recent years of constraining career paths and time-consuming committees. By the end of the century he felt that 'we are running into a situation where no scientific discoveries will be possible.'[66] In 2001 he told a journalist:

> I had the good luck to enter a climate where everything was open to challenge, in the wake of the quantum mechanics revolution . . . . In that atmosphere the only way for a young man to make a career was to find something that no-one believed and to show it was true. Today, it's exactly the opposite. To be successful, you need to prove something that everyone wants to believe.[67]

Like Hoyle, Burbidge worries for the future of science that does not tolerate dissent:

If you do unpopular things you don't get supported. Fred and I, I call us the elderly radicals, and we are a very small population. It's very hard because the young people, why would they want to work with us? They don't see any support. Arp is one of the great elderly radicals. Margaret has been recruited, she now believes most of this – not in the early days, she sat on the fence. She's done her own observations . . . and the data are coming out. . . . I am sure we are on the right track with these things, and Fred was convinced of that also.[68]

Fred Hoyle suffered a severe stroke in July 2001, and died on 20 August, at the age of 86. He was working, and walking, until this last illness. His final scientific communication was news from his friends in India, where high-altitude experiments were detecting organic material that could have come from space.[69]

Hoyle's friend Bernard Lovell, writing in the *Guardian*, described him as 'one of the most distinguished and controversial scientists of the 20th century.'[70] Lovell highlighted steady-state theory, and called its rivalry with big-bang theory 'one of the bitterest scientific divisions of the century.' He noted Hoyle's long commitment to continuous creation, and how 'long after the "big bang" universe had become conventional scientific wisdom he continued to probe its defects.' He described Hoyle's contributions to popular science and public life, and his prominent role in scientific affairs. But for Lovell:

> . . . there is little doubt that his most lasting and significant contribution to science concerns the origin of the elements. This theory of [nucleosynthesis] (the build-up of the elements in the hot interiors of stars) was an outstanding scientific landmark of the 1950s. . . . the theory has had a cardinal influence on astrophysics.[71]

In *Physics Today*, Astronomer Royal Martin Rees, a long-time member of the Institute of Astronomy, picked out steady-state theory ('a serious contender for 15 years'), the evolution of stars, stellar nucleosynthesis and the quasar work to support his claim that Hoyle was 'pre-eminent among astrophysicists in the range and influence of his contributions.' Rees regretted Hoyle's exclusion from Fowler's Nobel Prize, and his departure from Cambridge: 'His consequent isolation from the broad academic community was probably detrimental to his own science; it was certainly a sad deprivation for the rest of us.'[72] Rees also noted Hoyle's extensive committee work, and the great success of both the Anglo-Australian Telescope and the Institute. To those who, in the early 1960s, suspected that Hoyle wanted an institute to foster steady-state theory, Rees was able to say, forty years later, that 'Hoyle was supportive to us all, even when our researches were orthogonal (or even contradictory) to his own.' He also remarked on Hoyle's 'engaging wit and relish for controversy', his 'wide following as a popularizer of science and as a successful writer of science fiction,' and his Yorkshireman's 'agreeable manner and accent.'[73]

Walter Sullivan of the *New York Times* called Hoyle 'one of the most creative and provocative astrophysicists of the last half century' and 'a versatile scientist brimming with ideas and a lifelong rebel eager for intellectual combat.'[74] In *Nature*, John Maddox described him as 'the most imaginative of men, a kind of Leonardo. He made monumental contributions to astrophysics and cosmology, and was a brilliant popularizer (and science-fiction writer and occasional playwright). He also put his name to much rubbish and was embroiled in controversy for most of his life.' Hoyle was also a Yorkshireman to the end: 'directly spoken, combative, careful with money and self-reliant.'[75]

Leon Mestel commended Hoyle's popular work, particularly *The Nature of the Universe* and *Frontiers of Astronomy*, for impressing readers 'old and young, and many who later achieved scientific distinction have acknowledged that his books led them to make astronomy their vocation.'[76] In this long appreciation, Mestel also considered Hoyle critically:

> It must be admitted that at times his style was unnecessarily polemical and could be embarrassing to his friends and admirers, e.g. in his interactions with the Cambridge radio astronomers (where indeed there was provocation), and even more so in his forays into other areas, especially when commenting on what he thought was facile theorising. . . . But maybe one should not jib at his willingness to court controversy. One hears it said that academic institutions in which there is no internal rivalry, not to say in-fighting, are also significantly less productive.[77]

The acrimony of the cosmological debates of the 1960s was remembered in many obituaries. Other colleagues also offered qualified criticisms: John C. Brown, Astronomer Royal for Scotland, described the great range of Hoyle's achievements, and Hoyle himself as 'a fascinating man . . . if a shade gruff and not the world's best listener.' He noted that:

> It is true that at times, Hoyle flaunted caution over data analysis in pursuit of support for his ideas, but I distrust any scientist – especially a cosmologist or life theorist – who claims to be squeaky-clean in that regard. Contemporary

cosmology is riddled with dogmatic paradigms, albeit threaded with intense efforts to test them, and anyone who attacks Hoyle for lack of scientific objectivity must be suspected of sour grapes at the fertility of his mind.[78]

The *Daily Telegraph* was in a small minority of publications that highlighted Hoyle's more controversial work, and it did so at length. Calling him an 'outrageous mischief-maker who took a delight in enraging his academic colleagues', the obituary emphasized the *Archaeopteryx* episode, and diseases from space:

> He and his close associate, Prof Chandra Wickramasinghe ... used to make other scientists so angry that some even wrote a special sub-program for their word processors which, by pressing a single key, caused the words 'Contrary to the views of Hoyle and Wickramasinghe ...' to appear on the screen.

The *Telegraph* then reproduced some of the most stridently negative reviews of Hoyle's work before acknowledging that 'Hoyle made many genuine and significant contributions to physics and astronomy.'[79]

Wickramasinghe's appreciation of Hoyle also dwelt on the biological work, though with rather more optimism.[80] Science-fiction fans mourned the passing of the author of *The Black Cloud*; and Creationists commended Hoyle, though 'not a Biblical creationist or even a Christian', for his challenges to Darwinism and big-bang theory.[81] Local media noted Hoyle's passing: in California, the *Santa Monica Mirror*, with its headline, 'Good-bye, Sir Fred Hoyle!' spoke for Hoyle's fans among non-scientists. Describing him as 'One of the most prominent astrophysicists of the 20th century, as well as one of the most colorful personalities,' it concluded that 'we cannot blame Sir Fred for his ideas: he certainly was a very gifted, influential, and original thinker, and the road to understanding the world around us has always been paved by people like him.'[82] The *Yorkshire Post* reminded its readers that in 2000 they had voted Hoyle one of the Great Yorkshiremen of the Millennium.[83] The ultimate local appreciation came from *Matterdale Matters*, in the Lake District, which remembered Hoyle as a resident of Cockley Moor and as the author of *Ice*, and who was 'one of the greatest popularisers of science in the 20th century.'[84]

Since Hoyle's death, even those colleagues who were not among his friends have found plenty to celebrate about his life and career. Radio-astronomer Antony Hewish, who had borne the brunt of one of the darker episodes of Hoyle's career, valued his contributions to astrophysics and to the community of astronomers in Cambridge. He thought that Hoyle deserved thanks for:

. . . producing brilliant ideas. The steady-state theory was a very nice theory and it was good in the sense that you could disprove it quickly. Theories that you can't disprove quickly are not much good. . . . And his ideas about the resonance in carbon which enables nucleosynthesis to get past the hurdle . . . that's a brilliant insight. His talks were always fun to go to in Cambridge, he'd always come out with new stimulating ideas. He used to give talks at the Old Observatory Club, a marvellous occasion with a real English tea, which collected people – thinly sliced bread and butter done by staff wives, and fruit cake, and all that sort of thing, followed by a nice talk – very popular occasions. Whenever Fred was speaking the place would always be pretty full – people were always interested in what he would say. He was a very stimulating chap, and all his students say the same.[85]

In a contrast which perhaps explains Hoyle's broad appeal – he was many things to many people – Geoffrey Burbidge valued the more contrary side of his friend's character, and the qualities that set Hoyle apart:

I never met a man who was so original and had so much perception about problems. On the other hand we used to have some rare old goes and I would say, 'Fred you're nuts about this.' But he was prepared to try things.

I remember one day at Caltech 30 years ago. We were all having lunch [and] some very well known people were talking about astronomy – Dick Feynman, Maarten Schmidt, Fred, Margaret, I was there, Willy was there. . . . Fred made the remark that 'Geoff and I reckon we're doing well if we bat 500,' meaning, we think we're doing very well if we are right half of the time. And Maarten Schmidt . . . looked really serious – he looked aghast – and he said to Fred, 'do you mean to tell me that you and Geoff are prepared to be wrong half the time?'

That's a very different approach.[86]

References

References beginning 'PRO' are to papers held at the National Archive, Kew (formerly the Public Record Office). References to the Hoyle papers are to the collection of Hoyle's personal papers, which is held at St John's College, University of Cambridge.

Chapter 1: Coming to light

1. The talks ran for six weeks from 18 November 1930. The lecture was the Rede lecture, given on 4 November 1930. The editorial comments are from the *Daily Mail*, 15 November 1930.
2. Jeans, J., 1930, *The Mysterious Universe* (Cambridge University Press).
3. Editorial, *Church Times*, 7 November 1930.
4. Letter to *The Times* from William Schooling, 18 November 1930.
5. *Nature*, **126**(3184), 732.
6. Dingle, H., 1930, Physics and reality. *Nature*, **126**(3186), 799–800.
7. Hoyle, F, 1977, *Ten Faces of the Universe* (San Francisco: Freeman), p. 4.
8. Hoyle's early life is recorded in his two autobiographies: Hoyle, F., 1986, *The Small World of Fred Hoyle* (London: Joseph); and Hoyle, F., 1993, *Home is Where the Wind Blows: Chapters from a Cosmologist's Life* (Mill Valley, CA: University Science Books).
9. Hoyle, F., 1986, *The Small World of Fred Hoyle* (London: Joseph), p. 10.
10. Hoyle papers 34/4.
11. Hoyle, F., 1999, *Mathematics of Evolution* (Acorn Enterprises: Memphis, TN), xiv–xv.
12. At the time of writing, this building is used as a wine bar, and the library is in a modern building in a shopping precinct not far away.
13. Hoyle papers 19/3.
14. Narlikar, J.V., 2001, Fred Hoyle: scientist of multifaceted talents. *Current Science*, **81**(7), 843–845.
15. Interview by Brig Klyce with Fred Hoyle, Cambridge, 5 July 1996, http://www.panspermia.org/hoylintv.htm.
16. Hoyle, F., 1986, *The Small World of Fred Hoyle* (London: Joseph), p. 102.
17. Hoyle papers 22/12.
18. Hoyle, F., 1986, *The Small World of Fred Hoyle* (London: Joseph), p. 152.
19. Hoyle, F., 1986, *The Small World of Fred Hoyle* (London: Joseph), p. 159.
20. Hoyle, F., 1986, *The Small World of Fred Hoyle* (London: Joseph), p. 164.
21. Hoyle papers, 34/4.
22. Hoyle, F., 1986, *The Small World of Fred Hoyle* (London: Joseph), p. 165.
23. Hoyle, F., 1986, *The Small World of Fred Hoyle* (London: Joseph), p. 169.

24. Hoyle, F., 1937, The generalised Fermi interaction. *Proceedings of the Cambridge Philosophical Society*, **33**, 227.

25. Hoyle, F., 1986, *The Small World of Fred Hoyle* (London: Joseph), p. 146.

26. Hoyle, F., 1981, *Ice* (London: Hutchinson), pp. 169–170.

27. Hoyle, F., 1981, *Ice* (London: Hutchinson), p. 11.

28. Interview by Brig Klyce with Fred Hoyle, Cambridge, 5 July 1996, http://www.panspermia.org/hoylintv.htm.

29. Hoyle, F., 1994, *Home is Where the Wind Blows: Chapters from a Cosmologist's Life* (Mill Valley, CA: University Science Books), p. 130.

30. Hoyle, F., undated, probably late 1980s, The accretion affair, p. 4.4. MS chapter, probably early draft of his second autobiography. Hoyle papers 2/3.

31. Hoyle, F., undated, probably late 1980s, The accretion affair, p. 4.6. MS chapter, probably early draft of his second autobiography. Hoyle papers 2/3.

32. Hoyle, F., undated, probably late 1980s, The accretion affair, p. 4.6. MS chapter, probably early draft of his second autobiography. Hoyle papers 2/3.

33. Hoyle, F., undated, probably late 1980s, The accretion affair, p. 4.7. MS chapter, probably early draft of his second autobiography. Hoyle papers 2/3.

34. Hoyle, F., undated, probably late 1980s, The accretion affair, p. 4.11. MS chapter, probably early draft of his second autobiography. Hoyle papers 2/3.

35. Hoyle, F., and Lyttleton, R.A., 1939, The effect of interstellar matter on climatic variation. *Proceedings of the Cambridge Philosophical Society*, **35**, 405–418.

36. Hoyle, F., and Lyttleton, R.A., 1939, The evolution of the stars. *Nature*, **144**, 1019.

37. Chown, M., 2000, *The Magic Furnace: the Search for the Origin of Atoms* (London: Vintage), pp. 153–155.

38. Hoyle, F., 1986, *The Small World of Fred Hoyle* (London: Joseph), p. 185.

39. Domb, C., 2002, Fred and naval radar (1941–1945), talk given at Fred Hoyle's Universe: a conference celebrating Fred Hoyle's extraordinary contributions to science, 24–26 June, Cardiff University.

Chapter 2: Hut no. 2

1. French, A.P., 1974, The Cavendish in wartime. *A Hundred Years of Cambridge Physics* (Cambridge University Physics Society).

2. http://www.penleyradararchives.org.uk/history/introduction.htm.

3. http://www.penleyradararchives.org.uk/history/introduction.htm.

4. http://homepages.westminster.org.uk/hooke/issue12/tizard.html.

5. http://homepages.westminster.org.uk/hooke/issue12/tizard.html.

6. http://www.penleyradararchives.org.uk/history/introduction.htm.

7. http://www.penleyradararchives.org.uk/history/introduction.htm.

8. http://murray.newcastle.edu.au/users/staff/eemf/ELEC351/SProjects/Calligeros/invent_radar.htm.

9. Domb, C., 2002, Fred and naval radar (1941–1945), talk given at Fred Hoyle's Universe: a conference celebrating Fred Hoyle's extraordinary contributions to science, 24–26 June, Cardiff University.

10. Domb, C., 2002, Fred and naval radar (1941–1945), talk given at Fred Hoyle's Universe: a conference celebrating Fred Hoyle's extraordinary contributions to science, 24–26 June, Cardiff University.

11. Domb, C., 2002, Fred and naval radar (1941–1945), talk given at Fred Hoyle's Universe: a conference celebrating Fred Hoyle's extraordinary contributions to science, 24–26 June, Cardiff University.

12. Domb, C., 2002, Fred and naval radar (1941–1945), talk given at Fred Hoyle's Universe: a conference celebrating Fred Hoyle's extraordinary contributions to science, 24–26 June, Cardiff University.

13. Hoyle, F., 1989, Sir Hermann Bondi – Seventieth Birthday: Reminiscences from the impressionable years. *Bulletin of the Institute of Mathematics and its Applications*, **25**(12), 282–284.

14. Bondi, H., 1990, *Science, Churchill and Me: the Autobiography of Hermann Bondi* (Oxford: Pergamon Press), p. 13.

15. Bondi, H., 1990, *Science, Churchill and Me: the Autobiography of Hermann Bondi* (Oxford: Pergamon Press), p. 21.

16. Kragh, H., 1996, *Cosmology and Controversy: the Historical Development of Two Theories of the Universe* (Princeton University Press), pp. 166–167.

17. Bondi, H., 2002, Work with Fred (1942–1949), talk given at Fred Hoyle's Universe: a conference celebrating Fred Hoyle's extraordinary contributions to science, 24–26 June, Cardiff University.

18. Domb, C., 2002, Fred and naval radar (1941–1945), talk given at Fred Hoyle's Universe: a conference celebrating Fred Hoyle's extraordinary contributions to science, 24–26 June, Cardiff University.

19. Bondi, H., 1990, *Science, Churchill and Me: the Autobiography of Hermann Bondi* (Oxford: Pergamon Press), p. 39.

20. Bondi, H., 2002, Work with Fred (1942–1949), talk given at Fred Hoyle's Universe: a conference celebrating Fred Hoyle's extraordinary contributions to science, 24–26 June, Cardiff University.

21. Kragh, H., 1996, *Cosmology and Controversy: the Historical Development of Two Theories of the Universe* (Princeton University Press), pp. 166–167.

22. Domb, C., 2002, Fred and naval radar (1941–1945), talk given at Fred Hoyle's Universe: a conference celebrating Fred Hoyle's extraordinary contributions to science, 24–26 June, Cardiff University.

23. Bondi, H., 1990, *Science, Churchill and Me: the Autobiography of Hermann Bondi* (Oxford: Pergamon Press), p. 41.

24. Bondi, H., 1990, *Science, Churchill and Me: the Autobiography of Hermann Bondi* (Oxford: Pergamon Press), p. 43.

25. Domb, C., 2002, Fred and naval radar (1941–1945), talk given at Fred Hoyle's Universe: a conference celebrating Fred Hoyle's extraordinary contributions to science, 24–26 June, Cardiff University.

26. Hoyle, F., 1993, *Home is Where the Wind Blows: Chapters from a Cosmologist's Life* (Mill Valley, CA: University Science Books), p. 178.

27. Domb, C., 2002, Fred and naval radar (1941–1945), talk given at Fred Hoyle's Universe: a conference celebrating Fred Hoyle's extraordinary contributions to science, 24–26 June, Cardiff University; and Hoyle, F., 1993, *Home is Where the Wind Blows: Chapters from a Cosmologist's Life* (Mill Valley, CA: University Science Books), p. 179.

28. PRO ADM 1/25225.

29. Domb, C., 2002, Fred and naval radar (1941–1945), talk given at Fred Hoyle's

Universe: a conference celebrating Fred Hoyle's extraordinary contributions to science, 24–26 June, Cardiff University.

30. Bondi, H., 2002, Work with Fred (1942–1949), talk given at Fred Hoyle's Universe: a conference celebrating Fred Hoyle's extraordinary contributions to science, 24–26 June, Cardiff University.

31. Bondi, H., 1990, *Science, Churchill and Me: the Autobiography of Hermann Bondi* (Oxford: Pergamon Press), p. 49.

32. Domb, C., 2002, Fred and naval radar (1941–1945), talk given at Fred Hoyle's Universe: a conference celebrating Fred Hoyle's extraordinary contributions to science, 24–26 June, Cardiff University.

33. Bondi, H., 2002, Work with Fred (1942–1949), talk given at Fred Hoyle's Universe: a conference celebrating Fred Hoyle's extraordinary contributions to science, 24–26 June, Cardiff University. The paper was Bondi, H. and Hoyle, F., 1944, *Monthly Notices of the Royal Astronomical Society*, **104**, 273.

34. Kragh, H., 1996, *Cosmology and Controversy: the Historical Development of Two Theories of the Universe* (Princeton University Press), p. 168.

35. Hoyle, F., 1993, *Home is Where the Wind Blows: Chapters from a Cosmologist's Life* (Mill Valley, CA: University Science Books), pp. 208, 219–233.

36. Domb, C., 2002, Fred and naval radar (1941–1945), talk given at Fred Hoyle's Universe: a conference celebrating Fred Hoyle's extraordinary contributions to science, 24–26 June, Cardiff University.

37. Hoyle, F., 1945, On the structure of disk-shaped extra-galactic nebulae, I–IV, *Monthly Notices of the Royal Astronomical Society*, **105**, 287–302, 302–319, 345–362, 363–381. He published 30 papers between 1945 and 1949.

Chapter 3: Into the limelight

1. Letter from F. Hoyle to Sir Harold Spencer Jones, undated, probably summer 1952. Hoyle papers, 22/20.

2. Hoyle, F., 1988, An assessment of the evidence against the steady-state theory. Hoyle papers, 3/2. This paper subsequently appeared in Bertotti, B., Balbinot, R., Bergia, S., and Messina, A., 1990, Modern Cosmology in Retrospect (Cambridge University Press). See also Barrow, J.D., *The Origin of the Universe* (London: Weidenfeld and Nicholson), p. 31. Kragh is sceptical about the historical accuracy of Hoyle's recollections on this point: see Kragh, H., 1993, Steady-state theory. *Cosmology: Historical, Literary, Philosophical, Religious, and Scientific Perspectives*, edited by Norriss S. Hetherington (New York, London: Garland), p. 393. However, the correspondence with Spencer-Jones in 1952 support's Hoyle's later published account.

3. Lemaître, G., 1931, The beginning of the world from the point of view of quantum theory. *Nature*, **127**(3210), 706.

4. Eddington, A., 1926, *Stars and Atoms* (Cambridge University Press), p. 27.

5. Lemaître, G., 1931, The beginning of the world from the point of view of quantum theory. *Nature*, **127**(3210), 706. Like Hoyle, Lemaître had studied under Eddington at Cambridge.

6. For an account of Gamow's career see Gamow, G., 1970, *My World Line: an Informal Autobiography* (New York: Viking); and Kragh, H., 1996, *Cosmology and Controversy: the Historical Development of Two Theories of the Universe* (Princeton University Press).

7. Kragh, H., 1996, *Cosmology and Controversy: the Historical Development of Two Theories of the Universe* (Princeton University Press), p. 106.

8. Gamow, G., 1946, Expanding universe and the origin of the elements. *Physical Review*, **70**, 572–573.

9. Alpher, R.A., Bethe, H., and Gamow, G., 1948, The origin of chemical elements. *Physical Review*, **76**, 803–804.

10. Bondi, H., undated, probably 1988, The cosmological scene 1945–1952. Hoyle papers, 3/2. This paper subsequently appeared in Bertotti, B., Balbinot, R., Bergia, S., and Messina, A., 1990, Modern Cosmology in Retrospect (Cambridge University Press).

11. Bondi, H., undated, probably 1988, The cosmological scene 1945–1952. Hoyle papers, 3/2.

12. Overbye, D., 1991, *Lonely Hearts of the Cosmos* (New York: HarperCollins), p. 39.

13. Hoyle, F., 1988, An assessment of the evidence against the steady-state theory. Hoyle papers, 3/2.

14. Bondi, H., Gold, T., and Hoyle, F., 1995, The origins of steady-state theory. *Nature*, **373**, 10.

15. Bondi, H., undated, probably 1988, The cosmological scene 1945–1952. Hoyle papers, 3/2.

16. Bondi, H., undated, probably 1988, The cosmological scene 1945–1952. Hoyle papers, 3/2.

17. Bondi, H., Gold, T., and Hoyle, F., 1995, The origins of steady-state theory. *Nature*, 373, 10.

18. Letter from F. Hoyle to Sir Harold Spencer Jones, undated, probably summer 1952. Hoyle papers, 22/20.

19. Hoyle, F., undated, probably late 1980s, An assessment of the evidence against the steady-state theory. Hoyle papers, 3/2.

20. Bondi, H., undated, probably 1988, The cosmological scene 1945–1952. Hoyle papers, 3/2.

21. Gold, T., and Bondi, H., 1948, The steady-state theory of the expanding universe. *Monthly Notices of the Royal Astronomical Society*, **108**, 252–270; Hoyle, F., 1948, A new model for the expanding universe, *Monthly Notices of the Royal Astronomical Society*, **108**, 372–382.

22. Letter from H.H. Hopkins to F. Hoyle, 1948, Hoyle papers 22/20.

23. Kragh, H., 1996, *Cosmology and Controversy: the Historical Development of Two Theories of the Universe* (Princeton University Press), p. 178.

24. Interview with Bernard Lovell, Jodrell Bank, 31 October 2003.

25. Addison, P., 1985, *Now that the War is Over* (London: BBC Books/Jonathan Cape).

26. http://www.joh.cam.ac.uk/publications/eagle97/Eagle97-Professo-2.html.

27. http://www.joh.cam.ac.uk/publications/eagle97/Eagle97-Professo-2.html.

28. Hoyle, F., 1989, Sir Hermann Bondi – Seventieth Birthday: Reminiscences from the impressionable years. *Bulletin of the Institute of Mathematics and its Applications*, **25**(12), 282–284.

29. Bondi, H., undated, probably 1988, The cosmological scene 1945–1952. Hoyle papers, 3/2.

30. Dingle, H., 1953, Modern theories of the origin of the universe (1949 address to the British Association, Birmingham). *The Scientific Adventure* (New York: Philosophical Library), p. 167.

31. See Pirani, F., 1991, The crisis in cosmology. *New Left Review*, **191**, 72; and Pirani, F.P., and Roche, C., 1993, *The Universe for Beginners* (Cambridge: Icon), p. 58.

32. Bondi, H., 2002, Work with Fred (1942–1949), talk given at Fred Hoyle's Universe: a conference celebrating Fred Hoyle's extraordinary contributions to science, 24–26 June, Cardiff University.

33. Interview with Felix Pirani, London, 23 September 1997.

34. Domb, C., 2002, Fred and naval radar (1941–1945), talk given at Fred Hoyle's Universe: a conference celebrating Fred Hoyle's extraordinary contributions to science, 24–26 June, Cardiff University.

35. Hoyle, F., 1949, *Listener*, **41**, 567.

36. *The Star*, 15 August 1949, Are there men on other planets? by Fred Hoyle, Cambridge lecturer in mathematics. *The Star* was a London evening paper: it closed in 1960.

37. Briggs, A., 1985, *The BBC: the First Fifty Years* (Oxford University Press), pp. 249–251.

38. For a history of this interest see Seifer, M.J., 1996, *Wizard: the Life and Times of Nikola Tesla, Biography of a Genius* (Secaucus, NJ: Birch Lane), Chap. 17.

39. See Hoyle, F., 1986, *The Small World of Fred Hoyle* (London: Michael Joseph), p. 94. The 'big fuss' about Jeans was among 'philosophers and religious believers.' Hoyle, F., 1993, *Home is Where the Wind Blows: Chapters from a Cosmologist's Life* (Mill Valley, CA: University Science Books), p. 216.

40. Jeans, J., 1931, *The Mysterious Universe* (2nd edn) (Cambridge University Press), p. 3.

41. Jeans, J., 1931, *The Mysterious Universe* (2nd edn) (Cambridge University Press), p. 6.

42. Wells, H.G., Huxley, J., and Wells, G.P., 1931, *The Science of Life* (London, Toronto, Melbourne, Sydney: Cassell), pp. 8–9.

43. Hoyle, F., 1994, *Home is Where the Wind Blows: Chapters from a Cosmologist's Life* (Mill Valley, CA: University Science Books), p. 159.

44. Butterfield, H., 1949, *The Origins of Modern Science* (London: Bell & Hyman), vii.

45. Interview with Fred Hoyle, Bournemouth, 11 August 1993.

46. Hoyle, F., The invention of Big Bang. Souvenir programme from the Big Bang City Ball, 26 September 1986.

47. Interview with Fred Hoyle, Bournemouth, 11 August 1993.

48. Recordings of the broadcasts are still held by the BBC. They are faithfully presented in the book of the series, *The Nature of the Universe*.

49. Horgan, J., 1995, The return of the maverick. *Scientific American*, March, p. 47.

50. Hoyle, F., 1960, *The Nature of the Universe* (3rd edn) (Oxford: Basil Blackwell), p. 93.

51. BBC Listener Research Report, LR/50/217, 271, 377 and 427.

52. BBC Listener Research Report, LR/50/427.

53. BBC Listener Research Report, LR/50/271.

54. This claim was made in the *Daily Graphic*, 22 November 1950, Fred Hoyle is First, by Walter Hayes.

55. Hoyle papers 15/1.

56. *Daily Graphic*, 22 November 1992, Fred Hoyle is First, by Walter Hayes.

57. Hoyle, F., The invention of Big Bang. Souvenir programme from the Big Bang City Ball, 26 September 1986.

58. Hoyle, F., 1960, *The Nature of the Universe* (third edn) (Oxford: Basil Blackwell), pp. 100–101.

59. BBC Listener Research Report, LR/50/427.

60. Muggeridge, M., 1950, A scientist rewrites Genesis: of the making of stars there is no end. *Daily Telegraph*, 1 May.

61. Kragh, H., 1993, Big bang cosmology. *Encyclopedia of Cosmology: Historical, Philosophical and Scientific Foundations of Modern Cosmology*, edited by N.S. Hetherington (New York, London: Garland), p. 632; Kragh, H., 1993, Steady state theory. *Encyclopedia of Cosmology: Historical, Philosophical and Scientific Foundations of Modern Cosmology*, edited by N.S. Hetherington (New York, London: Garland), p. 397.

62. For a recent attack on J.B.S. Haldane, see Dingle, H., 1945, Kinematical relativity and the nebular redshift. *Nature*, **155**, 511.

63. Lance Day, personal communication, December 1990.

64. See Hoyle, F., 1960, *The Nature of the Universe* (second edn) (Oxford: Basil Blackwell), p. 87.

65. Dingle, H., 1950, *Nature*, **166**, 15 July.

66. Dingle, H., 1950, *Nature*, **166**, 15 July.

67. Interview with John Faulkner, Santa Cruz, 5 November 2002

68. *Yorkshire Post*, 31 May 1950, Mathematician on cricket. The *Yorkshire Post* is the regional newspaper of Hoyle's home county. At this time, regional papers of the quality of the *Yorkshire Post* had considerably more influence nationally than they do today. See Seymour-Ure, Colin, 1991, *British Press and Broadcasting since 1945* (Oxford: Blackwell), p. 21.

69. *Sunday Express*, 27 August 1950, How will the world end? 'We shall all be scorched', by Peter Dacre.

70. *Evening Standard*, 13 September 1950, Fred Hoyle tonight.

71. Interview with Charles Barnes, Pasadena, 20 September 2000.

72. Interview with Charles Barnes, Pasadena, 20 September 2000.

73. Hoyle, F., 1994, *Home is Where the Wind Blows: Chapters from a Cosmologist's Life* (Mill Valley, CA: University Science Books), p. 255.

74. PRO WORK 25/24: FB. Science/Physics (50(4)) revised; Destined for the 'dome of discovery', section on the Nature of Matter; based on Nature of the Universe, Fred Hoyle; written by A.J. Garratt of the Science Directorate.

75. National Gallery, 1993, *An Introduction to Boris Anrep's Mosaics at the National Gallery London* (London: National Gallery). The mosaic of Hoyle is in the south-east corner of the North Vestibule at the Gallery. The *Daily Telegraph* mentioned the mosaic in its coverage of Hoyle's appointment to the Plumian Chair: see *Daily Telegraph*, 28 January 1958, Stegophilist.

76. Hoyle papers 27/8.

77. Martin, B.R., 1976, The origins, development and capitulation of steady-state cosmology: a sociological study of authority and conflict in science. MSc thesis, Department of Liberal Studies in Science, University of Manchester, p. 99.

78. Burbidge, E. M., 1994, *Annual Reviews of Astronomy and Astrophysics*, **32**, 1–36.

79. Interview with Geoffrey Burbidge, La Jolla, 8 November 2002.

80. Interview with Geoffrey Burbidge, La Jolla, 8 November 2002.

81. Sullivan, W.T., 1990, Radio stars and Martin Ryle's 2C survey. *Modern Cosmology in Retrospect*, edited by Bertotti, B., Balbinot, R., Bergia, S., and Messina, A. (Cambridge University Press), p. 315.

82. Interview with Thomas Gold, Ithaca, 3 April 1996.
83. Interview with Thomas Gold, Ithaca, 3 April 1996.
84. See, for example, *Time and Tide*, Astronomers with mud on their boots, by John Glover, 17 February 1961; and unidentified magazine articles from the Cavendish Laboratory scrapbook.
85. Unless otherwise attributed, these observations on Ryle are from interviews with his colleagues John Baldwin and Dennis Bly of the radioastronomy group at the Cavendish Laboratory, Cambridge University, 8 March 1995.
86. Interview with Bernard Lovell, Jodrell Bank, 31 October 2003.
87. Interview with Antony Hewish, Cambridge, 26 May 2004.
88. *Sunday Dispatch*, 'Creation' professors in bitter row, by Gerald McKnight, 12 February 1961.
89. Edge, D., 1984, Styles of research in three early radio astronomy groups. *The Early Years of Radioastronomy: Reflections Fifty Years after Jansky's Discovery*, edited by W.T. Sullivan (Cambridge University Press), p. 362.
90. Interview with Thomas Gold, Ithaca, 3 April 1996.
91. Interview with Antony Hewish, Cambridge, 26 May 2004.
92. Interview with Geoffrey Burbidge, La Jolla, 8 November 2002.
93. Greenstein, J.L., 1984, Optical and radio astronomers in the early years. *The Early Years of Radioastronomy: Reflections Fifty Years after Jansky's Discovery*, edited by W.T. Sullivan (Cambridge University Press), pp. 76–77.
94. Interview with Thomas Gold, Ithaca, 3 April 1996.
95. Hoyle, F., 1994, *Home is Where the Wind Blows: Chapters from a Cosmologist's Life* (Mill Valley, CA: University Science Books), p. 270.

Chapter 4: New world

1. Hoyle published 11 papers in 1949, and five in 1950, but only three in 1951 and only one in 1952. His citations also dropped, to a mere seven in 1954 of a total of 63 published papers – the lowest rate of citations per published paper in the whole of his career.
2. Interview with Fred Hoyle, Bournemouth, 11 August 1993.
3. For a discussion of the sociology of popularization, see Gregory, J., and Miller, S., 1998, *Science in Public: Communication, Culture and Credibility* (New York: Plenum), Chap. 4.
4. Interview with Fred Hoyle, Bournemouth, 11 August 1993.
5. Letter from Jesse Greenstein to Fred Hoyle, 30 November 1951, Greenstein Papers, folder 16.1, California Institute of Technology Archives.
6. Greenstein Papers, folder 16.1, California Institute of Technology Archives.
7. Letter from Fred Hoyle to Jesse Greenstein, 18 January 1952, Greenstein Papers, folder 16.1, California Institute of Technology Archives.
8. Interview with Fred Hoyle, Bournemouth, 11 August 1993.
9. Personal diary, December 1952, Hoyle papers 35/2.
10. Christmas 1952. MS, Hoyle papers 20/7.
11. Personal diary, December 1952, Hoyle papers 35/2.
12. California Institute of Technology News Bureau release, Friday 26 December 1952, H section Z, folder 219, California Institute of Technology Archives.

13. Two decades of collaboration with WAF, by Fred Hoyle, Fowler Papers, folder 164.15, California Institute of Technology Archives.
14. Two decades of collaboration with WAF, by Fred Hoyle, Fowler Papers, folder 164.15, California Institute of Technology Archives.
15. Two decades of collaboration with WAF, by Fred Hoyle, Fowler Papers, folder 164.15, California Institute of Technology Archives.
16. Two decades of collaboration with WAF, by Fred Hoyle, Fowler Papers, folder 164.15, California Institute of Technology Archives.
17. Transcript of an interview by Charles Wiener for the American Institute of Physics History Centre, 6 February 1973, Fowler Papers, folder 199.26, California Institute of Technology Archives.
18. Transcript of an interview by Charles Wiener for the American Institute of Physics History Centre, 6 February 1973, Fowler Papers, folder 199.26, California Institute of Technology Archives. This is Fowler's own copy, and has been amended by hand (not in Fowler's handwriting): 'funny little man' has been changed to 'opinionated Yorkshireman', and 'young fellow' has been changed to 'theorist'.
19. Interview with Charles Barnes, Pasadena, 20 September 2002.
20. Interview with Charles Barnes, Pasadena, 20 September 2002.
21. Interview with Charles Barnes, Pasadena, 20 September 2002.
22. Hoyle, F., How old is the universe? Undated MS, probably early draft of autobiography, late 1980s. Hoyle papers 1/3.
23. Hoyle, F., 1994, *Home is Where the Wind Blows: Chapters from a Cosmologist's Life* (Mill Valley, CA: University Science Books), p. 286.
24. Sharov, A.S., and Novikov, I.D., 1993, *Edwin Hubble: Discoverer of the Big Bang Universe* (New York: Cambridge), p. 146.
25. Hoyle, F., 1994, *Home is Where the Wind Blows: Chapters from a Cosmologist's Life* (Mill Valley, CA: University Science Books), p. 404.
26. McCrea, W.H., 1955, Jubilee of relativity theory, conference at Berne. *Nature*, **176**(4477), 331.
27. Letter from Thomas Gold to Hiram Caton of the University of Virginia, posted by Caton on the SCIFRAUD internet bulletin board, July 1994.
28. Interview with Anthony Hewish, Cambridge, 26 May 2004.
29. Contributed by a member of the audience after a talk by the author at the Science Policy Research Unit, Sussex University, in 1995.
30. Personal communication from Dr Geoffrey Thomas.
31. Hoyle papers 22/19.
32. Sir Harold's lecture was reported in the *New York Times* on 24 May 1952. The correspondence with Hoyle followed shortly thereafter. Hoyle papers, 22/20.
33. Interview by Ben Martin of Hermann Bondi, 5 July 1976. Martin, B.R., 1976, The origins, development and capitulation of steady-state cosmology: a sociological study of authority and conflict in science. MSc thesis, Department of Liberal Studies in Science, University of Manchester.
34. Kragh, H., 1996, *Cosmology and Controversy: the Historical Development of Two Theories of the Universe* (Princeton University Press), p. 208. See also Gregory, J., 1998, Fred Hoyle and the popularisation of cosmology. PhD thesis, University of London.
35. Dingle quotes from, and cites '*M.N.*, **108**, 372, 1948', Hoyle's paper announcing steady-state theory. Bondi's 1952 book *Cosmology* also receives much criticism along similar lines.

36. Dingle, H., 1953, Address. *Monthly Notices of the Royal Astronomical Society*, **113**, 394.

37. Dingle, H., 1953, Address. *Monthly Notices of the Royal Astronomical Society*, **113**, 393–394.

38. Dingle, H., 1953, Address. *Monthly Notices of the Royal Astronomical Society*, **113**, 403.

39. Hoyle, F., 1994, *Home is Where the Wind Blows: Chapters from a Cosmologist's Life* (Mill Valley, CA: University Science Books), pp. 401, 404.

40. Overbye, D., 1991, *Lonely Hearts of the Cosmos* (New York: HarperCollins), p. 39.

41. Eddington. A.S., 1931, *Nature*, 21 March, p. 447; Lemaître, G., 1931, The beginning of the world from the point of view of quantum theory. *Nature*, **127**(3210), 706.

42. Interview with Thomas Gold, Ithaca, 3 April 1996.

43. Letter from Jesse Greenstein to Fred Hoyle, 17 January 1961, Greenstein Papers, folder 16.1, California Institute of Technology Archives.

44. McCrea, W.H., 1984, The influence of radio astronomy on cosmology. *The Early Years of Radioastronomy: Reflections Fifty Years after Jansky's Discovery*, edited by W.T. Sullivan (Cambridge University Press), p. 370.

45. Bauer, M., Durant, J., Ragnarsdottir, A., and Rudolphsdottir, A., 1995, *Science and Technology in the British Press, 1946–1990* (London: Science Museum).

46. See, for example, Sullivan, J.W.N., 1931, The physical nature of the universe. *An Outline of Modern Knowledge*, 1931, edited by W. Rose (London: Victor Gollanz), p. 109.

47. Interview with Antony Hewish, Cambridge, 26 May 2004.

48. Dingle, H., 1953, Address. *Monthly Notices of the Royal Astronomical Society*, **113**, 398.

49. Martin, B.R., 1976, The origins, development and capitulation of steady-state cosmology: a sociological study of authority and conflict in science. MSc thesis, Department of Liberal Studies in Science, University of Manchester, p. 63.

50. See Kragh, H., 1996, *Cosmology and Controversy: the Historical Development of Two Theories of the Universe* (Princeton University Press), pp. 39–40, 251–259; Miller, S., 1994, Wrinkles, ripples and fireballs: cosmology on the front page. *Public Understanding of Science*, **3**, 445–466; Smoot, G., and Davidson, K., 1993, *Wrinkles in Time: the Imprint of Creation* (London: Little, Brown), p. 18.

51. Weinberg, S., 1993, *The First Three Minutes: a Modern View of the Origin of the Universe* (London: HarperCollins), p. 148.

52. Interview with Thomas Gold, Ithaca, 3 April 1996.

53. Alpher, R.A., and Herman, R., 1988, *Reflections on early work on 'big bang' cosmology. Physics Today*, **41**(8), 26.

54. See the drawings in Scientific American in 1956: Gamov, G., 1956, The evolutionary universe. *Scientific American*, **195**(3), 136; and the group photograph of the 1958 Solvay conference, in Hoyle, F., 1994, *Home is Where the Wind Blows: Chapters from a Cosmologist's Life* (Mill Valley, CA: University Science Books), plate 24.

55. Kragh, H., 1993, Big bang cosmology. *Encyclopedia of Cosmology: Historical, Philosophical and Scientific Foundations of Modern Cosmology*, edited by N.S. Hetherington (New York, London: Garland), p. 38.

56. Lemaître, G., 1958, The primaeval atom hypothesis and the problem of the clusters of galaxies. *La Structure et L'Evolution de L'Univers* (Proceedings of the Eleventh Solvay Conference) (Brussels: Stoops), p. 7.

57. Gamow, G., 1970, *My World Line: An Informal Autobiography* (New York: Viking), Chap. 1.

58. For examples of the satirical writing see Gamow, G., 1970, *My World Line: an Informal Autobiography* (New York: Viking), p. 127. The Pope is mentioned in Gamow, G., 1946, Expanding universe and the origin of the elements. *Physical Review*, **70**, 572–573. See also Kragh, H., 1993, Big bang cosmology. *Encyclopedia of Cosmology: Historical, Philosophical and Scientific Foundations of Modern Cosmology*, edited by N.S. Hetherington (New York, London: Garland), p. 38.

59. Gamow, G., 1970, *My World Line: an Informal Autobiography* (New York: Viking); Gamow, G., 1965, *Mr Tompkins in Paperback* (Cambridge University Press), p. 58.

60. Hoyle, F., 1958, A scientist in Russia. *Observer*, 7 September.

61. See Hoyle, F., 1994, *Home is Where the Wind Blows: Chapters from a Cosmologist's Life* (Mill Valley, CA: University Science Books), pp. 116–121; and Hoyle, F., 1986, *The Small World of Fred Hoyle* (London: Joseph), pp. 158–163.

62. Kragh, H., 1996, *Cosmology and Controversy: the Historical Development of Two Theories of the Universe* (Princeton University Press), pp. 166–167.

63. Interview with Felix Pirani, London, 23 September 1997.

64. Interview with Felix Pirani, London, 23 September 1997.

65. Kevles, D.J., 1995, *The Physicists: the History of a Scientific Community in Modern America* (Cambridge, Mass., London: Harvard University Press), p. 379.

66. Kevles, D., 1995, *The Physicists: the History of a Scientific Community in Modern America* (Harvard University Press), pp. 386–387; interview with Felix Pirani, London, 23 September 1997.

67. See, for example, Pugh, S., and McLeave, H., 1961, Spaceship to Venus. *Daily Mail*, 13 February, p. 1, and in the same edition McLeave, H., 1961, A hop and a skip to the Morning Star: as astronomers wrangle over their starry theories of creation the Russians leapfrog out 26,000,000 miles to see for themselves . . . and get the facts to clinch all arguments. *Daily Mail*, 13 February, p. 8.

68. Hollis, C., 1961, Rearranging the universe: how the world started is a party political game. *Punch*, 22 February.

69. Gamow, G., 1970, *My World Line: an Informal Autobiography* (New York: Viking), p. 126.

70. Kragh, H., 1993, Steady state theory. *Cosmology: Historical, Literary, Philosophical, Religious, and Scientific Perspectives*, edited by N.S. Hetherington (New York, London: Garland), pp. 396–397; Martin, B.R., 1976, The origins, development and capitulation of steady-state cosmology: a sociological study of authority and conflict in science. MSc thesis, Department of Liberal Studies in Science, University of Manchester.; Gamov, G., 1956, The evolutionary universe. *Scientific American*, **195**(3), 136.

71. Gamow, G., 1956, The evolutionary universe. *Scientific American*, **195**(3), 136.

72. See Hoyle, Fred, 1986, *The Small World of Fred Hoyle* (London: Joseph), p. 146; Hoyle, F., 1993, *Home is Where the Wind Blows: Chapters from a Cosmologist's Life* (Mill Valley, CA: University Science Books), p. 127.

73. Alpher, R.A., and Herman, R., 1988, Reflections on early work on 'big bang' cosmology. *Physics Today*, **41**(8), 26. For the role of strong disciplinary communities in the support of controversial theories see Collins, H.M., 1985, *Changing Order: Replication and Induction in Scientific Practice* (London, Beverly Hills, CA: Sage), p. 141.

74. See, for example, Weart, S., 1988, *Nuclear Fear: A History of Images* (London,

Cambridge, Mass.: Harvard University Press), p. 387; and Bauer, M., Durant, J., Ragnarsdottir, A., and Rudolphsdottir, A., 1995, *Science and Technology in the British Press, 1946–1990* (London: Science Museum).

75. *Observer*, 30 July 1950, uncertain cosmos.

76. Morris, D., 1992, *The Masks of Lucifer: Technology and the Occult in Twentieth Century Popular Literature* (London: B.T. Batsford).

77. Martin, B.R., 1976, The origins, development and capitulation of steady-state cosmology: a sociological study of authority and conflict in science. MSc thesis, Department of Liberal Studies in Science, University of Manchester, p. 90.

78. Dingle, H., 1956, Cosmology and science. *Scientific American*, **195**(3), 224.

79. Hoyle, F., 1994, *Home is Where the Wind Blows: Chapters from a Cosmologist's Life* (Mill Valley, CA: University Science Books), pp. 401, 404.

80. Hoyle, F., 1955, *Frontiers of Astronomy* (London: Heinemann), xv.

81. Interview with Felix Pirani, London, 23 September 1997.

82. On the segregation of conflicting ideas see Collins, H.M., and Pinch, T., 1979, The construction of the paranormal: nothing unscientific is happening. *On The Margins of Science: the Social Construction of Rejected Knowledge*, edited by Roy Wallis, Sociological Review Monograph No. 27 (University of Keele Press), pp. 237–270.

83. Hoyle, F., 1994, *Home is Where the Wind Blows: Chapters from a Cosmologist's Life* (Mill Valley, CA: University Science Books), p. 130.

84. Interview with Thomas Gold, Ithaca, 3 April 1996.

85. Hoyle, F., 1956, The steady-state universe. *Scientific American*, **195**(3), 157.

86. Hoyle, F., and Narlikar, J.V., 1962, Mach's principle and the creation of matter (discussion meeting). *Proceedings of the Royal Society (A)*, **270**, 341.

87. *Daily Express*, 12 September 1965.

88. *Sunday Times*, 14 June 1964, Proving the apple can't fall upwards.

89. Hoyle, F., 1960, *The Nature of the Universe* (third edn) (Oxford: Basil Blackwell), p. 89.

90. Stephen Hawking, however, describes steady-state theory as 'simple': see Hawking, S., 1988, *A Brief History of Time* (London: Bantam), p. 47.

91. Hoyle papers 37/3.

92. Hoyle, F., 1953, *A Decade of Decision* (Melbourne, London, Toronto: Heinemann).

93. Hoyle. F., 1953, *A Decade of Decision* (Melbourne, London, Toronto: William Heinemann Ltd), x.

94. *Daily Telegraph*, 2 October 1953, An astronomer gets down to earth, by J. C. Johnstone.

95. *Economist*, 10 October 1953, Radical surgery.

96. *Sunday Times*, 11 October 1953, A 25-year plan, by C.P. Snow.

97. Snow, C.P., 1959, *The New Men* (Harmondsworth: Penguin), p. 137.

Chapter 5: Under fire

1. Transcript of an interview with Willy Fowler by Charles Wiener for the American Institute of Physics History Centre, 6 February 1973, Fowler Papers, folder 199.26, California Institute of Technology Archives.

2. Transcript of an interview with Willy Fowler by Charles Wiener for the American Institute of Physics History Centre, 6 February 1973, Fowler Papers, folder 199.26, California Institute of Technology Archives.

3. Burbidge, E.M., 1994, Watcher of the skies. *Annual Reviews in Astronomy and Astrophysics*, **32**, 1–36.
4. Burbidge, E.M., 1994, Watcher of the skies. *Annual Reviews in Astronomy and Astrophysics*, **32**, 1–36.
5. Transcript of an interview with Willy Fowler by Charles Wiener for the American Institute of Physics History Centre, 6 February 1973, Fowler Papers, folder 199.26, California Institute of Technology Archives.
6. Transcript of an interview with Willy Fowler by Charles Wiener for the American Institute of Physics History Centre, 6 February 1973, Fowler Papers, folder 199.26, California Institute of Technology Archives.
7. Hoyle, F., 1954, *Astrophysical Journal Suppl.* **1**, 121.
8. Burbidge, E.M., 1994, Watcher of the skies. *Annual Reviews in Astronomy and Astrophysics*, **32**, 17.
9. Hoyle, F., 1955, *Daily Telegraph*, February, New worlds to conquer? The planets and their origin.
10. Martin, B.R., 1976, The origins, development and capitulation of steady-state cosmology: a sociological study of authority and conflict in science. MSc thesis, Department of Liberal Studies in Science, University of Manchester.
11. Smith, F.G., 1984, Early work on radio stars in Cambridge. *The Early Years of Radioastronomy: Reflections Fifty Years after Jansky's Discovery*, edited by W.T. Sullivan (Cambridge University Press), p. 247.
12. Interview with Antony Hewish, Cambridge, 26 May 2004.
13. McCrea, W.H., 1984, The influence of radio astronomy on cosmology. *The Early Years of Radioastronomy: Reflections Fifty Years after Jansky's Discovery*, edited by W.T. Sullivan (Cambridge University Press), p. 371.
14. Interview with Bernard Lovell, Jodrell Bank, 31 October 2003.
15. Interview with Bernard Lovell, Jodrell Bank, 31 October 2003.
16. Interview with Antony Hewish, Cambridge, 26 May 2004.
17. Interview with Francis Graham-Smith, Jodrell Bank, 31 October 2003.
18. Interview with Antony Hewish, Cambridge, 26 May 2004.
19. Martin, B.R., 1976, The origins, development and capitulation of steady-state cosmology: a sociological study of authority and conflict in science. MSc thesis, Department of Liberal Studies in Science, University of Manchester, p. 79.
20. See, for example, Narlikar, J.V., 1988, *The Primeval Universe* (Oxford University Press), pp. 217–225; McCrea, W.H., 1984, The influence of radioastronomy on cosmology. *The Early Years of Radioastronomy: Reflections Fifty Years after Jansky's Discovery*, edited by W.T. Sullivan (Cambridge University Press), pp. 365–384; and Graham-Smith, F., 1974, *Radio Astronomy* (Harmondsworth: Penguin), pp. 166–169.
21. Interview with Geoffrey Burbidge, La Jolla, 8 November 2002.
22. *The Times*, 1955, Nearly 2,000 'radio stars' found: results of Cambridge survey, from our science correspondent, 14 May.
23. Interview with Francis Graham-Smith, Jodrell Bank, 31 October 2003.
24. See, for example, Kragh, H., 1996, *Cosmology and Controversy: the Historical Development of Two Theories of the Universe* (Princeton University Press), p. 245; Bondi, H., 1973, Setting the scene. *Cosmology Now*, edited by Laurie John (London: BBC), p. 11; Bondi, H., 1991, Arrogance of certainty. *The Times*, 20 May.
25. Bondi, H., undated, probably 1988, The cosmological scene. Hoyle papers, 3/2.

26. Other participants were Stanley Dreser, Richard Feynman, Felix Pirani, Leon Rosenfeld, Alfred Schild, Dennis Sciama and John Wheeler.

27. Cecilé M. de Witt, 1957, *Proceedings of the Conference on the Role of Gravitation in Physics*, University of North Carolina, Chapel Hill, 18–23 January 1957, WADC Technical Report 57–216, Astia Document No. AD 118180 (Wright Patterson Air Force Base, Ohio: Air Research and Development Command, United States Air Force), p. 53.

28. Letter from Fred Hoyle to Jesse Greenstein, 30 May 1955, Greenstein Papers, folder 16.1, California Institute of Technology Archives.

29. See Mills, B., 1984, Radio sources and the log N – log S controversy. *The Early Years of Radioastronomy: Reflections Fifty Years after Jansky's Discovery*, edited by W.T. Sullivan (Cambridge University Press), p. 155.

30. Interview with Antony Hewish, Cambridge, 26 May 2004.

31. Mills, B., and Slee, O.B., 1957, A preliminary survey of radio sources in a limited region of the sky at a wavelength of 3.5m. *Australian Journal of Physics*, **10**, 162–194.

32. Mills, B., 1984, Radio sources and the log N – log S controversy. *The Early Years of Radioastronomy: Reflections Fifty Years after Jansky's Discovery*, edited by W.T. Sullivan (Cambridge University Press), p. 163.

33. Interview with Antony Hewish, Cambridge, 26 May 2004.

34. Interview with Antony Hewish, Cambridge, 26 May 2004.

35. Interview with Francis Graham-Smith, Jodrell Bank, 31 October 2003.

36. Interview with John Faulkner, Santa Cruz, 5 November 2002.

37. Interview with Brian Robinson by Carolyn Little, recorded in Little, C., 2002, 'Conflict in Radioastronomy', unpublished paper; and personal communication with Carolyn Little.

38. Little, C., 2002, Conflict in radioastronomy (unpublished paper), and personal communication with Carolyn Little, September 2002.

39. Personal communication from David Edge, 17 December 2002.

40. Interview with Antony Hewish, Cambridge, 26 May 2004.

41. Letter from Jesse Greenstein to Fred Hoyle, 17 January 1955, Greenstein Papers, folder 16.1, California Institute of Technology Archives.

42. See, for example, Hughes, D., 1997, In retrospect. *Nature*, **387**, 364.

43. Spencer Jones, H., 1955, Stargazing. *Sunday Times*, 10 July 1955. *The Times* was also positive: see *The Times*, 30 June 1955, Starry amorist.

44. McCrea's Spectator review is quoted on the cover of the 1963 reprint of *Frontiers of Astronomy* (London: Mercury).

45. D.W.M., 1955, Shorter notices. *Financial Times*, 4 July.

46. Hoyle, F., 1955, *Frontiers of Astronomy* (London: Heinemann), pp. 100–104.

47. Hoyle, F., 1955, *Frontiers of Astronomy* (London: Heinemann), pp. 100–103.

48. *Scientific American* had a circulation outside the USA of 5700 in 1956, of which, if current patterns held then, probably 1300–1400 would have been in Britain. Personal communication from John Rennie, editor-in-chief of *Scientific American*, to Michael Rodgers, 7 February 1996.

49. Gamow, G., 1956, The evolutionary universe. *Scientific American*, **195**(3), 136. This quote is from a subheading that was probably written by an editor rather than by Gamow himself.

50. Hoyle, F., 1956, The steady-state universe. *Scientific American*, **195**(3), 157. As for

Gamow's article which preceded this one, this quote is from a subheading that was probably written by an editor rather than by Hoyle.

51. Sandage, A., 1956, The red-shift. *Scientific American*, **195**(3), 171.

52. Ryle, M., 1956, Radio galaxies. *Scientific American*, **195**(3), 205.

53. Dingle, H., 1956, Cosmology and science. *Scientific American*, **195**(3), 224.

54. Hoyle, F., 1960, *The Nature of the Universe* (third edn) (Oxford: Basil Blackwell), p. 86.

55. Hoyle, F., 1994, *Home is Where the Wind Blows: Chapters from a Cosmologist's Life* (Mill Valley, CA: University Science Books), p. 407, emphasis added.

56. Kragh, H., 1993, Steady state theory. *Cosmology: Historical, Literary, Philosophical, Religious, and Scientific Perspectives*, edited by Norriss S. Hetherington (New York, London: Garland), p. 397.

57. Gamow, G., 1964, Evolutionary and steady state cosmologies. *The Universe and its Origin*, edited by H. Messel and S.T. Butler (New York: St Martin's), p. 35.

58. Interview with Felix Pirani, London, 23 September 1997.

59. Martin, B.R., 1976, The origins, development and capitulation of steady-state cosmology: a sociological study of authority and conflict in science. MSc thesis, Department of Liberal Studies in Science, University of Manchester, p. 63. For Bondi's reaction, see Bondi, H. [quoted in discussion], 1958, Discussion of Hoyle's report. *La Structure et L'Evolution de L'Univers* (Proceedings of the Eleventh Solvay Conference) (Brussels: Stoops), p. 78.

60. Hoyle, F., 1994, *Home is Where the Wind Blows: Chapters from a Cosmologist's Life* (Mill Valley, CA: University Science Books), p. 151.

61. Kragh, H., 1996, *Cosmology and Controversy: the Historical Development of Two Theories of the Universe* (Princeton University Press), p. 345.

62. Hoyle, F., 1982, *The Anglo-Australian Telescope* (University College Cardiff Press), p. 16.

Chapter 6: New Genesis

1. Suess, H.E., and Urey, H.C., 1956, Abundances of the elements. *Reviews in Modern Physics*, **28**, 53–74.

2. Burbidge, E.M., 1994, Watcher of the skies. *Annual Reviews in Astronomy and Astrophysics*, **32**, 1–36.

3. For more technical details of this story see Chown, M., 1999, *The Magic Furnace: the Search for the Origin of Atoms* (London: Jonathan Cape).

4. Faulkner, J., 2002, After dinner speech at Fred Hoyle's Universe: a conference celebrating Fred Hoyle's extraordinary contributions to science, 24–26 June, Cardiff University.

5. Two decades of collaboration with WAF, by Fred Hoyle, Fowler Papers, folder 164.15, California Institute of Technology Archives.

6. Transcript of an interview with Willy Fowler by Charles Wiener for the American Institute of Physics History Centre, 6 February 1973, Fowler Papers, folder 199.26, California Institute of Technology Archives.

7. Hoyle, F., Fowler, W.A., Burbidge, G.R., and Burbidge, E.M., 1956, Origin of the elements in stars. *Science*, **124**, 611–614; Burbidge, E.M., Burbidge, G.R., Fowler, W.A., and Hoyle, F., 1957, Synthesis of elements in stars. *Reviews in Modern Physics*, **29**, 547–650.

8. See, for example, Smoot, G., and Davidson, K., 1993, *Wrinkles in Time: the Imprint of*

Creation (London: Little, Brown), p. 74; Gribbin, J., and Rees, M., 1991, *Cosmic Coincidences* (London: Black Swan), pp. 244–247; Lynden-Bell, D., 1973, Sources of cosmic power. *Cosmology Now* (London: BBC), p. 44.

9. See Gamow, G., 1970, *My World Line: an Informal Autobiography* (New York: Viking), p. 127. According to Alpher and Herman, Gamow's jokes prevented some people from taking big-bang theory seriously. See Alpher, R.A., and Herman, R., 1988, Reflections on early work on 'big bang' cosmology. *Physics Today*, **41**(8), 26.

10. Manchester *Guardian*, 22 March 1957, Mr Fred Hoyle an F.R.S.: cosmological theory.

11. Hoyle, F., 1993, *Home is Where the Wind Blows: Chapters from a Cosmologist's Life* (Mill Valley, CA: University Science Books), p. 244.

12. Etienne, J., 2001, *Space News International*, http://www.spacenews.be/art2001/hoyle_230801.html, 7323-3901-2631EC, 23 August, from an article by J.M. Bonnet-Bidaud and other sources. Translated from the French by JG.

13. O'Connell, D.J.K., 1958, *Stellar Populations: Proceedings of the Conference Sponsored by the Pontifical Academy of Science and the Vatican Observatory* (Amsterdam: North Holland/Interscience: New York).

14. Two decades of collaboration with WAF, by Fred Hoyle, Fowler Papers, folder 164.15, California Institute of Technology Archives.

15. 'New Exodus' by William A. Fowler, Greenstein Papers, folder 11.10, California Institute of Technology Archives.

16. Domb, C., 2002, Fred and naval radar. Talk at Fred Hoyle's Universe: a conference celebrating Fred Hoyle's extraordinary contributions to science, 24–26 June, Cardiff University.

17. Hoyle, F., 1957, *The Black Cloud* (London, Melbourne, Toronto, Cape Town, Auckland, The Hague: Heinemann).

18. Thompson, J., 1958, *Evening Standard*, 12 February.

19. *Daily Telegraph*, 16 September 1957, Fred Hoyle's frolic.

20. Interview with Hermann Bondi, Swansea, August 1990.

21. Hoyle, F., 1957, *The Black Cloud* (London, Melbourne, Toronto, Cape Town, Auckland, The Hague: Heinemann), p. 183.

22. Hoyle, F., 1957, *The Black Cloud* (London: Heinemann), pp. 202–203.

23. Hoyle, F., 1957, *The Black Cloud* (London: Heinemann), p. 246.

24. See, for example, Stevenson, J., 1975, Nobel Prize star-spotter gets rocket from Hoyle. *Daily Mail*, 22 April; Dover, C., 1975, Sir Fred Hoyle starts Cambridge science furore. *Daily Telegraph*, 23 April.

25. Hoyle papers 37/3.

26. *Daily Telegraph*, 1 December 1957, Hoyle's nimbus.

27. Hoyle papers 23/1.

28. *Spectator*, quoted on the dust jacket of the first reprint (1957) of *The Black Cloud*.

29. Thompson, J., 1958, *Evening Standard*, 12 February.

30. Hoyle, F., and Lyttleton, R. A., 1939, The effect of interstellar matter on climatic variation. *Proceedings of the Cambridge Philosophical Society*, **35**, 405–418.

31. See, for example, Heiles, C., 1969, *Astrophysical Journal*, **157**, 123; Berendze, R., 1972, *Annals of the New York Academy of Sciences*, **198**, 114; Dyson, F., 1979, *Reviews in Modern Physics*, **51**, 447; Talbot, R.J., 1980, *Interdisciplinary Science*, **5**, 102; Miles, J., 1984, *Advances in Applied Mechanics*, **24**, 189.

32. Interview with Fred Hoyle, Bournemouth, 11 August 1993.
33. Interview with Felix Pirani, London, 23 September 1997.
34. Hoyle, F., 1957, *The Black Cloud* (London, Melbourne, Toronto, Cape Town, Auckland, The Hague: Heinemann), pp. 107–108.
35. Introduction by W.A. Fowler to the Friends of the YMCA, The Athenaeum, 26 February 1975. Fowler Papers, folder 147.11, California Institute of Technology Archives.
36. Dawkins, R., 1996, Rereadings: Look, up in the sky. . . . *Financial Times*, 11–12 May, xiii.
37. Hoyle, F., 1993, *Home is Where the Wind Blows: Chapters from a Cosmologist's Life* (Mill Valley, CA: University Science Books), p. 309.
38. Note from Adrian to Hoyle, 25 January 1958, Hoyle papers 23/1.
39. Transcript of an interview with Willy Fowler by Charles Wiener for the American Institute of Physics History Centre, 6 February 1973, Fowler Papers, folder 199.26, California Institute of Technology Archives.
40. *Guardian*, 27 January 1958, Professorship for Mr Fred Hoyle: Cambridge election. See also *Daily Telegraph*, 27 January 1958.
41. *The Times*, 2 March 1956, letter from Fred Hoyle.
42. *Daily Telegraph*, 27 January 1958.
43. *Evening Standard*, 12 February 1958, by John Thompson.
44. Hoyle papers 23/1.
45. Hoyle papers 23/1.
46. Hoyle papers 34/9.
47. See Fösling, A., 1997, *Albert Einstein, A Biography* (London: Viking), p. 284.
48. Gamow, G., 1970, *My World Line: An Informal Autobiography* (New York: Viking), p. 126.
49. Martin, B.R., 1976, The origins, development and capitulation of steady-state cosmology: a sociological study of authority and conflict in science. MSc thesis, Department of Liberal Studies in Science, University of Manchester, p. 102.
50. Schücking, E., and Heckmann, O., 1958, World models. *La Structure et L'Evolution de L'Univers* (Proceedings of the Eleventh Solvay Conference) (Brussels: Stoops), pp. 149–150.
51. Shapley, H., 1958, Acknowledgement. *La Structure et L'Evolution de L'Univers* (Proceedings of the Eleventh Solvay Conference) (Brussels: Stoops), ix.
52. Hoyle, F., 1994, *Home is Where the Wind Blows: Chapters from a Cosmologist's Life* (Mill Valley, CA: University Science Books), pp. 309–311.
53. Hoyle, F., 1994, *Home is Where the Wind Blows: Chapters from a Cosmologist's Life* (Mill Valley, CA: University Science Books), p. 313.
54. Hoyle papers 34/4.
55. Hoyle papers 15/1.
56. *Sunday Times*, 1 May 1958, Astronomers in Moscow.
57. *Observer*, 7 September 1958, Astronomical occasion: a scientist in Russia, by Fred Hoyle.
58. *The Times*, 14 November 1958, Cosmic rays from the Sun: satellite research theory.
59. *The Times*, 15 April 1959, Moon trips 'lunatic fringe stuff': astronomer urges use of television camera.
60. *The Times*, 16 May 1959, Still deploring?, letter from Fred Hoyle; see also a response: *The Times*, 18 May 1959, Still deploring?, letter from John Hayward.

61. *Sunday Times,* June 1959, The new world of science – 2: what the Universe is made of, by Fred Hoyle.

62. Letter from J. de Grouchy, Compagnie Francaise Cinématographique, to F. Hoyle, 9 July 1959, Hoyle papers 23/1.

63. Mestel, L., 2001, Professor Sir Fred Hoyle, 1915–2001, *The Johnian*. This obituary is an edited version of the original that appeared in the October 2001 issue of *Astronomy & Geophysics*, the Journal of the Royal Astronomical Society.

64. Letter from Jesse Greenstein to Fred Hoyle, 29 April 1959, Greenstein Papers, folder 16.1, California Institute of Technology Archives.

65. Hoyle, F., 1994, *Home is Where the Wind Blows: Chapters from a Cosmologist's Life* (Mill Valley, CA: University Science Books), p. 314.

66. See Hoyle, F., 1994, *Home is Where the Wind Blows: Chapters from a Cosmologist's Life* (Mill Valley, CA: University Science Books), Chap. 22.

67. Burbidge, E.M., Burbidge, G.R., Fowler, W.A., and Hoyle, F., 1957, Synthesis of elements in stars. *Reviews in Modern Physics*, **29**, 641.

68. *Guardian*, 4 March 1960, Support for new theories on origin of universe: no specific beginning in time?, by our Scientific Correspondent.

69. Letter to Hoyle from Norman Tucker, director of Sadler's Wells Theatre, 24 August 1960, and letter to Hoyle from Helen O'Neill, 31 July 1960, Hoyle papers 23/1.

Chapter 7: Eclipsed

1. Conducted by the Science Service, Washington DC, January 1959, and cited in Gamow, G., 1964, Evolutionary and steady state cosmologies. *The Universe and its Origin*, edited by H. Messel and S.T. Butler (New York: St Martin's), p. 35.

2. *Evening Standard*, 11 February 1961, Away from space, by John London.

3. Interview with Antony Hewish, Cambridge, 26 May 2004.

4. *Guardian*, 3 March 1960. A large government grant was awarded to radio-astronomy in 1962, but the lion's share went to Ryle's group at the Cavendish. See Martin, B.R., 1976, The origins, development and capitulation of steady-state cosmology: a sociological study of authority and conflict in science. MSc thesis, Department of Liberal Studies in Science, University of Manchester, p. 100.

5. The report is discussed in Crowther, J.G., 1967, *Science in Modern Society* (London: Cresset), section 16.

6. Crowther, J.G., 1967, *Science in Modern Society* (London: Cresset), p. 50.

7. *Observer*, January 1961, Science: a vocation in the sixties, by Professor Fred Hoyle.

8. See, for example, *The Times*, 3 August 1960, Research into space: how much should Britain spend?, letter from Fred Hoyle; *Economist*, 6 August 1960, Space research: a question of proportion; *The Times*, 10 August 1960, Research into space: part Britain must play, letter from G.B.B.M. Sutherland; *The Times*, 12 August 1960, Space research, letter from Fred Hoyle; *The Times*, 11 August 1960, Space research, letter from Austen Albu.

9. Crowther, J.G., 1967, *Science in Modern Society* (London: Cresset), p. 54.

10. *The Times*, 13 September 1962, Efforts 'almost worthless', from our correspondent. See also *Daily Telegraph*, 14 September 1962, Space race 'not worth cost': Prof. Hoyle's view.

11. 'The needs of British astronomy'. Memo to the Advisory Council on Science Policy from R. Woolley, 1 January 1960, PRO: CAB 124/2152.

12. This note, written by Hoyle, is dated 24 February 1960. A very similar document was distributed in September as a report from the Royal Society's British National Committee for Astronomy. PRO: CAB 124/2152.

13. PRO: CAB 124/2152.

14. PRO: CAB 124/2152.

15. Interview with Geoffrey Burbidge, La Jolla, 8 November 2002.

16. PRO: CAB 124/2152.

17. PRO: CAB 124/2152. Bondi's proposal was dated 2 January 1961.

18. PRO: CAB 124/2152.

19. Hoyle papers 7/8.

20. Interview with Geoffrey Burbidge, La Jolla, 8 November 2002.

21. Hoyle, F., 1994, *Home is Where the Wind Blows: Chapters from a Cosmologist's Life* (Mill Valley, CA: University Science Books), p. 236.

22. Interview with John Faulkner, Santa Cruz, 5 November 2002.

23. Interview with John Faulkner, Santa Cruz, 5 November 2002.

24. PRO: CAB 124/2152. The BNC meeting was on 13 January 1961.

25. PRO: CAB 124/2152. Quirk contacted Turnbull on 18 January 1961, after meeting Massey.

26. PRO: CAB 124/2152.

27. Ryle's paper is dated 28 January 1961. Ryle, M., and Clarke. R.W., 1961, An examination of the steady-state model in the light of some recent observations of radio sources. *Monthly Notices of the Royal Astronomical Society*, **122**, 349–362.

28. Hoyle, F., 1994, *Home is Where the Wind Blows: Chapters from a Cosmologist's Life* (Mill Valley, CA: University Science Books), p. 409.

29. Ryle, M., and Clarke, R.W., 1961, An examination of the steady-state model in the light of some recent observations of radio sources. *Monthly Notices of the Royal Astronomical Society*, **122**, 361.

30. See, for example, Smoot, G., and Davidson, K., 1993, *Wrinkles in Time: the Imprint of Creation* (London: Little, Brown), pp. 77–79.

31. *Evening News* Science Reporter, 1961, The Bible was right. *Evening News and Star*, 10 February, p. 1.

32. Fairley, P., 1961, 'How it all began' fits in with Bible story. *Evening Standard*, 10 February, p. 1.

33. See, for example, 'Pertinax', 1961, Fleet Street's spacemen: who said *Genesis* began with a bang? *Time and Tide*, p. 230.

34. Hoyle, F., 1994, *Home is Where the Wind Blows: Chapters from a Cosmologist's Life* (Mill Valley, CA: University Science Books), pp. 409–410.

35. Interview with Jayant Narlikar, Cardiff, 26 June 2002.

36. Interview with Jayant Narlikar, Cardiff, 26 June 2002.

37. Interview with Jayant Narlikar, Cardiff, 26 June 2002.

38. Interview with Jayant Narlikar, Cardiff, 26 June 2002.

39. See, for example, Ryle, M., and Clarke, R.W., 1961, An examination of the steady-state model in the light of some recent observations of radio sources. *Monthly Notices of the Royal Astronomical Society*, **122**, 361.

40. Interview with Francis Graham-Smith, Jodrell Bank, 21 February 1996.

41. Hoyle said this at a conference in Italy in 1988, according to Smoot, G., and Davidson, K., 1993, *Wrinkles in Time: the Imprint of Creation* (London: Little, Brown), p. 79.

42. Interview with Antony Hewish, Cambridge, 26 May 2004.

43. *Evening Standard*, 12 February 1958, by John Thompson.

44. *The Times*, 11 February 1961, New theory of universe: expanding and changing: evidence from radioastronomy, from our Science Correspondent.

45. For examples of coverage of Ryle as 'personality' see *Evening Standard*, 11 February 1961, Away from space, by John London; *Punch*, Rearranging the universe, by Christopher Hollis, 22 February 1961; *Time and Tide*, Astronomers with mud on their boots, by John Glover, 17 February 1961; and unidentified magazine articles from the Cavendish Laboratory scrapbook, 'Scientific research in Britain,' and 'The world's most romantic men.' See, for example, *Evening Standard*, 10 February 1961; *Evening News and Star*, 10 February 1961; *Daily Telegraph*, 11 February 1961; *Daily Sketch*, 11 February 1961; *Observer*, 12 February 1961.

46. Hoyle, F., 1994, *Home is Where the Wind Blows: Chapters from a Cosmologist's Life* (Mill Valley, CA: University Science Books), p. 408.

47. Heimer, M., 1961, Brouillés a cause de la creation du monde. *Paris Match*, 4 March.

48. Unidentified magazine article from the Cavendish Laboratory scrapbook, 'Scientific research in Britain.'

49. Unidentified magazine article from the Cavendish Laboratory scrapbook, 'The world's most romantic men.'

50. Martin, B.R., 1976, The origins, development and capitulation of steady-state cosmology: a sociological study of authority and conflict in science. MSc thesis, Department of Liberal Studies in Science, University of Manchester, p. 76.

51. *Daily Express*, 13 February 1961, 'Sorry' – but the big bang row is still on, by *Express* Staff Reporter.

52. *Evening News and Star*, Universe theory: more tests are needed, by *Evening News* Reporter, 11 February 1961.

53. *Daily Telegraph*, 11 February 1961, Star watch gives clue to start of universe: 'big bang' theory backed, by Anthony Smith, *Daily Telegraph* science correspondent.

54. For Ryle's attitudes to the media see Martin, B.R., 1976, The origins, development and capitulation of steady-state cosmology: a sociological study of authority and conflict in science. MSc thesis, Department of Liberal Studies in Science, University of Manchester, p. 76. The Mullard press agent was Peter Wymer. Interviews with Francis Graham-Smith, 21 February 1996, and John Baldwin and Dennis Bly, 8 March 1996.

55. See, for example, the *Evening Standard*, 'How it all began' fits in with Bible story, by Peter Fairley, 10 February 1961; *Evening News and Star*, 'The Bible was right', by *Evening News* Science Reporter, 10 February 1961; *The Times*, New theory of universe: expanding and changing: evidence from radioastronomy, from our Science Correspondent, 11 February 1961.

56. See *Time and Tide*, Fleet Street's spacemen, by 'Pertinax', 17 February 1961.

57. See *Time and Tide*, Fleet Street's spacemen, by 'Pertinax', 17 February 1961.

58. *Daily Express*, 'Sorry' – but the big bang row is still on, by *Express* Staff Reporter, 13 February 1961.

59. *Sunday Dispatch*, Creation professors in bitter row: Hoyle slams Ryle on cheap success, by Gerald McKnight, 12 February 1961.

60. *Evening Standard*, And now a look at the edge of beyond, by Peter Fairley, 11 February 1961.

61. *Evening Standard*, 'How it all began' fits in with Bible story, by Peter Fairley, 10 February 1961; *Evening News and Star*, 'The Bible was right', by *Evening News* Science Reporter, 10 February 1961.

62. *The Times*, 11 February 1961, New theory of universe: expanding and changing: evidence from radioastronomy, from our Science Correspondent.

63. *Sunday Dispatch*, Creation professors in bitter row: Hoyle slams Ryle on cheap success, by Gerald McKnight, 12 February 1961.

64. Unidentified newspaper cutting from the Cavendish Laboratory scrapbook, A quarrel of cosmologists, by David Luytens, 12 February 1961.

Chapter 8: Fighting for space

1. *Daily Express*, Sorry – but the big bang row is still on, by *Express* Staff Reporter, 13 February 1961.

2. See *The Times*, New theory of the universe, from our science correspondent, 11 February 1961; the unidentified newspaper cutting from the Cavendish scrapbook, A quarrel of cosmologists, by David Luytens, 12 February 1961; *Sunday Times*, a setback for the big bang theory, 17 February 1963; *Daily Express*, 13 February 1961.

3. Martin, B.R., 1976, The origins, development and capitulation of steady-state cosmology: a sociological study of authority and conflict in science. MSc thesis, Department of Liberal Studies in Science, University of Manchester. See also Wright, P., 1980, Business Diary profile: Sir Hermann Bondi, enfant terrible of science. *The Times*, 18 August.

4. See, for example, *Evening Standard*, How it all began fits in with Bible story, by Peter Fairley, 10 February 1961; and *Sunday Dispatch*, Creation professors in bitter row: Hoyle slams Ryle on cheap success, by Gerald McKnight, 12 February 1961.

5. See, for example, *Evening Standard*, How it all began fits in with Bible story, by Peter Fairley, 10 February 1961, and *Daily Mail*, Scientists challenge world with new theory, by Angus MacPherson and Keith Thompson, 11 February 1961.

6. See, for example, *Evening News and Star*, 10 February 1961; *Observer*, 12 February 1961; *Sunday Times News Magazine* (Johannesburg), 19 February 1961; *Listener*, 9 March 1961.

7. See, for example, the *Daily Mail*, 13 February 1961.

8. McLeave, H., 1961, A hop and a skip to the Morning Star: as astronomers wrangle over their starry theories of creation the Russians leapfrog out 26,000,000 miles to see for themselves ... and get the facts to clinch all arguments. *Daily Mail*, 13 February, p. 8. See also Pugh, S., and McLeave, H., 1961, Spaceship to Venus. *Daily Mail*, 13 February, p. 1.

9. See *New Statesman*, London Diary, by Charon, 17 February 1961.

10. *Daily Mail*, 13 February 1961.

11. *Guardian*, Universe constantly evolving, by John Maddox, 11 February 1961. For examples of quality newspaper articles which do not mention God, see *Daily Telegraph*, Universe began 'with a bang', by Dr Tom Margerison, 11 February 1961;

Observer, The universe: how Ryle solved the riddle, by John Davy, 12 February 1961.

12. *Evening News and Star*, The Bible was right, by *Evening News* Science Reporter, 10 February 1961; *Evening Standard*, How it all began fits in with Bible story, by Peter Fairley, 10 February 1961; *Daily Express*, Telescope shows Genesis was right, by Chapman Pincher, 11 February 1961.

13. *Evening News and Star*, 'The Bible was right', by *Evening News* Science Reporter, 10 February 1961; *Daily Express*, Telescope shows Genesis was right, by Chapman Pincher, 11 February 1961.

14. *Sunday Express*, The divine quest, 12 February 1961.

15. *Daily Sketch*, 11 February 1961.

16. Interview with Francis Graham-Smith, 21 February 1996. See also *Sunday Dispatch*, 'Creation' professors in bitter row, by Gerald McKnight, 12 February 1961.

17. *Daily Mirror*, How it all started, by Ronald Bedford, 11 February 1961.

18. *Sunday Dispatch*, Creation professors in bitter row: Hoyle slams Ryle on cheap success, by Gerald McKnight, 12 February 1961.

19. *Daily Mirror*, How it all started, by Ronald Bedford, 11 February 1961; and *Daily Express*, Telescope shows 'Genesis was right', by Chapman Pincher, 11 February 1961.

20. Unidentified newspaper cutting, *Stars – 1*, by Malcolm Muggeridge from the Cavendish Laboratory scrapbook.

21. *Sunday Dispatch*, 'Creation' professors in bitter row, by Gerald McKnight, 12 February 1961.

22. *Daily Mail*, Comment, According to Hoyle, 13 February 1961.

23. Art Buchwald, *New York Herald Tribune*, February 1961.

24. Letter from William Fowler to Fred Hoyle, 21 February 1961, Fowler Papers, folder 12.16, California Institute of Technology Archives.

25. *New Statesman*, London Diary, by Charon, 17 February 1961.

26. Letter from Fred Hoyle to William Fowler, 27 March 1961, Fowler Papers, folder 183.8, California Institute of Technology Archives.

27. Hoyle, F., and Narlikar, J.V., 1961, Meeting of the Royal Astronomical Society. *Observatory*, **81**, 86–90.

28. Letter from Willliam Fowler to Fred Hoyle, 12 May 1961, Fowler Papers, folder 12.16, California Institute of Technology Archives.

29. Hoyle, F., and Narlikar, J.V., 1961, On the counting of radio sources in the steady-state cosmology. *Monthly Notices of the Royal Astronomical Society*, **123**, 133–166.

30. Letter from Fred Hoyle to Jesse Greenstein, 8 May 1961, Greenstein Papers, folder 16.1, California Institute of Technology Archives.

31. Hoyle, F., and Narlikar, J.V., 1962, On the counting of radio sources in the steady state cosmology, II. *Monthly Notices of the Royal Astronomical Society*, **125**, 13–20.

32. See McCrea, W.H., 1984, The influence of radio astronomy on cosmology. *The Early Years of Radioastronomy: Reflections Fifty Years after Jansky's Discovery*, edited by W.T. Sullivan (Cambridge University Press), p. 381.

33. *Yorkshire Post*, Where the universe stands now, by J.N. Buxton, 2 March 1961.

34. PRO: CAB 124/2152.

35. PRO: CAB 124/2152.

36. Interview with Geoffrey Burbidge, La Jolla, 8 November 2002.

37. PRO: CAB 124/2152.
38. PRO: CAB 124/2152.
39. Letter from Fred Hoyle to Jesse Greenstein, 8 May 1961, Greenstein Papers, folder 16.1, California Institute of Technology Archives.
40. PRO: CAB 124/2152.
41. PRO: CAB 124/2152.
42. PRO: CAB 124/2152.
43. Letter from W.A. Fowler to Fred Hoyle, 22 January 1960, Fowler Papers, folder 12.16, California Institute of Technology Archives.
44. See, for example, the correspondence with Greenstein, May 1960, Greenstein Papers, folder 16.1, California Institute of Technology Archives.
45. PRO: CAB 124/2152.
46. Faulkner, J., 2002, After dinner speech at Fred Hoyle's Universe: a conference celebrating Fred Hoyle's extraordinary contributions to science, 24–26 June, Cardiff University.
47. Faulkner, J., 2002, After dinner speech at Fred Hoyle's Universe: a conference celebrating Fred Hoyle's extraordinary contributions to science, 24–26 June, Cardiff University.
48. PRO: CAB 124/2152. Quirk's note is dated 27 June 1961.
49. PRO: CAB 124/2152. Quirk's note is dated 20 June 1962.
50. Hoyle, F., and Elliot, J., 1962, *A for Andromeda* and *Andromeda Breakthrough* (London: Souvenir Press).
51. These endorsements are from the dust jacket of *A for Andromeda*.
52. *The Times*, 4 October 1961, Science fiction serial starts well.
53. Gander, L.M., 1961, Television: strange call from space. *Daily Telegraph*, 4 October.
54. Letter from Fred Hoyle to Jesse Greenstein, 3 October 1961, Greenstein Papers, folder 16.1, California Institute of Technology Archives.
55. Letter from the BBC responding to a telephone request from Barbara Hoyle, Hoyle papers 9/10.
56. McLeave, H., 1961, Can Fred keep ahead of the test tubes . . .? Hoyle's TV baby isn't so out of this world as it seems. *Daily Mail*, November.
57. *Observer*, 10 October 1961, The *Observer* profile: Fred Hoyle.
58. See Low, R., 1991, *The Observer Book of Profiles* (London: W.H. Allen), xv.
59. *Sunday Times*, 4 March 1962, Professor Hoyle writes a play, by our theatre correspondent.
60. *Daily Express*, 11 April 1962.
61. *Daily Express*, 11 April 1962.
62. *Daily Telegraph*, 12 April 1962, Only 2 cheers for Mermaid: space play is too earthbound, by Eric Shorter. *The Times*, 12 April 1962, Played according to Hoyle: Mermaid Theatre, *Rockets in Ursa Major*.
63. *Daily Telegraph*, 27 December 1962, Slightly gayer space war: occasional thrills, by E.S. [Eric Shorter].
64. *The Times*, 28 December 1962, Science fiction that works: Mermaid Theatre, *Rockets in Ursa Major*.
65. Hoyle papers 22/19.
66. PRO: CAB 124/2152.

67. PRO: CAB 124/2152. The memo is dated 4 July 1962.
68. PRO: CAB 124/2152.
69. PRO: CAB 124/2152. Quirk's reply to Woolley is dated 19 July 1962.
70. PRO: CAB 124/2152. The Department of Scientific and Industrial Research's letter to Quirk is dated 27 September 1962.
71. PRO: CAB 124/2152. Quirk's note of his conversation with Woolley is dated 27 September 1962.
72. PRO: CAB 124/2152.

Chapter 9: Storm clouds

1. *The Times*, 1 November 1962, The sky's the limit.
2. *Sunday Times*, 17 February 1963, Science: A setback for Big Bang theory.
3. *Daily Telegraph*, 13 March 1963, Radio stars mystery to end soon, by L. Marsland Gander.
4. *Daily Mail*, 21 October 1963, TV dons expect a million to turn up over breakfast, by Roy Nash.
5. Freud, C., 1963, $x + y + z$ = no breakfast. *Daily Herald*, 22 October. The *Daily Herald* was relaunched as the *Sun* in 1964.
6. Hammerton, M., 1964, A case of an inappropriate model. *Nature*, **203**(4940), 63–64.
7. *The Times*, 25 May 1963, Slur on Lord Hailsham 'monstrous', Dons criticized by Prof. Hoyle, Cambridge degree withdrawal, from our University Correspondent.
8. Letter from Lord Hailsham to Hoyle, 26 March 1963, Hoyle papers 34/9.
9. Faulkner, J., 2002, After-dinner speech at Fred Hoyle's Universe: a conference celebrating Fred Hoyle's extraordinary contributions to science, 24–26 June, Cardiff University.
10. Interview with Sverre Aarseth, Cambridge, 26 May 2004.
11. Interview with John Faulkner, Santa Cruz, 5 November 2002.
12. Interview with John Faulkner, Santa Cruz, 5 November 2002.
13. Lavington, S.H., 1980, *Early British Computers: the Story of Vintage Computers and the People who Built Them* (Manchester University Press).
14. http://www.bobjanes.com/modules/sections/index.php?op=viewarticle&artid=123
15. *Financial Times*, 8 December 1962, World's most powerful computer commissioned.
16. *Hansard*,1964, Guided Missiles, 29 April, pp. 407–551.
17. Command Paper 2428, July 1964, First report into the Pricing of Ministry of Aviation Contracts, under Sir John Lang.
18. *The Times*, 10 December 1964, Ferranti want time to pay.
19. *Daily Sketch*, 3 July 1962, and personal communication from Geoffrey Hoyle, 18 February 2005. See also *Evening Standard*, 20 July 1963, Hoyle the Second.
20. Faulkner, J., 2002, After-dinner speech at Fred Hoyle's Universe: a conference celebrating Fred Hoyle's extraordinary contributions to science, 24–26 June, Cardiff University.
21. Hoyle, F., and Hoyle, G., 1963, *Fifth Planet* (New York: Harper and Row/Fawcett Crest), p. 60.

22. Hoyle, F., and Hoyle, G., 1963, *Fifth Planet* (New York: Harper and Row/Fawcett Crest), p. 18.

23. Hoyle, F., and Hoyle, G., 1963, *Fifth Planet* (New York: Harper and Row/Fawcett Crest), p. 12.

24. Hoyle, F., and Hoyle, G., 1963, *Fifth Planet* (New York: Harper and Row/Fawcett Crest), p. 45.

25. Hoyle, F., and Hoyle, G., 1963, *Fifth Planet* (New York: Harper and Row/Fawcett Crest), p. 123.

26. Hoyle, F., and Hoyle, G., 1963, *Fifth Planet* (New York: Harper and Row/Fawcett Crest), p. 58.

27. Hoyle, F., and Fowler, W.A., 1963, On the nature of strong radio sources. *Monthly Notices of the Royal Astronomical Society*, **125**, 169–176.

28. *Sunday Telegraph*, 17 March 1963, Implosion – a star's big bang, by John Delin, *Sunday Telegraph* science correspondent.

29. Interview with Maarten Schmidt, Pasadena, 12 November 2002.

30. Interview with Maarten Schmidt, Pasadena, 12 November 2002.

31. Interview with Maarten Schmidt, Pasadena, 12 November 2002.

32. Hoyle, F., undated, probably 1972, Two decades of collaboration with WAF. Fowler Papers, folder 164.15, California Institute of Technology Archives.

33. Interview with Maarten Schmidt, Pasadena, 12 November 2002.

34. Interview with Maarten Schmidt, Pasadena, 12 November 2002.

35. *Daily Express*, 18 December 1963, from Robin Stafford: New York, Tuesday.

36. Becker, B.J., 1989, The redshift controversy: exposing the boundaries of acceptable research. Teaching notes for 'Exploring the Cosmos: An Introduction to the History of Astronomy'. http://eee.uci.edu/clients/bjbecker/ExploringtheCosmos/week10e.html

37. Hoyle, F., 1968, Welcome to Slippage City. *Element 79* (New York: Signet), pp. 54–55.

38. Letter from James T. Barkelew to Fowler, 26 December 1962, Hoyle papers 39/7 and letter from Fowler to Hoyle, February 1963, Hoyle papers 9/10.

39. See, for example, Narlikar, J., *The Structure of the Universe* (Oxford University Press), pp. 133–137.

40. Hoyle, F., and Narlikar, J.V., 1962, Mach's principle and the creation of matter. *Proceedings of the Royal Society (A)*, **270**, 334–341.

41. Hoyle, F., and Narlikar, J.V., 1963, Mach's principle and the creation of matter. *Proceedings of the Royal Society (A)*, **273**, 1–11.

42. Hoyle, F., and Narlikar, J.V., 1964, Time symmetric electrodynamics and the arrow of time in cosmology. *Proceedings of the Royal Society (A)*, **277**, 1–23; Hoyle, F., and Narlikar, J.V., 1964, On the avoidance of singularities in *C*-field cosmology. *Proceedings of the Royal Society (A)*, **278**, 465; Hoyle, F., and Narlikar, J.V., 1964, The *C*-field as a direct particle field. *Proceedings of the Royal Society (A)*, **282**, 178; Hoyle, F., and Narlikar, J.V., 1964, On the gravitational influence of direct particle fields. *Proceedings of the Royal Society (A)*, **282**, 184.

43. Hoyle, F., and Narlikar, J.V., 1964, A new theory of gravity. *Proceedings of the Royal Society (A)*, **282**, 191.

44. Brown, P., 1964, After 300 years better than Newton. *Daily Express*, 4 June. See also *Evening Standard*, 1 June 1964, No apples for Fred.

45. *Evening Standard*, 1 June 1964, No apples for Fred.
46. *Evening Standard*, 1 June 1964, No apples for Fred.
47. *Sunday Times*, 14 June 1964, Proving the apple can't fall upwards.
48. Interview with Sverre Aarseth, Cambridge, 26 May 2004.
49. *Daily Mail*, 15 June 1964, Newton, Einstein and now Fred Hoyle, by Angus McPherson.
50. Personal communication from John Faulkner, 28 January 2003.
51. Interview with Fred Hoyle, 11 August 1993.
52. *New York Times*, 12 June 1964, Britons revise gravity theory – present new concept based on 'totality' of matter, by John Hillaby, special to the *New York Times*.
53. *Sunday Times*, 14 June 1964, Proving the apple can't fall upwards.
54. See, for example, *Daily Mail*, 15 June 1964, Newton, Einstein and now Fred Hoyle, by Angus Macpherson. For other coverage see *New York Times*, 12 and 23 June 1964; *Guardian*, 12 and 16 June 1964; *Daily Mail*, 15 June 1964; *Daily Telegraph*, 15 June 1964.
55. *Guardian*, 12 June 1964, Back to Newton: Hoyle's theory of the Universe, by John Maddox, our Science Correspondent.
56. *Guardian*, 18 June 1964, Science today: Gravitational revival, by John Maddox.
57. Personal communication from Donald Gillies, 2004.
58. *Guardian*, 12 June 1964, Back to Newton: Hoyle's theory of the Universe, by John Maddox, our Science Correspondent.
59. Hetherington, N.S., 1993, Hoyle–Narlikar theory. *Encyclopedia of Cosmology: Historical, Philosophical and Scientific Foundations of Modern Cosmology* (New York, London: Garland).
60. *New Scientist*, 18 June 1964, Notes and Comments: Natural philosophers; Narlikar, J.V., 1964, A new look at gravitation. *New Scientist*, 18 June, 730–732.
61. *New Scientist*, 18 June 1964, Notes and Comments: Natural philosophers.
62. Hoyle, F., and Tayler, R.J., 1964, *Nature*, **203**, 1108–1110.
63. Kragh, H., 1996, *Cosmology and Controversy: the Historical Development of Two Theories of the Universe* (Princeton, NJ: Princeton University Press), pp. 339–340.

Chapter 10: 'Dear Mr Hogg'

1. Draft letter from Lyttleton to Davenport, Hoyle papers 8/1.
2. Letter from Fred Hoyle to W.J. Sartain, undated, Spring 1964, Hoyle papers 8/1.
3. Letter from R.A. Lyttleton to H. Davenport, 27 April 1964, Hoyle papers 15/7.
4. Letter from Mestel, Sciama, Tayler et al. to Hoyle, 6 May1964, Hoyle papers 8/1.
5. MS note by Hoyle, Hoyle papers 8/1.
6. *Newsweek*, 25 May 1964, Life and Death of the Universe, p. 64.
7. Hoyle to Sartain, 26 May 1964, Hoyle papers 7/8.
8. Letter from George Batchelor to Hoyle, 25 May 1964, Hoyle papers 7/8.
9. Hoyle papers 7/8.
10. Tomalin, N., 1964, *Sunday Times*, 19 July.
11. Hoyle papers 8/1 and 8/9.
12. See, for example, *Financial Times*, 15 February 1964, Prof. Hoyle threatens to go to the US; *Daily Express*, 20 July 1964, 'Skyman' Hoyle may join Atlantic brain drain;

The Times, 14 February 1964, Migrating scientists; *New York Herald Tribune*, 16 February 1964, Astronomer–author Hoyle: UK frustrates scientists; *Sunday Times*, 19 July 1964, Fred Hoyle in a dilemma; *Daily Express*, 20 July 1964.

13. *Sunday Times*, 19 July 1964, Fred Hoyle's dilemma; *Daily Express*, 20 July 1964, 'Skyman' Hoyle may join Atlantic brain drain.

14. Hoyle papers 22/25.

15. Undated MS by Hoyle to 'Geoff', signed 'D', attached to the diary. Hoyle papers 24/3.

16. PRO: CAB 124/2152, 7 March 1963.

17. *The Times*, Government back astronomy plan, 13 March 1963.

18. Letter from Roger Blin-Stoyle to Hoyle, 10 January 1964, Hoyle papers 7/7.

19. PRO: CAB 124/2152.

20. Godwin, M., 2005, Skylark and the European Space Research Organisation, 1957–1972. PhD thesis, University of London.

21. PRO: CAB 124/2152.

22. Godwin, M., 2005, Skylark and the European Space Research Organisation, 1957–1972. PhD thesis, University of London.

23. Undated draft in Hoyle's hand, Hoyle papers 8/2.

24. PRO: CAB 124/2152.

25. PRO: CAB 124/2152.

26. PRO: CAB 124/2152.

27. PRO: CAB 124/2152.

28. Blaker to Beaven, 26 August 1964. CAB 124/2152.

29. PRO: CAB 124/2152.

30. PRO: CAB 124/2152.

31. PRO: CAB 124/2152. Blaker to Turnbull, 9 September 1964.

32. PRO: CAB 124/2152, 11 September 1964.

33. PRO: CAB 124/2152.

34. PRO: CAB 124/2152.

35. PRO: CAB 124/2152.

36. PRO: CAB 124/2152, Turnbull's letter to Melville is dated 2 October 1964.

37. Interview with Geoffrey Burbidge, La Jolla, 8 November 2002; Hoyle, F., 1994, *Home is Where the Wind Blows: Chapters from a Cosmologist's Life* (Mill Valley, CA: University Science Books), p. 236.

38. PRO: CAB 124/2152.

39. Hoyle papers 7/7, letter from Herzig to Hoyle, 7 October 1964.

40. PRO: CAB 124/2152, 7 October 1964.

41. Interview with Geoffrey Burbidge, La Jolla, 8 November 2002.

42. See, for example, letter to Francis, 15 October 1964, PRO:CAB 124/2152.

43. PRO: CAB 124/2152, 8 October 1964.

44. PRO: CAB 124/2152.

45. Hoyle, F., 1994, *Home is Where the Wind Blows: Chapters from a Cosmologist's Life* (Mill Valley, CA: University Science Books), p. 302.

46. Interview with Geoffrey Burbidge, La Jolla, 8 November 2002.

47. Interview with Geoffrey Burbidge, La Jolla, 8 November 2002.

48. PRO: CAB 124/2152.

49. PRO: CAB 124/2152; Dean wrote to the Vice-Chancellor on 18 October 1964.
50. PRO: CAB 124/2152, 12 October 1964.
51. PRO: CAB 124/2152, 13 October 1964.

Chapter 11: His institute

1. PRO: CAB 124/2152; Note by Sir Maurice Dean, 15 October 1964.
2. PRO: CAB 124/2152.
3. PRO: CAB 124/2152; Background Notes, undated but probably October 1964.
4. PRO: CAB 124/2152.
5. PRO: CAB 124/2152, letter dated 24 October 1964.
6. PRO: CAB 124/2152.
7. PRO: CAB 124/2152.
8. PRO: CAB 124/2152; Confidential note B/167/03 file CAB 124/2152.
9. PRO: CAB 124/3036, 1 December 1964.
10. PRO: CAB 124/3036, 29 December 1964.
11. PRO: CAB 124/3036, 31 December 1964.
12. PRO: CAB 124/3036, 6 January 1965.
13. PRO: CAB 124/3036, Bowden to Melville, 11 January 1965.
14. PRO: CAB 124/3036, 27 January 1965.
15. PRO: CAB 124/3036.
16. Letter from Fred Hoyle to William Fowler, 25 February 1965, Fowler Papers, folder 12.16, California Institute of Technology Archives.
17. Letter from Geoffrey Burbidge to Fred Hoyle, 9 March 1965, Fowler Papers, folder 12.16, California Institute of Technology Archives.
18. Letter from Geoffrey Burbidge to Fred Hoyle, 9 March 1965, Fowler Papers, folder 12.16, California Institute of Technology Archives.
19. Letter from William Fowler to Fred Hoyle, 12 March 1965, Fowler Papers, folder 12.16, California Institute of Technology Archives.
20. Tomalin, N., 1965, Prof Hoyle may join the 'brain drain': row over work. *The Times*, 7 March 1965.
21. *The Times*, 8 March 1965, A special grant for Cambridge University? Professor Hoyle asks for new institute.
22. *Daily Mail*, 1965, Why I may go, by Fred Hoyle by *Daily Mail* reporter, 9 March. See also *Daily Telegraph*, 1965, Grants sought for Hoyle's project, by our science correspondent, 20 March.
23. *Guardian*, 1965, Prof Hoyle angry over lack of computers, by our science correspondent, 8 March.
24. Hoyle papers, 8/9, undated MS. It is not clear whether this letter was actually sent. It was not filed with government records.
25. PRO: CAB 124/3036, 13 January 1965.
26. PRO: CAB 124/3036, Minister of State's meeting with Sir Harry Melville, 9 March 1965.
27. Hoyle, F., 1965, Remarks on the Institute of Theoretical Astronomy, typescript, 14 March, Hoyle papers, 8/2.
28. PRO: CAB 124 3036.

29. PRO: CAB 124 3036.
30. PRO: CAB 124 3036 Francis to Boys Smith, 24 May 1965.
31. Hoyle, F., 1965, *Of Men and Galaxies* (London: Heinemann), p. 31.
32. Hoyle, F., 1965, But can you be sure? *Daily Express*, 14 June.
33. Weatherall, R., 1965, Review of *Of Men and Galaxies*. *Observer*, 20 June.
34. Minutes of the General Board, 9 June 1965.
35. Science Research Council draft press release 24 June 1965; letter from Turnbull to Francis 28 June 65. PRO: CAB 124 3036.
36. *Daily Express*, 1965, Britain will give him his new institute, by Chapman Pincher, 29 June.
37. Letter from H. Davenport to Hoyle, 28 June 1965, Hoyle papers 8/2.
38. Note submitted to the Faculty Board of Mathematics, 10 July 1965, by John Dougherty, Hoyle papers 6/13.
39. Untitled document headed Faculty of Mathematics – University of Cambridge, dated 10 July 1965. Hoyle papers 6/13.
40. Minutes of the meeting on 12 July 1965 of the Council of the School of Physical Sciences, dated 6 December 1965. Hoyle papers 8/2.
41. Letter from J. F. Baker to W.F. Sartain, 15 July 1965, Hoyle papers 8/2.
42. Minutes of the General Board, 21 July 1965.
43. *Radio Times*, 24–30 July 1965. *Fred Hoyle's Universe* was scripted by Gordon Rattray Taylor and first shown on 27 December 1964. This screening was on 25 July 1965.
44. Kragh, H., 1996, *Cosmology and Controversy: the Historical Development of Two Theories of the Universe* (Princeton University Press), pp. 344–345.
45. Wilson's Nobel speech: http://www.nobel.se/physics/laureates/1078/wilson-lecture.pdf.
46. Kragh, H., 1996, *Cosmology and Controversy: the Historical Development of Two Theories of the Universe* (Princeton University Press), p. 351.
47. Sullivan, W., 1965, Signals imply a 'big bang' universe. *New York Times*, 21 May.
48. Kragh, H., 1996, *Cosmology and Controversy: the Historical Development of Two Theories of the Universe* (Princeton University Press), p. 350.
49. http://www.nobel.se/physics/laureates/1078/wilson-lecture.pdf
and http://www.nobel.se/physics/laureates/1078/penzias-lecture.pdf
50. *The Times*, 7 September 1965, New look at theory of universe.
51. Hoyle, F., 1965, Recent developments in cosmology. *Nature*, **208**(5006), 111. The lecture was delivered on 6 September 1965 and the paper published on 9 October.
52. *Daily Mirror*, 9 October 1965, Hoyle says: I was wrong about creation, by Arthur Smith.
53. *Daily Telegraph*, 9 October 1965, Hoyle retracts 20-yr-old theory of universe: 'steady state' abandoned by Dr Anthony Michaelis, science correspondent.
54. *Sunday Telegraph*, 10 October 1965, Welcome for Hoyle rethink, by John Delin, science correspondent.
55. Interview with Felix Pirani, London, 23 September 1997.
56. *Sunday Telegraph*, 10 October 1965, Welcome for Hoyle rethink, by John Delin, science correspondent.
57. *Daily Express*, 12 September 1965.
58. Letter from Hoyle to John Cockcroft, 28 October 1965. Hoyle papers 8/2.
59. Letter from Hoyle to the Vice Chancellor, 24 January 1966, Hoyle papers 8/2.

60. Faculty of Mathematics Board, 1965, Confidential: Institute of Theoretical Astronomy, 2 December.
61. Report 7723 to the General Board.
62. Letter from Bullard to Hoyle, 14 December 1965, Hoyle papers 8/2.
63. Minutes of the General Board, 12 January 1966.
64. Report 7761 to the General Board, 7 January 1966.
65. Untitled MS in Hoyle's hand, probably September 1965, Hoyle papers 8/2.
66. Untitled, undated MS in Hoyle's hand, probably winter 65/66, Hoyle papers 7/7.
67. Untitled, undated MS in Hoyle's hand, probably winter 65/66, Hoyle papers 7/7.
68. Untitled, undated MS in Hoyle's hand, probably winter 65/66, Hoyle papers 7/7. Bondi had written to Hoyle on 25 May 1965, Hoyle papers 15/7.
69. Untitled MS in Hoyle's hand, probably September 1965, Hoyle papers 8/2.
70. Letter from Hoyle to the Vice-Chancellor of Cambridge University, 20 January 1966, Hoyle papers 8/2.
71. Letter from Hoyle to the Vice-Chancellor, 24 January 1966, Hoyle papers 8/2.
72. Draft Letter from Hoyle to Francis, MS, 24 January 1966, Hoyle papers 8/2.
73. Letter from Hoyle to Francis, 24 January 1966, Hoyle papers 8/2.
74. Report 7799 to the General Board, 8 February 1966.
75. See, for example, Reports 7697 and 7727 to the General Board, 1965–6.
76. Hoyle papers 39/10.
77. Report 7799 to the General Board, 8 February 1966.
78. Letter from Hoyle to the Royal Society, 16 March 1966, Hoyle papers 39/10.
79. *Daily Telegraph*, 1966, University support for Prof. Hoyle, by the *Daily Telegraph* education reporter, 17 March.
80. *The Times*, 1966, Plan for Cambridge astronomy centre, from our university correspondent, 17 March.
81. *The Times*, 6 June 1966, Prof Hoyle heads new institute.

Chapter 12: The end of the beginning

1. Letter from Fred Hoyle to William Fowler, 9 June 1966, Fowler Papers, folder 12.16, California Institute of Technology Archives.
2. Undated MS, Hoyle papers, 8/3.
3. Letter from Neil Clarke, 24 March 1965, Hoyle papers 8/3.
4. Letter from Fowler to Hoyle, Fowler Papers, folder 12.16, California Institute of Technology Archives.
5. Interview with John Faulkner, Santa Cruz, 5 November 2002.
6. Faulkner, J., 2002, After-dinner talk at Fred Hoyle's Universe: a conference celebrating Fred Hoyle's extraordinary contributions to science, 24–26 June, Cardiff University.
7. Interview with Sverre Aarseth, Cambridge, 26 May 2004.
8. Letters from Bondi to Strittmatter, 10 January 1966 and Strittmatter to Hoyle, 4 February 1966, Hoyle papers, St John's, 9/3; letter from Faulkner to Hoyle, 26 January 1966, Hoyle papers 15/7.
9. Interview with John Faulkner, Santa Cruz, 5 November 2002.
10. Faulkner, J., 2002, After-dinner talk at Fred Hoyle's Universe: a conference

celebrating Fred Hoyle's extraordinary contributions to science, 24–26 June, Cardiff University.

11. Interview with John Faulkner, Santa Cruz, 5 November 2002.

12. Interview with John Faulkner, Santa Cruz, 5 November 2002.

13. Interview with Sverre Aarseth, Cambridge, 26 May 2004.

14. Letter dated 8 December 1966 from A.A.L. Rylance at Cambridge University to Jim Hosie at the Science Research Council, Hoyle papers 6/13.

15. Hoyle papers 9/3.

16. Minutes of the General Board, 18 January 1967.

17. Interview with John Faulkner, Santa Cruz, 5 November 2002.

18. Interview with Wallace Sargent, Pasadena, 12 November 2002.

19. Interview with Sverre Aarseth, Cambridge, 26 May 2004.

20. Interview with Sverre Aarseth, Cambridge, 26 May 2004.

21. Interview with Sverre Aarseth, Cambridge, 26 May 2004.

22. Hoyle papers 9/3 and 15/7.

23. Interview with John Faulkner, Santa Cruz, 5 November 2002.

24. Interview with Wallace Sargent, Pasadena, 12 November 2002.

25. Interview with Sverre Aarseth, Cambridge, 26 May 2004.

26. Hoyle papers 9/1.

27. *Sunday Times*, 24 April 1966.

28. Hoyle, F., 1966, *October the First is Too Late* (New York: Fawcett Crest), p. 147.

29. Hoyle, F., 1966, *October the First is Too Late* (New York: Fawcett Crest), p. 158.

30. Letter to Hoyle from Donald Menzel at Harvard University, Hoyle papers 9/4.

31. Hoyle, F., 1966, *Galaxies, Nuclei and Quasars* (London: Heinemann); for a review see, for example, *Observer*, 1 May 1966.

32. Hoyle, F., and Burbidge, G., 1966, Relation between the redshifts of quasi-stellar objects and their radio and optical magnitudes. *Nature*, **210**, 1346.

33. See, for example, Delin, J., 1966, Space theories in the melting pot, by our science correspondent. *Sunday Telegraph*, 10 July.

34. Becker, B.J., 1989, The redshift controversy: exposing the boundaries of acceptable research. Teaching notes for 'Exploring the Cosmos: An Introduction to the History of Astronomy'. http://eee.uci.edu/clients/bjbecker/ExploringtheCosmos/week10e.html

35. Letter from George Gamow to Jesse Greenstein, 4 February 1968, Greenstein Papers, folder 11.8, California Institute of Technology Archives.

36. Arp, H., 1987, *Quasars, Redshifts and Controversies* (Berkeley, CA: Interstellar Media), pp. 7–8.

37. Arp, H., 1967, *Astrophysical Journal*, **148**, 321.

38. Lynden-Bell, D., Cannon, R.D., Penston, M.V., and Rothman, V.C.A., 1966, *Nature*, **211**, 838.

39. Interview with Charles Barnes, Pasadena, 20 September 2000.

40. Interview with Fred Hoyle, Bournemouth, 11 August 1993.

41. Interview with Geoffrey Burbidge, La Jolla, 8 November 2002.

42. Hoyle, F., 1968, The Bakerian Lecture 1968: Review of recent developments in cosmology. *Proceedings of the Royal Society (A)*, **308**, 16.

43. Letter from Fred Hoyle to William Fowler, 9 June 1966, Fowler Papers, folder 12.16, California Institute of Technology Archives; letter from William Fowler to

Fred Hoyle, 14 September 1966, Fowler Papers, folder 12.16, California Institute of Technology Archives.

44. Karl G. Jansky lecture, September 1969, Hoyle papers 22/11.

45. Uppsala Symposium, Sweden, August 1970, Hoyle papers 22/10.

46. See the report in the *Observer*, 31 July 1968, Hoyle backs Stonehenge as a space clock, by John Davy.

47. Cited in Perfect, F.W., 1966, Stonehenge 'built for astronomy.' *Daily Telegraph*, 5 December.

48. Hoyle, F., 1966, *Antiquity*, **40**, 262.

49. Perfect, F.W., 1966, Stonehenge 'built for astronomy.' *Daily Telegraph*, 5 December.

50. Hoyle's popular books were *From Stonehenge to Modern Cosmology* (W.H. Freeman, 1972) and *On Stonehenge* (Heinemann, 1977). For coverage in 1966 see, for example, Davy, J., 1966, Astronomy by John Davy: Hoyle backs Stonehenge as a space clock. *Observer*, 31 July; *Sunday Times*, 31 July 1966, Archaeology: Fred Hoyle's theory of Stonehenge; Perfect, F.W., 1966, Stonehenge built for astronomy: Hoyle's theory, by our archaeological correspondent. *Daily Telegraph*, 5 December. *On Stonehenge* was reviewed in *Nature*, 21 January 1978.

51. Gascoigne, S.C.B., Proust, K.M., and Robins, M.O., 1990, *The Creation of the Anglo-Australian Telescope* (Cambridge University Press), p. 29.

52. Gascoigne, S.C.B., Proust, K.M., and Robins, M.O., 1990, *The Creation of the Anglo-Australian Telescope* (Cambridge University Press), p. 29.

53. Minutes of the BNCA meeting, 14 July 1959, quoted in Lovell, B., 1985, The early history of the Anglo-Australian Telescope. *Quarterly Journal of the Royal Astronomical Society*, **26**, 393–455.

54. Gascoigne, S.C.B., Proust, K.M., and Robins, M.O., 1990, *The Creation of the Anglo-Australian Telescope* (Cambridge University Press), p. 31.

55. Lovell, B., 1985, The early history of the Anglo-Australian (150-inch) Telescope (AAT). *Quarterly Journal of the Royal Astronomical Society*, **26**, 408.

56. Lovell, B., 1985, The early history of the Anglo-Australian (150-inch) Telescope (AAT). *Quarterly Journal of the Royal Astronomical Society*, **26**, 416–417.

57. Gascoigne, S.C.B., Proust, K.M., and Robins, M.O., 1990, *The Creation of the Anglo-Australian Telescope* (Cambridge University Press), p. 35.

58. Lovell, B., 1985, The early history of the Anglo-Australian (150-inch) Telescope (AAT). *Quarterly Journal of the Royal Astronomical Society*, **26**, 423.

59. Lovell, B., 1985, The early history of the Anglo-Australian (150-inch) Telescope (AAT). *Quarterly Journal of the Royal Astronomical Society*, **26**, 426.

60. Gascoigne, S.C.B., Proust, K.M., and Robins, M.O., 1990, *The Creation of the Anglo-Australian Telescope* (Cambridge University Press), p. 19.

61. Hoyle, F., 1982, *The Anglo-Australian Telescope* (University College Cardiff Press), p. 5.

62. Hoyle, F., 1982, *The Anglo-Australian Telescope* (University College Cardiff Press), p. 6.

63. Hoyle, F., and Lyttleton, R.A., 1939, The effect of interstellar matter on climatic variation. *Proceedings of the Cambridge Philosophical Society*, **35**, 405–418.

64. Wickramasinghe, N.C., 9 December 1963, Graphite particles as interstellar grains, PhD thesis, University of Cambridge.

65. Hoyle, F., and Wickramasinghe, C., 1984, *From Grains to Bacteria* (University College Cardiff Press).

66. Hoyle, F., and Wickramasinghe, N.C., 1962, *Monthly Notices of the Royal Astronomical Society*, **125**(5), 417–433.
67. Hoyle, F., and N.C. Wickramasinghe, 1963, *Monthly Notices of the Royal Astronomical Society*, **126**(4), 401–404.
68. Hoyle, F., and Wickramasinghe, C., 1967, Impurities in interstellar grains. *Nature* **214**, 969–971.
69. Letter from George Gamow to Jesse Greenstein, 4 February 1968, Greenstein Papers, folder 11.8, California Institute of Technology Archives.
70. Hoyle, F., and Wickramasinghe, C., 1968, Condensation of the planets. *Nature* **217**(5127), 415; Hoyle, F., and Wickramasinghe, C., 1968, Condensation of dust in galactic explosions. *Nature*, **218**(5147), 1126; Hoyle, F., and Wickramasinghe, C., 1970, Dust in supernova explosions. *Nature*, **226**(5240), 62.
71. Hoyle, F., and Wickramasinghe, C., 1984, *From Grains to Bacteria* (University College Cardiff Press), p. 49.
72. Hoyle, F., and Wickramasinghe, C., 1969, Interstellar grains. *Nature*, **223**(5205), 459.
73. Hoyle, F., and Wickramasinghe, C., 1984, *From Grains to Bacteria* (University College Cardiff Press), p. 56.
74. Wickramasinghe, N.C., 1974, Formaldehyde polymers in interstellar space. *Nature*, **252**(5483), 462.

Chapter 13: The Astronomer Hoyle

1. Correspondence in *The Times*, 10 and 11 July 1966.
2. Letter from Dael Wolfle to William Fowler, 12 April 1966, Fowler Papers, folder 12.19, California Institute of Technology Archives.
3. Interview with Sverre Aarseth, Cambridge, 26 May 2004.
4. See, for example, *Daily Telegraph*, 1968, Science Prize for Prof. Hoyle, 3 October.
5. Acceptance speech at the Kalinga Prize ceremony, New Delhi, 22 February 1969, Hoyle papers 25/4.
6. Hoyle, F., and Hoyle, G., 1970, *Seven Steps to the Sun* (New York: Fawcett Crest), p. 144.
7. *Daily Mail*, 1970, How Hoyle found love beyond the Milky Way, by Vincent Mulchrone, 2 February.
8. Interview with Sverre Aarseth, Cambridge, 26 May 2004.
9. Interview with John Faulkner, Santa Cruz, 5 November 2002.
10. Interview with John Faulkner, Santa Cruz, 5 November 2002.
11. Hoyle papers 8/15.
12. Interview with John Faulkner, Santa Cruz, 5 November 2002.
13. Hoyle papers 8/15.
14. Hoyle papers 8/15.
15. Hoyle papers 8/15.
16. Interview with John Faulkner, Santa Cruz, 5 November 2002.
17. Interview with John Faulkner, Santa Cruz, 5 November 2002.
18. Interview with John Faulkner, Santa Cruz, 5 November 2002.
19. Hoyle papers 9/1.

20. Hoyle papers 9/1.
21. Cover page to the final report. Hoyle papers 7/12.
22. Interview with Wallace Sargent, Pasadena, 12 November 2002.
23. Letter from Wallace Sargent to Hoyle, 19 October 1964, Hoyle papers 34/9.
24. Letter to *The Times* from Fred Hoyle, 10 July 1966.
25. Notebook, Hoyle papers 7/6.
26. Hoyle papers 9/4.
27. Hoyle papers 9/4.
28. Hoyle papers 9/4.
29. Letter from Vincent Reddish to Hoyle, 9 October 1969, Hoyle papers 9/4.
30. Interview with Geoffrey Burbidge, La Jolla, 8 November 2002.
31. Minutes of the meeting on 31 July 1969, Northerm Hemisphere Review Committee, 31 July 1969, Hoyle papers, 9/4.
32. Letter from Jim Hosie to Hoyle, 18 September 1969, Hoyle papers 9/4.
33. Letter from G.A. Winbow to Hoyle, 3 May 1970, Hoyle papers 9/2.
34. Letter to G.A. Winbow, 18 June 1970, Hoyle papers 9/2.
35. Lovell, B., 1991, The genesis of the Northern Hemisphere Observatory. *Quarterly Journal of the Royal Astronomical Society*, **32**, 1–16.
36. Interview with Wallace Sargent, Pasadena, 12 November 2002.
37. Interview with Bernard Lovell, Jodrell Bank, 31 October 2003.
38. Final version of the Northern Hemisphere Review report, sent to Hoyle by Jim Hosie on 21 September 1970, Hoyle papers 45/5.
39. Interview with Geoffrey Burbidge, La Jolla, 8 November 2002.
40. Interview with Geoffrey Burbidge, La Jolla, 8 November 2002.
41. Letter from Brian Flowers to Bernard Lovell, October 1993, cited in Lovell, B., 1994, The Royal Society, the Royal Greenwich Observatory and the Astronomer Royal. *Notes and Records of the Royal Society of London*, **48**(2), 283–297.
42. PRO: CAB 124/3011.
43. Interview with Bernard Lovell, Jodrell Bank, 31 October 2003.
44. Interview with Bernard Lovell, Jodrell Bank, 31 October 2003.
45. Letter from Maarten Schmidt to Bernard Lovell, May 1993, cited in Lovell, B., 1994, The Royal Society, the Royal Greenwich Observatory and the Astronomer Royal. *Notes and Records of the Royal Society of London*, **48**(2), 283–297.
46. Letter from Thomas Gold to Bernard Lovell, 14 February 1983, cited in Lovell, B., 1994, The Royal Society, the Royal Greenwich Observatory and the Astronomer Royal. *Notes and Records of the Royal Society of London*, **48**(2), 283–297.
47. Lovell, B., 1994, The Royal Society, the Royal Greenwich Observatory and the Astronomer Royal. *Notes and Records of the Royal Society of London*, **48**(2), 294.
48. Interview with Bernard Lovell, Jodrell Bank, 31 October 2003.
49. Interview with Francis Graham-Smith, Jodrell Bank, 31 October 2003.
50. Letter from Margaret Burbidge to Brian Flowers, 22 August 1972, Hoyle papers 9/4.
51. Burbidge, E.M., 1994, Watcher of the skies. *Annual Reviews in Astronomy and Astrophysics*, **32**, 1–36.
52. Interview with Geoffrey Burbidge, La Jolla, 8 November 2002.
53. Letter from Margaret Burbidge to Bernard Lovell, 27 August 1972, Hoyle papers 9/4.

54. Burbidge, E.M., 1994, Watcher of the skies. *Annual Reviews in Astronomy and Astrophysics*, **32**, 1–36.
55. Interview with Geoffrey Burbidge, La Jolla, 8 November 2002.
56. Letter from Jacquetta Priestly to Barbara Hoyle, 20 September 1971, Hoyle papers 23/1.
57. Hoyle F., and Hoyle, G., 1971, *The Molecule Men* (New York: Harper & Row), p. 33.
58. Hoyle, F., and Hoyle, G., 1971, *The Molecule Men* (New York: Harper & Row), p. 5.
59. Hoyle papers 8/4.
60. Hoyle papers 8/4.
61. Letter to Hoyle from the Ford Foundation, 1 February 1971, Hoyle papers 9/6.
62. Minutes of the Executive Board, 21 June 1971, Hoyle papers 9/6.
63. Document dated 21 November 1971, Hoyle papers 9/1.
64. Domb, C., 2002, Fred and naval radar (1941–1945), talk given at Fred Hoyle's Universe: a conference celebrating Fred Hoyle's extraordinary contributions to science, 24–26 June, Cardiff University.
65. Western Union telefax signed Kellogg and Wagoner, 5 January 1972, Hoyle papers 6/6.

Chapter 14: The beginning of the end

1. Interview with John Faulkner, Santa Cruz, 5 November 2002.
2. Interview with Wallace Sargent, Pasadena, 12 November 2002.
3. Letter from F. Hoyle to Roger Tayler, c.1970, Hoyle papers 18/6.
4. Interview with Margaret Burbidge, 8 November 2002.
5. Hoyle, F., 1993, *Home is Where the Wind Blows: Chapters from a Cosmologist's Life* (Mill Valley, CA: University Science Books), p. 374.
6. Interview with Geoffrey Burbidge, La Jolla, 8 November 2002.
7. Letter from F. Hoyle to the Vice-Chancellor, University of Cambridge, 14 February 1972, Hoyle papers 24/3.
8. Statement headed 'Institute of Theoretical Astronomy, University of Cambridge', and signed Fred Hoyle, 14 February 1972, Hoyle papers 24/3.
9. Letter from F. Hoyle to Barbara Hoyle, 14 February 1972, Hoyle papers 24/3.
10. Letter from Fred Hoyle to Bernard Lovell, cited in Lovell's obituary for Hoyle in the *Guardian*, 23 August 2001.
11. Letter from Ray Lyttleton to Hoyle, 19 July 1972, Hoyle papers 7/11.
12. Interview with John Faulkner, Santa Cruz, 5 November 2002.
13. Letter from Donald Lynden-Bell to Fred Hoyle, 18 February 1972, Hoyle papers 24/3.
14. Interview with Antony Hewish, Cambridge, 26 May 2004.
15. Letter from Donald Lynden-Bell to Hoyle, 19 July 1972.
16. Letter from Roderick Redman to Hoyle, 15 May 1972, Hoyle papers 7/11.
17. *Sunday Times*, 1972, Hoyle quits in row over bureaucracy, by Bryan Silcock, 23 April.
18. *Daily Telegraph*, 1972, Sir Fred Hoyle quits Cambridge professorship, by Anthony Michaelis, 24 April; and *Daily Telegraph*, 1972, Hoyle denies he is taking a job in California, by Ian Ball, 25 April.

19. *The Times*, 1972, Astronomers back protest by Hoyle, by Pearce Wright, 24 April.
20. *The Times*, 1972, letter from W.A. Deer, 26 April.
21. *Guardian*, 1972, Black cloud over Cambridge, letter from Brian Pippard, 27 April.
22. Letter from Hoyle to Bernard Lovell, undated MS, Hoyle papers, 7/11.
23. *Nature*, 1972, Departure of Professor Hoyle, **236**(5348), 417.
24. *Nature*, 1972, Cambridge astronomy without Sir Fred, **236**(5348), 419.
25. Readhead, A.C.S., and Hewish, A., 1972, Galactic structure and the apparent size of radio sources. *Nature*, **236**(5348), 440–443.
26. Interview with Antony Hewish, Cambridge, 26 May 2004.
27. Interview with Sverre Aarseth, Cambridge, 26 May 2004.
28. Letter from Mrs E.P. Hubble to Sir Fred and Lady Barbara Hoyle, Hoyle papers 7/10.
29. Letter from Don Clayton to Fred Hoyle, 22 April 1972, Hoyle papers 7/10.
30. Postcard from the Radcliffe Observatory, Hoyle papers 7/10.
31. Letters from Donald Blackwell to Hoyle, 24 April 1972 and 19 May 1972, Hoyle papers 7/11.
32. Letter from Asa Briggs to Hoyle, 8 May 72, Hoyle papers 7/11.
33. Letter from Neville J. Woolf to Hoyle, 27 April 1972, Hoyle papers 7/11.
34. Letter from Peter Strittmatter to Hoyle, 31 May 1972, Hoyle papers 7/11.
35. Telegram from the Institute of Theoretical Astronomy, June 1972, Hoyle papers 7/11.
36. Letter from Mrs Warren C. Thompson to Hoyle, 17 May 1972, Hoyle papers 7/11.
37. Letter from William Fowler to R.F. Christy and R.B. Leighton, 9 March 1972, Fowler Papers, folder 12.17, California Institute of Technology Archives.
38. Interview with Charles Barnes, Pasadena, 20 September 2000.
39. Interview with Charles Barnes, Pasadena, 20 September 2000.
40. Fowler's transcript of the AIP History Center Interview conducted by Charles Weiner, Fowler Papers, folder 199.26, California Institute of Technology Archives.
41. Interview with Wallace Sargent, Pasadena, 12 November 2002.
42. Letter from Kip Thorne to R.B. Leighton, 3 April 1972, Fowler Papers, folder 12.17, California Institute of Technology Archives.
43. Letter from William Fowler to R.B. Leighton, 12 April 1972, Fowler Papers, folder 12.17, California Institute of Technology Archives.
44. See, for example, *Observer*, 1972, Fred Hoyle will go to Manchester, 9 July; *Sunday Times*, 1972, New job for Hoyle, 9 July; *Guardian*, 1972, Hoyle goes North, by Peter Hildrew, 10 July.
45. Interview with Bernard Lovell, Jodrell Bank, 31 October 2003.
46. Document dated 8 January 1972, Hoyle papers 9/1.
47. Hoyle papers 8/12.
48. Letter from Wickramasinghe to Hoyle, undated, quoting Lynden-Bell's letter to Wickramasinghe of 14 April 1972, Hoyle papers 7/11.
49. Letter from Wickramasinghe to Hoyle, undated, quoting Lynden-Bell's letter to Wickramasinghe of 14 April 1972, Hoyle papers 7/11.
50. Letter from the Master of Gonville and Caius to the Provost of King's, 5 June 1972, Hoyle papers 7/11.

51. Letter from G.S. Hankinson, General Board of the Faculties, to F. Hoyle, 12 October 1972, Hoyle papers 22/25.
52. Letter from Hoyle to Lady Barbara Hoyle, probably September 1972, Hoyle papers 7/10.
53. Interview with Antony Hewish, Cambridge, 26 May 2004.
54. Hoyle, F., and Hoyle, G., 1973, *The Inferno* (London: Heinemann), p. 4.
55. Hoyle, F., and Hoyle, G., 1973, *The Inferno* (London: Heinemann), p. 210.
56. Haynes, R., Haynes, R., Malin, D., and McGee, R., 1996, *Explorers of the Southern Sky: a History of Australian Astronomy* (Cambridge University Press), p. 179; Suntzeff, N., 1998, A remembrance of Olin Eggen. NOAO Newsletter, December, no. 56. http://www.noao.edu/noao/noaonews/dec98/node6.html.
57. *Daily Express*, 13 February 1961, 'Sorry' – but the big bang row is still on, by *Express* Staff Reporter.
58. Little, C., 2002, Conflict in radioastronomy (unpublished paper), and personal communication with Carolyn Little, September 2002.
59. Gascoigne, S.C.B., Proust, K.M., and Robins, M.O., 1990, *The Creation of the Anglo-Australian Telescope* (Cambridge University Press), pp. 135–6.
60. Hoyle, F., 1982, *The Anglo-Australian Telescope* (university College Cardiff Press), pp. 13–14.
61. Quoted in Gascoigne, S.C.B., Proust, K.M., and Robins, M.O., 1990, *The Creation of the Anglo-Australian Telescope* (Cambridge University Press), p. 130.
62. Gascoigne, S.C.B., Proust, K.M., and Robins, M.O., 1990, *The Creation of the Anglo-Australian Telescope* (Cambridge University Press), p. 134.
63. Gascoigne, S.C.B., Proust, K.M., and Robins, M.O., 1990, *The Creation of the Anglo-Australian Telescope* (Cambridge University Press), p. 136.
64. Hoyle, F., 1994, *Home is Where the Wind Blows: Chapters from a Cosmologist's Life* (Mill Valley, CA: University Science Books), p. 385; Hoyle, F., 1982, *The Anglo-Australian Telescope* (University College Cardiff Press), p. 15.
65. Hoyle, F., 1982, *The Anglo-Australian Telescope* (University College Cardiff Press), p. 3.
66. Interview with Fred Hoyle, Bournemouth, 11 August 1993.
67. Hoyle, F., 1994, *Home is Where the Wind Blows: Chapters from a Cosmologist's Life* (Mill Valley, CA: University Science Books), p. 384.
68. Hoyle, F., 1994, *Home is Where the Wind Blows: Chapters from a Cosmologist's Life* (Mill Valley, CA: University Science Books), p. 386.
69. Hoyle, F., 1982, *The Anglo-Australian Telescope* (University College Cardiff Press), p. 22.
70. Gascoigne, S.C.B., Proust, K.M., and Robins, M.O., 1990, *The Creation of the Anglo-Australian Telescope* (Cambridge University Press), p. 146.
71. Haynes, R., Haynes, R., Malin, D., and McGee, R., 1996, *Explorers of the Southern Sky: a History of Australian Astronomy* (Cambridge University Press), p. 402.
72. Hoyle, F., 1982, *The Anglo-Australian Telescope* (University College Cardiff Press), p. 17.
73. Letter from Joe Wampler to Fred Hoyle, 30 January 1975, Hoyle papers 7/20.
74. Hoyle, F., 1982, *The Anglo-Australian Telescope* (University College Cardiff Press), p. 24.
75. Gascoigne, S.C.B., Proust, K.M., and Robins, M.O., 1990, *The Creation of the Anglo-Australian Telescope* (Cambridge University Press), p. 282.

76. Gascoigne, S.C.B., Proust, K.M., and Robins, M.O., 1990, *The Creation of the Anglo-Australian Telescope* (Cambridge University Press), p. 23.
77. Gascoigne, S.C.B., Proust, K.M., and Robins, M.O., 1990, *The Creation of the Anglo-Australian Telescope* (Cambridge University Press), p. 161.
78. Interview with Margaret Burbidge, La Jolla, 8 November 2002.
79. Hoyle, F., 1982, *The Anglo-Australian Telescope* (University College Cardiff Press), p. 24.

Chapter 15: On the loose

1. *Cambridge Evening News*, 8 September 1972, p. 19.
2. For replies see Hoyle papers 7/10.
3. Hoyle, F., 1994, *Home is Where the Wind Blows: Chapters from a Cosmologist's Life* (Mill Valley, CA: University Science Books), p. 383.
4. Interview with Bernard Lovell, Jodrell Bank, 31 October 2002.
5. Hoyle, F., 1974, *Nicolaus Copernicus: an Essay on his Life and Work* (London: Heinemann).
6. *Observer*, 7 December 1975.
7. Hoyle, F., and Narlikar, J.V., 1974, *Action at a Distance in Physics and Cosmology* (San Francisco: W.H. Freeman).
8. Lovell, B., 2001, Professor Sir Fred Hoyle. *Guardian*, 23 August.
9. Hoyle, F., and Hoyle, G., 1975, *Into Deepest Space* (Harmondsworth: Penguin), p. 176.
10. Hoyle, F., 1975, *Astronomy and Cosmology: A Modern Course* (San Francisco: W.H. Freeman), Preface.
11. Hoyle, F., 1975, *Astronomy and Cosmology: A Modern Course* (San Francisco: W.H. Freeman), p. 673
12. Etienne, J., 2001, *Space News International*, http://www.spacenews.be/art2001/hoyle_230801.html, 7323–3901–2631EC, 23 August, from an article by J.M. Bonnet-Bidaud and other sources. Translated from the French by JG.
13. Fowler's transcript of the AIP History Center Interview conducted by Charles Weiner, Fowler Papers, folder 199.26, California Institute of Technology Archives.
14. Mulchrone, V., 1974, The totally happy man . . . Professor Sir Fred, in tune with his mountains. *Daily Mail*, 28 August.
15. The promise of science fiction: prophetic or profane? Ray Bradbury and Fred Hoyle, moderated by Bruce Murray, 26 February 1975. Typescript, Hoyle papers 22/5.
16. Letter from William Fowler to C.A. Barnes, 25 July 1974, Fowler Papers, folder 147.20, California Institute of Technology Archives.
17. Letter from William Fowler to Don Clayton, 23 May 1974, Fowler Papers, folder 147.20, California Institute of Technology Archives.
18. Letter from Don Clayton to William Fowler, 30 May 1974, Fowler Papers, folder 147.20, California Institute of Technology Archives.
19. Letter from Ray Lyttleton to Geoffrey Burbidge, 24 February 1975, Fowler Papers, folder 147.20, California Institute of Technology Archives.
20. Letter from Martin Ryle to Geoffrey Burbidge, 26 February 1975, Fowler Papers, folder 147.20, California Institute of Technology Archives.
21. Press release: The Nobel Prize for Physics, 1974. Swedish Academy of Sciences, 15 October 1974.

22. Presentation speech by Professor Hans Wilhelmsson, http://www.nobel.se/physics/laureates/1974/presentation-speech.html

23. Interview with Antony Hewish, Cambridge, 26 May 2004.

24. Faulkner, J., 2002, After dinner talk at Fred Hoyle's Universe: a conference celebrating Fred Hoyle's extraordinary contributions to science, 24–26 June, Cardiff University.

25. Faulkner, J., 2002, After dinner talk at Fred Hoyle's Universe: a conference celebrating Fred Hoyle's extraordinary contributions to science, 24–26 June, Cardiff University.

26. S. Jocelyn Bell Burnell, 1977, Petit four. *Annals of the New York Academy of Science*, **302**, 685–689.

27. Faulkner, J., 2002, After dinner talk at Fred Hoyle's Universe: a conference celebrating Fred Hoyle's extraordinary contributions to science, 24–26 June, Cardiff University.

28. Interview with Thomas Gold, Ithaca, 3 April 1996.

29. Stevenson, J., 1975, Nobel Prize star-spotter gets rocket from Hoyle. *Daily Mail*, 22 April.

30. Hoyle, F., 1994, *Home is Where the Wind Blows: Chapters from a Cosmologist's Life* (Mill Valley, CA: University Science Books), p. 245.

31. Wright, P., 1975, Sir Fred Hoyle attacks Nobel prize winner. *The Times*, 22 April.

32. *Montreal Gazette*, Astronomical award scandal bared, by Andre Potworowski, 21 March 1975, p. 41. Hoyle papers 7/2.

33. Interview with Antony Hewish, Cambridge, 26 May 2004.

34. *Montreal Star*, 22 March 1975, Student says that she doesn't feel 'squeezed out' of Nobel prize, by Linda Cahill.

35. *Cambridge Evening News*, 24 March 1975.

36. Anonymous handwritten note dated 21 March 1975, Fowler Papers, folder 147.21, California Institute of Technology Archives.

37. *New York Times*, 21 March 1975.

38. Anonymous typed note dated 21 March 1975, Fowler Papers, folder 147.21, California Institute of Technology Archives.

39. Anonymous handwritten note dated 21 March 1975, Fowler Papers, folder 147.21, California Institute of Technology Archives.

40. A handwritten note on American quarto paper, Hoyle papers 6/8.

41. Interview with Antony Hewish, Cambridge, 26 May 2004.

42. Letter from Gold to Hoyle, 8 April 1975, Hoyle papers 7/2.

43. Interview with Thomas Gold, Ithaca, 3 April 1996.

44. Dover, C., 1975, Sir Fred Hoyle starts Cambridge science furore. *Daily Telegraph*, 22 April.

45. Letter from A. Hewish to *The Times*, 10 April 1975.

46. Letter from Fred Hoyle to *The Times*, 8 April 1975.

47. Letter from F. Hoyle to *The Times*, 8 April 1975.

48. Letter from Martin Ryle to *The Times*, 12 April 1975.

49. Interview with Antony Hewish, Cambridge, 26 May 2004.

50. Interview with Antony Hewish, Cambridge, 26 May 2004.

51. Interview with Antony Hewish, Cambridge, 26 May 2004.

52. Interview with Antony Hewish, Cambridge, 26 May 2004.
53. Interview with Antony Hewish, Cambridge, 26 May 2004.
54. Interview with Antony Hewish, Cambridge, 26 May 2004.
55. Interview with Antony Hewish, Cambridge, 26 May 2004.
56. Interview with Antony Hewish, Cambridge, 26 May 2004.
57. S. Jocelyn Bell Burnell, 1977, Petit four. *Annals of the New York Academy of Science*, **302**, 685–689.
58. Interview with Antony Hewish, Cambridge, 26 May 2004.
59. S. Jocelyn Bell Burnell, 1977, Petit four. *Annals of the New York Academy of Science*, **302**, 685–689.
60. Letter from Jesse Greenstein to William Fowler, 1 April 1975, Fowler Papers, folder 147.21, California Institute of Technology Archives.
61. Letter from Wallace Sargent to Geoffrey Burbidge, 1 April 1975, Fowler Papers, folder 147.21, California Institute of Technology Archives.
62. Letter from Don Clayton to Wallace Sargent, 7 April 1975, Fowler Papers, folder 147.21, California Institute of Technology Archives.
63. Note to Fowler attached to Letter from Don Clayton to Wallace Sargent, 7 April 1975, Fowler Papers, folder 147.21, California Institute of Technology Archives.
64. Letter from Don Clayton to Martin Rees, 12 May 1975, Fowler Papers, folder 147.21, California Institute of Technology Archives.
65. Hoyle papers, 8/4.
66. Hoyle papers, 8/4.
67. Transcript of 'Frontiers of Astronomy – 1975', *Scientifically Speaking*, broadcast on 30 July 1975, BBC Radio 3, produced by David Paterson.

Chapter 16: Apocalyptic visions

1. Taylor, J., 1977, H for Hoyle. *Radio Times*, 8 January.
2. ITV1 Wales this Week, 13 September 2003, http://www.htvwales.com/walesthisweek/pages/2003/space.shtml.
3. The most prominent example is the song *Woodstock*, by Joni Mitchell.
4. See Clayton, D., and Wickramasinghe, N.C., 1976, *Astrophysics and Space Science*, **42**, 463, and Clayton, D., 1975, *Nature*, **257**, 36.
5. Interview by Brig Klyce of Fred Hoyle, 5 July 1996, http://www.panspermia.org/hoylintv.htm.
6. See Hoyle, F., and Wickramasinghe, N.C., 1977, Polysaccharides and infrared spectra of galactic sources. *Nature*, **268**(5621), 610; Hoyle, F., and Wickramasinghe, N.C., 1977, Prebiotic polymers and infrared spectra of galactic sources. *Nature*, **269**(5630), 674; Hoyle, F., and Wickramasinghe, N.C., 1977, Identification of interstellar polysaccharides and related hydrocarbons. *Nature*, **271**(5642), 229; Hoyle, F., and Wickramasinghe, N.C., Dry polysaccharides and the infrared spectrum of OH 26.5 + 0.6. *Astrophysics and Space Science*, **72**, 247.
7. Hoyle, F., and Wickramasinghe, C., 1984, *From Grains to Bacteria* (University College Cardiff Press), p. 60.
8. Hoyle, F., 1957, *The Black Cloud* (London: Heinemann), pp. 170–171. See also

Gregory, J., 2003, Popularization and excommunication of Fred Hoyle's 'life-from-space' theory. *Public Understanding of Science*, **12**, 25–46.

9. Hoyle, F., and Wickramasinghe, C., 1984, *From Grains to Bacteria* (University College Cardiff Press), p. 61.

10. Davidson, K., 1999, *Carl Sagan: a Life* (New York: Wiley), pp. 158–161.

11. Hoyle, F., and Wickramasinghe, C., 1984, *From Grains to Bacteria* (University College Cardiff Press), p. 66.

12. Interview with Bernard Lovell, Jodrell Bank, 31 October 2002.

13. Interview with Bernard Lovell, Jodrell Bank, 31 October 2002.

14. Letter from N.C. Wickramasinghe to Fred Hoyle, 26 October 1976, Hoyle papers 6/4.

15. Letter from N.C. Wickramasinghe to Fred Hoyle, 5 November 1976, Hoyle papers 6/4.

16. Dixon, B., personal communication, September 2002; Hoyle, F., and Wickramasinghe, C., 1977, Does epidemic disease come from space? *New Scientist*, 17 November.

17. Dixon, B., 1977, Celestial pathology. *New Scientist*, 17 November, 396.

18. Hoyle, F., and Wickramasinghe, C., 1977, Does epidemic disease come from space? *New Scientist*, 17 November.

19. Ridley, H.B., 1977, Epidemics from space? *New Scientist*, 1 December, 593; Hardy, D., 1977, Epidemics from space? *New Scientist*, 1 December, 593; Penman, J.M., 1977, Epidemics from space? *New Scientist*, 8 December.

20. Hoyle, F., and Wickramasinghe, C., 1978, Diseases from space, *New Scientist*, 5 January.

21. Hoyle, F., 1977, *Energy or Extinction: the Case for Nuclear Energy* (London: Heinemann). For reviews, see: *Guardian*, 14 September 1977, Sir Fred and fast reaction; *Daily Telegraph*, 14 September 1977, Ecological dupes.

22. Everitt, A., 1977, A nuclear attack from Hoyle. *Birmingham Post*, 8 September.

23. Hoyle, F., and Hoyle, G., 1980, *Commonsense in Nuclear Energy* (London: Heinemann). For coverage see Hoyle, F., 1980, Should Britain take the nuclear route? *Now!*, 11 January, pp. 62–63; Berry, A., 1988, In praise of nuclear power. *Daily Telegraph*, 28 January; Lampitt, L.F., 1980, Energy and the nuclear family. *Evening News*, 18 February; Hoyle, F., 1980, Stop this nonsense, nuclear energy IS safe! *Daily Mail*, 17 June.

24. Oakley, C., 1977, Bulldozing logic of nuclear energy argument. *Yorkshire Post*, September.

25. For Hoyle's campaign, see, for example, Berry, A., 1980, In praise of nuclear power. *Daily Telegraph*, 28 January; Lampitt, L.F., 1980, Energy and the nuclear family. *Evening News*, 18 February; and Hoyle, F., 1980, Stop this nonsense, nuclear energy IS safe! *Daily Mail*, 17 June; for Bondi's, see, for example, Fishlock, D., 1979, Protests would hurt 'third world.' *Financial Times*, 16 October; Davies, T., 1980, Assessing the nuclear risks. *Observer*, 6 January.

26. Hoyle, F., and Hoyle, G., 1978, *The Incandescent Ones* (New York: Signet), p. 25.

27. Hoyle, F., and Hoyle, G., 1978, *The Incandescent Ones* (New York: Signet), p. 135.

28. Hoyle, F., and Hoyle, G., 1978, *The Incandescent Ones* (New York: Signet), p. 169.

29. Hoyle, F., and Hoyle, G., 1978, *The Incandescent Ones* (New York: Signet), p. 121.

30. Hoyle, F., 1977, *Ten Faces of the Universe* (San Francisco, CA: W.H. Freeman).

31. Stubbs, P., 1977, Review: *Ten Faces of the Universe* by Fred Hoyle. *New Scientist*, 1 December, p. 578.

32. Fairhall, J., 1978, 'Life on comets' comes down to earth. *Guardian*, 14 April.

33. McKie, R., 1978, Disease from outer space research 'obstructed' claim. *The Times*, 14 April.

34. PRO: FD 23/4849.

35. Hoyle, F., and Wickramasinghe, C., 1984, *From Grains to Bacteria* (University College Cardiff Press), pp. 116–120. The book refers to the SERC, the Science and Engineering Research Council, which came into existence 1981, when it replaced the Science Research Council.

36. Berry, A., 1978, Bugs from space spread flu, says Fred Hoyle. *Daily Telegraph*, 9 August.

37. Berry, A., 1978, Space flu vaccines a problem. *Daily Telegraph*, August.

38. Hoyle, F., 1978, The invading viruses from space (letter). *Daily Telegraph*, 28 August.

39. Dixon, B., 1977, Celestial pathology. *New Scientist*, 17 November, 396.

40. Coleman, T., 1978, *Guardian*, 13 September.

41. Macpherson, A., 1978, The cosmic connection. *Daily Mail*, 16 September.

42. Hoyle, F., and Wickramasinghe, C., 1984, *From Grains to Bacteria* (University College Cardiff Press), p. 116.

43. Hoyle, F., and Wickramasinghe, N.C., 1976, Primitive grain clumps and organic compounds in carbonaceous chondrites. *Nature*, **264**(5581), 45; Hoyle, F., and Wickramasinghe, N.C., 1977, Identification of the λ2,200A interstellar absorption feature. *Nature*, **270**(5635), 323; Sakata, A., Nakagawa, N., Iguchi, T., Isobe, S., Morimoto, M., Hoyle, F., and Wickramasinghe, N.C., 1977, Spectroscopic evidence for interstellar grain clumps in meteoric inclusions. *Nature*, **266**(5599), 241.

44. Hoyle, F., and Wickramasinghe, N.C., 1978, *Lifecloud* (London: Dent), p. 2.

45. This and the following extracts are from Hoyle, F., and Wickramasinghe, C., 1984, *From Grains to Bacteria* (University College Cardiff Press), pp. 96–98. See also the papers Hoyle, F., and Wickramasinghe, N.C., 1979, On the nature of interstellar grains. *Astrophysics and Space Science*, **66**, 77–90; Hoyle, F., and Wickramasinghe, N.C., 1979, Organic grains in space. *Astrophysics and Space Science*, **69**, 511.

46. Interview by Brig Klyce of Fred Hoyle, 5 July 1996, http://www.panspermia .org/ hoylintv.htm.

47. Interview by Brig Klyce of Fred Hoyle, 5 July 1996, http://www.panspermia .org/ hoylintv.htm.

48. Hoyle, F., and Wickramasinghe, C., 1984, *From Grains to Bacteria* (University College Cardiff Press), pp. 116–120.

49. The papers quoted are Duley, W.W., and Williams, D.A., 1979, *Nature*, **277**, 4 January and Whittet, D.C.B., 1979, *Nature*, **281**, 25 October.

50. Hoyle, F., Wickramasinghe, N.C., Al-Mufti, S., and Olavesen, A.H., 1982, Infrared spectroscopy of micro-organisms near 4.3 μm in relation to geology and astronomy. *Astrophysics and Space Science*, 81, 489–92; Hoyle, F., Wickramasinghe, N.C., Al-Mufti, S., Olavesen, A.H., and Wickramasinghe, D.T., 1982, Infrared spectroscopy over the 2.9–3.9 μm waveband in biochemistry and astronomy. *Astrophysics and Space Science*, **83**, 405–9; Hoyle, F., and Wickramasinghe, N.C., 1982, A model for interstellar extinction. *Astrophysics and Space Science*, **86**, 321–29.

51. Hoyle, F., 1993, *Home is Where the Wind Blows: Chapters from a Cosmologist's Life* (Mill Valley, CA: University Science Books), pp. 395–398.

52. See Gregory, J., 2003, Popularization and excommunication of Fred Hoyle's 'life-from-space' theory, *Public Understanding of Science*, **12**, 25–46.

53. Hoyle, F., and Wickramasinghe, C., 1980, *Influenza – A Genetic Virus with the External Trigger* (Department of Applied Mathematics and Astronomy, University College Cardiff); Berry, A., 1980, Invaders from space 'trigger evolution'. *Daily Telegraph*, 27 May.

54. Hoyle, F., 1980, *The Relation of Biology to Astronomy* (University College Cardiff Press), p. 13.

55. Hoyle on evolution, *Nature*, **294**(5837):105, November 12, 1981.

56. Hoyle, F., 1980, *The Relation of Biology to Astronomy* (University College Cardiff Press), p. 23.

57. Hoyle, F., and Wickramasinghe, C., 1980, *Origin of Life* (University College Cardiff Press), p. 3.

58. Hoyle, F., and Wickramasinghe, C., 1980, *Origin of Life* (University College Cardiff Press), p. 15.

59. Hoyle, F., and Wickramasinghe, C., 1980, *Origin of Life* (University College Cardiff Press), p. 18.

60. Hoyle, F., and Wickramasinghe, N.C., 1981, *Evolution from Space* (London: Dent), p. 143.

61. Levy, G., 1981, There *must* be a God. *Daily Express*, 14 August.

62. Smith, A., 1981, Lifestyles. *Sunday Telegraph*, 16 August; *New Scientist*, Extraterrestrial genes, 15 October 1981.

63. Hoyle, F., and Wickramasinghe, N.C., 1981, *Evolution from Space* (London: Dent), p. 149.

64. Hoyle, F., and Wickramasinghe, C., 1981, *Space Travellers: the Bringers of Life* (University College Cardiff Press), p. 30.

65. Hoyle, F., 1981, *Ice* (London: Hutchinson), p. 14.

66. Wright, P., 1981, Glaciers over Britain is the Hoyle forecast. *The Times*, 6 June.

67. Lord, G., 1981, Is this the chilling destiny of Earth? *Sunday Express*, 7 June.

68. Hoyle, F., 1981, The next ice age. *Sunday Times* magazine, 7 June; Hoyle, F., 1981, Race against the ice age. *Sunday Times* magazine, 14 June.

69. *New Scientist*, 1981, Hoyle's ice-age theory meets glacial response, 11 June.

70. Gribbin, J., 1981, Review of *Ice*, by Fred Hoyle. *New Scientist*, 11 June.

71. Interview with Geoffrey Burbidge, La Jolla, 8 November 2002.

72. Personal communication from Clark Friend, 17 November 2003.

73. Hoyle, F., 1981, *Ice* (London: Hutchinson), p. 168.

74. Alvarez, L.W., Alvarez, W., Asaro, F., Michel, H.V., 1980, Extraterrestrial cause for the Cretaceous-Tertiary extinction. *Science*, **208**, 1095–1108; and for example Clements, E., 1986, Of asteroids and dinosaurs: the role of the press in the shaping of scientific debate. *Social Studies of Science*, **16**, 421–456.

75. See *Nuclear War: the Aftermath*, 1982, a special edition of *Ambio*, **2**(2/3) (Swedish Academy of Sciences); Birks, J.W., and Crutzen, P.J., 1983, *Chemistry in Britain*, November, pp. 927–930; Turco, R.P., Toon, O.B., Ackerman, T.P., Pollack, J.B., and Sagan, C., 1983, Nuclear winter: global consequences of multiple nuclear explosions. *Science*, **222**, December, 1283–1297. Sagan published for the international relations community with his 1983 paper 'Nuclear war and climate catastrophe: some policy implications', *Foreign Affairs*, Winter 83/84.

Chapter 17: Evolution on trial

1. Hoyle, F., 1981, Will Darwin bite the dust in Little Rock? *The Times*, 7 December.
2. Letter from Willy Fowler to Chandra Wickramasinghe, 5 December 1981, Hoyle papers 31/10.
3. http://people.hofstra.edu/faculty/robert_l_hall/ISB1F01/ ScienceInCreationScience.html.
4. Wickramasinghe, N.C., quoted on http://www.panspermia.org/chandra.htm
5. Wickramasinghe, N.C., quoted on http://www.panspermia.org/chandra.htm
6. Wickramasinghe, N.C., quoted on http://www.panspermia.org/chandra.htm
7. http://people.hofstra.edu/faculty/robert_l_hall/ISB1F01/ ScienceInCreationScience.html.
8. *The Arkansas Democrat*, 17 December 1981.
9. See, for example, *The Times*, 23 December 1981.
10. Letter from Chandra Wickramasinghe to Willy Fowler, 22 December 1981, Hoyle papers 31/10.
11. The judge's decision was printed in *Science*, **215**, 934–943; see p. 940.
12. Timmins, N., 1982, Survivors of disaster in an earlier world: evolution according to Hoyle. *The Times*, 13 January.
13. Silcock, B., 1982, Hoyle's law: was man's ancestor a spore from outer space? *Sunday Times*, 17 January.
14. Hoyle, F., and Wickramasinghe, N.C., 1982, *Proofs that Life is Cosmic*. Memoirs of the Institute of Fundamental Studies, Sri Lanka, No. 1 (Sri Lanka: Government Press).
15. Wright, P., 1983, New claims over origin of life 'between the stars.' *The Times*, 8 March; and Tucker, J., 1983, Life in space (cont.). *Sunday Times*, 20 March.
16. Cherfas, J., 1983, The word according to Hoyle. *New Scientist*, 19 May.
17. Macpherson, A., 1983, In the beginning there *was* a creator. *Mail on Sunday*, 23 October.
18. *Awake!* 22 January 1996 (New York: Watchtower Bible and Tract Society).
19. Royal Swedish Academy of Science, 1983, Press release, The 1983 Nobel Prize for Physics, 19 October.
20. Royal Swedish Academy, 1983, Press release, The 1983 Nobel Prize for Physics, 19 October.
21. McKie, R., The maverick physicist catches the God bug. *Observer*, 30 October 1983.
22. Editorial, 1983, Where was Hoyle? *Nature*, **305**, 750.
23. Editorial, 1983, Where was Hoyle? *Nature*, **305**, 750.
24. Shallis, M., 1984, In the eye of a storm. *New Scientist*, 19 January.
25. Interview with Thomas Gold, Ithaca, 3 April 1996.
26. Interview with Thomas Gold, Ithaca, 3 April 1996.
27. Etienne, J., 2001, *Space News International*, http://www.spacenews.be/art2001/ hoyle_230801.html, 7323-3901-2631EC, 23 August, from an article by J.M. Bonnet-Bidaud and other sources. Translated from the French by JG.
28. Telegram from Antony Hewish to William Fowler, 20 October 1983, Fowler Papers, folder 198.2, California Institute of Technology Archives.
29. Letter from William Fowler to Fred Hoyle, 3 November 1983, Fowler Papers, folder 12.17, California Institute of Technology Archives.

30. Letter from Fred Hoyle to William Fowler, 8 November 1983, Fowler Papers, folder 12.17, California Institute of Technology Archives.
31. Fowler, W., 1983, Fred Hoyle, my friend and collaborator. *New Scientist*, 10 November.
32. Hoyle, F., and Wickramasinghe, C., 1984, *From Grains to Bacteria* (University College Cardiff Press), pp. 172, 197–199.
33. Campbell, P., 1983, Life in outer space: infrared data debugged. *Nature*, **306**, 218–219.
34. Hoyle, Sir Fred, 1984, Imagined worlds V: Doing battle with the universe. *Listener*, 6 August.
35. Hoyle, Sir Fred, 1984, Imagined worlds V: Doing battle with the universe. *Listener*, 6 August.
36. Shippey, T., 1985, review of *Comet Halley*. *Guardian*, 2 August.
37. Hoyle, F., 1985, *Comet Halley* (London: Michael Joseph), p. 23.
38. Hoyle, F., 1985, *Comet Halley* (London: Michael Joseph), pp. 43–44.
39. Hoyle, F., 1985, *Comet Halley* (London: Michael Joseph), p. 185.
40. Hoyle, F., 1985, *Comet Halley* (London: Michael Joseph), p. 397.
41. Hoyle, F., 1985, *Comet Halley* (London: Michael Joseph), pp. 197–201.
42. Wright, P., 1986, Comet 'confirms epidemic theory.' *The Times*, 2 April, and Scepticism on comet virus idea. *The Times*, 3 April.
43. Tucker, A., 1986, New evidence of organic material in Halley's comet. *Guardian*, 3 April.
44. Wright, P., 1986, 'Stardust' leads to scientific conflict. *The Times*, 8 August.
45. Veitch, A., 1986, If Halley gets up your nose. *Guardian*, 30 April.
46. See Hoyle, F., and Wickramasinghe, C., 1986, Does AIDS come from space? Letter to *Daily Telegraph*, 1 December; and Prentice, T., 1986, Aids virus blamed on comet. *The Times*, 2 December.
47. Gould, D., 1986, Well, where *does* AIDS come from? *Evening Standard*, 5 December.
48. Hoyle, F., and Wickramsinghe, C., Comets a possible source of BSE? *The Times*, 5 December 2000.
49. Letter from Lee Spetner to Hoyle, 5 September 1984, Hoyle papers 33/1.
50. Letter from Hoyle to Lee Spetner, 14 September 1984, Hoyle papers 33/1.
51. Watkins, R.S., Hoyle, F., Wickramasinghe, N.C., Watkins, J., Rabilizirov, R., and Spetner, L.M., 1985a, *Archaeopteryx* – a photographic study. *British Journal of Photography* **132**, 264–266; Watkins, R.S. et al., 1985b, *Archaeopteryx* – a further comment. *British Journal of Photography* **132**, 358–359, 367; Watkins, R.S. et al., 1985c, *Archaeopteryx* – more evidence. *British Journal of Photography* **132**, 468–470; Hoyle, F., Wickramasinghe, N.C., and Watkins, R.S., 1985, *Archaeopteryx*: Problems arise – and a motive. *British Journal of Photography* **132**(6516), 693–695, 703; and then in 1988, Spetner, L.M., Hoyle, F., Wickramasinghe, N.C. and Magaritz, M., 1988, *Archaeopteryx* – more evidence for a forgery. *British Journal of Photography* **135**, 14–17.
52. Letter from Lee Spetner to Alan Charig, 3 April 1985, Hoyle papers 33/1.
53. Correspondence between Spetner and Hoyle, June 1985, Hoyle papers 33/1.
54. Notes from a meeting on 23 May 1985 at the Natural History Museum, Hoyle papers 33/1.
55. Correspondence, June 1985, Hoyle papers 33/2.

56. Charig, A., Greenaway, F., Milner, A., Walker, C., and Whybrow, P., 1986, *Science*, **232**, 622–625.

57. Hoyle, F., and Wickramsinghe, C., 1986, *Archaeopteryx, the Primordial Bird: a Case of Fossil Forgery?* (London: Christopher Davies).

58. Kemp, T., 1986, Feathered flights of fancy. *Nature*, **324**, 185.

59. Whybrow, P., 1986, Rare controversy. Letter to *New Scientist*, 4 September, p. 62.

60. Whybrow, P., 1986, *Archaeopteryx*. Letter to *New Scientist*, 2 October, p. 60.

61. Letter from Alan Charig to Lee Spetner, 29 November 1986, Hoyle papers 33/2.

62. Natural History Museum, 1987, *The Feathers Fly!* (London: Natural History Museum), p. 7.

63. See Natural History Museum, 1987, *The Feathers Fly!* (London: Natural History Museum), and Kennedy, M., 1987, Museum gives the bird to Sir Fred's 'flight of fancy.' *Guardian*, 18 August.

64. Cohen, N., 1987, Feathers fly over fossil forgery claim. *Independent*, 18 August.

65. Connor, S., 1991, Hoyle still in a world of his own. *Independent on Sunday*, p. 4.

66. Maddox, J., 2001, Obituary: Fred Hoyle (1915–2001). *Nature* **413**, 270.

67. Woolley, B., 1988, Orbital dreams. *Guardian*, 22 July.

68. Hoyle, F. and Wickramasinghe, C., 1993, *Our Place in the Cosmos* (London: Dent).

69. Shapiro, R., 1993, Life, the Universe and anything goes. *Nature*, **363**, 124.

70. Atkins, P., 1993, 100 on the poppycock scale. *Science and Public Affairs*, Autumn, p. 58.

71. Interview with Ben Morgan, assistant editor, *Science and Public Affairs*, Autumn 1993.

72. Personal communication from Lady Barbara Hoyle, Autumn 1993.

73. See, for example, Wansell, G., 1985, Sir Fred Hoyle. *Sunday Telegraph*, 30 June.

74. Naughton, J., 1986, Astronomer extraordinary. *Observer*, 10 August; see also Lyttleton, R.A., 1986, Fred in transit. *Guardian*, 22 August.

75. Hoyle, F., 1986, *The Small World of Fred Hoyle* (London: Michael Joseph).

76. Hoyle, F., 1993, *Home is Where the Wind Blows: Chapters from a Cosmologist's Life* (Mill Valley, CA: University Science Books).

77. Hoyle papers 23/1.

78. Interview with Antony Hewish, Cambridge, 26 May 2003.

79. Interview with Geoffrey Burbidge, La Jolla, 8 November 2002; Interview with Bernard Lovell, Jodrell Bank, 31 October 2002.

80. Interview with Chandra Wickramasinghe by Walter Jayawardhana for LankaWeb News, http://www.lankaweb.com/news/items01/180801–3.html.

81. http://nai.arc.nasa.gov/institute/timeline.cfm.

82. Interview with Fred Hoyle, Bournemouth, 11 August 1993.

Chapter 18: A new cosmology

1. *Nature* **240**, 439 (signed M.R.-R. – the cosmologist Michael Rowan Robinson).

2. Hoyle, F., 1980, *The Steady State Cosmology Revisited* (University College Cardiff Press), p. 26.

3. Hoyle papers 23/8.

4. Script of Hoyle's Henry Norris Russell lecture at the meeting of the American Astronomical Society in Seattle, 8–12 April 1972, pp. 12–13. Orange Aid preprint

Series in Nuclear, Atomic and Relativistic Astrophysics, OAP-286, May 1972, Hoyle papers 23/8.

5. Hoyle, F., 1975, *Astronomy and Cosmology: A Modern Course* (San Francisco: W.H. Freeman), p. 673.

6. Hoyle, F., and Narlikar, J.V., 1980, *The Physics–Astronomy Frontier* (San Francisco: W.H. Freeman).

7. Hoyle, F., 1980, *The Steady State Cosmology Revisited* (University College Cardiff Press), pp. 56–57.

8. Hoyle, F., 1994, *Home is Where the Wind Blows: Chapters from a Cosmologist's Life* (Mill Valley, CA: University Science Books), p. 290.

9. Hoyle, F., Burbidge, G., and Narlikar, J.V., 1999, *A Different Approach to Cosmology: from a Static Universe through the Big Bang towards Reality* (Cambridge University Press), p. 93.

10. This survey was summarized in Durant, J.R., Evans, G.A., and Thomas, G.P., 1989, The public understanding of science. *Nature*, **340**, 11–14.

11. Interview with Thomas Gold, Ithaca, 3 April 1996.

12. Letter to the Hoyles from Hermann Bondi, 13 June 1988. Hoyle papers 1/8.

13. Hoyle papers 15/1.

14. Hoyle, F., and Wickramasinghe, N.C., 1988, The microwave background in steady-state cosmology. 22nd ESLAB Symposium on infra-red spectroscopy in astronomy, University of Salamanca, December.

15. Hoyle, F., Wickramasinghe, N.C., and Reddish, V.C., 1983, Solid hydrogen and the microwave background. *Nature*, **218**(5147), 1124.

16. Hoyle, F., and Wickramasinghe, N.C., 1988, The microwave background in steady-state cosmology. 22nd ESLAB Symposium on infra-red spectroscopy in astronomy, University of Salamanca, December.

17. Hoyle, F., 1988, An assessment of the evidence against steady state theory. Hoyle papers 3/2.

18. Hoyle papers 3/5.

19. Hoyle, F., 1988, An assessment of the evidence against steady state theory. Hoyle papers 3/2.

20. Hoyle, F., 1988, An assessment of the evidence against steady state theory. Hoyle papers 3/2.

21. Hoyle, F., and Wickramasinghe, N.C., 1988, The microwave background in steady-state cosmology. 22nd ESLAB Symposium on Infra-red spectroscopy in astronomy, University of Salamanca, December.

22. Arp, H., 1987, *Quasars, Redshifts and Controversies* (Berkeley, CA: Interstellar Media), p. 47.

23. Becker, B.J., 1989, The redshift controversy: exposing the boundaries of acceptable research. Teaching notes for 'Exploring the Cosmos: An Introduction to the History of Astronomy'.
http://eee.uci.edu/clients/bjbecker/ExploringtheCosmos/week10e.html.

24. Becker, B.J., 1989, The redshift controversy: exposing the boundaries of acceptable research. Teaching notes for 'Exploring the Cosmos: An Introduction to the History of Astronomy'.
http://eee.uci.edu/clients/bjbecker/ExploringtheCosmos/week10e.html.

25. Becker, B.J., 1989, The redshift controversy: exposing the boundaries of accept-

able research. Teaching notes for 'Exploring the Cosmos: An Introduction to the History of Astronomy'. http://eee.uci.edu/clients/bjbecker/ExploringtheCosmos/week10e.html.

26. Interview with Geoffrey Burbidge, La Jolla, 8 November 2002.

27. Hoyle, F., 1987, Review of *Quasars, Redshifts and Controversies*, by Halton Arp. MS, Hoyle papers 3/2.

28. Hoyle, F., 1987, Review of *Quasars, Redshifts and Controversies*, by Halton Arp. MS, Hoyle papers 3/2.

29. Arp, H., 1987, *Quasars, Redshifts and Controversies* (Berkeley, CA: Interstellar Media), p. 169.

30. Interview with Fred Hoyle, Bournemouth, 11 August 1993.

31. Letter from H. Arp to Fred Hoyle, 18 April 1988, Hoyle papers 3/6.

32. Carilli, C.L., van Gorkom, J.H., and Stocke, J.T., 1989, Disturbed neutral hydrogen in the galaxy NGC3067 pointing to the quasar 3C232. *Nature*, **338**, 134–136.

33. Letter from Halton Arp to *Nature*, 16 March 1989, Hoyle papers 3/6.

34. Letter from Halton Arp to Fred Hoyle, 17 March 1989, Hoyle papers 3/6.

35. Interview with Fred Hoyle, Bournemouth, 11 August 1993.

36. Personal communication from Philip Campbell, 10 December 2004.

37. Interview with Fred Hoyle, Bournemouth, 11 August 1993.

38. Horgan, J., 1995, Fred Hoyle: the return of the maverick. *Scientific American*, March, 46–47.

39. Interview with Geoffrey Burbidge, La Jolla, 8 November 2002.

40. Interview with Geoffrey Burbidge, La Jolla, 8 November 2002.

41. Hoyle, F., 1976, The future of physics and astronomy. *American Scientist*, **64**, 197.

42. Interview with Fred Hoyle, Bournemouth, 11 August 1993.

43. *Nature*, 1990, **346**, 780.

44. Arp, H.C., Burbidge, G., Hoyle, F., Narlikar, J.V. and Wickramasinghe, N.C., 1990, The extragalactic universe: an alternative view. *Nature*, **346**, 807–812.

45. Maddox, J., 1992, Hoyle statue unveiled. *Nature*, **358**, 269.

46. Hoyle, F., Burbidge, G., and Narlikar, J.V., 1993, A quasi-steady-state cosmology. *Astrophysical Journal*, **410**, 437.

47. Etienne, J., 2001, *Space News International*, http://www.spacenews.be/art2001/hoyle_230801.html, 7323-3901-2631EC, 23 August, from an article by J.M. Bonnet-Bidaud and other sources. Translated from the French by JG.

48. Etienne, J., 2001, *Space News International*, http://www.spacenews.be/art2001/hoyle_230801.html, 7323-3901-2631EC, 23 August, from an article by J.M. Bonnet-Bidaud and other sources. Translated from the French by JG.

49. See, for example, Matthews, R., 1993, *Sunday Telegraph*, 11 April.

50. Interview with Fred Hoyle, Bournemouth, 11 August 1993.

51. Etienne, J., 2001, *Space News International*, http://www.spacenews.be/art2001/hoyle_230801.html, 7323-3901-2631EC, 23 August, from an article by J.M. Bonnet-Bidaud and other sources. Translated from the French by JG.

52. Interview with Charles Barnes, Pasadena, 20 September 2000.

53. Horgan, J., 1995, Fred Hoyle: the return of the maverick. *Scientific American*, March, 46–47.

54. Hoyle, F., 1993, *The Origin of the Universe and the Origin of Religion* (Wakefield, Rhode Island: Moyer Bell).
55. Clube, V., and Napier, B., 1982, *The Cosmic Serpent* (Universe Publications); Hoyle cites Clube, V., and Napier, B., 1990, *The Cosmic Winter* (Oxford: Blackwell).
56. Interview with Sverre Aarseth, Cambridge, 26 May 2004.
57. Horgan, J., 1995, Fred Hoyle: the return of the maverick. *Scientific American*, March, 46–47.
58. Interview with Geoffrey Burbidge, La Jolla, 8 November 2002.
59. Interview with Geoffrey Burbidge, La Jolla, 8 November 2002.
60. Hoyle, F., Burbidge, G., and Narlikar, J.V., 1999, *A Different Approach to Cosmology: from a Static Universe through the Big Bang towards Reality* (Cambridge University Press), p. 336.
61. Interview with Geoffrey Burbidge, La Jolla, 8 November 2002.
62. Hoyle, F., Burbidge, G., and Narlikar, J.V., 1999, *A Different Approach to Cosmology: from a Static Universe through the Big Bang towards Reality* (Cambridge University Press), p. 93.
63. Coles, P., 2000, Universal rage. *New Scientist*, **166**(2233), 58.
64. Maddox, J., 2001, Obituary: Fred Hoyle (1915–2001). *Nature* **413**, 270.
65. Mestel, L., 2001, Professor Sir Fred Hoyle, 1915–2001, *The Johnian*. This obituary is an edited version of the original that appeared in the October 2001 issue of *Astronomy & Geophysics*.
66. Interview with Fred Hoyle, Bournemouth, 11 August 1993.
67. Etienne, J., 2001, *Space News International*, http://www.spacenews.be/art2001/hoyle_230801.html, 7323-3901-2631EC, 23 August, from an article by J.M. Bonnet-Bidaud and other sources. Translated from the French by JG.
68. Interview with Geoffrey Burbidge, La Jolla, 7 November 2002.
69. Interview with Sverre Aarseth, Cambridge, 26 May 2004.
70. Lovell, B., 2001, Professor Sir Fred Hoyle. *Guardian*, 23 August.
71. Lovell, B., 2001, Professor Sir Fred Hoyle. *Guardian*, 23 August.
72. Rees, M., 2001, *Physics Today*, **54**, 72.
73. Rees, M., 2001, *Physics Today*, **54**, 72.
74. Sullivan, W., 2001, Fred Hoyle dies at 86; opposed 'big bang' but named it. *New York Times*, 22 August 2001. This piece had been on file for some time: Sullivan died in 1996.
75. Maddox, J., 2001, Obituary: Fred Hoyle (1915–2001). *Nature* **413**, 270.
76. Mestel, L., 2001, Professor Sir Fred Hoyle, 1915–2001, *The Johnian*. This obituary is an edited version of the original that appeared in the October 2001 issue of *Astronomy & Geophysics*.
77. Mestel, L., 2001, Professor Sir Fred Hoyle, 1915–2001, *The Johnian*. This obituary is an edited version of the original that appeared in the October 2001 issue of *Astronomy & Geophysics*.
78. Brown, John C., *The Herald*, 23 August 2001.
79. *Daily Telegraph*, 22 August 2001.
80. Wickramasinghe, C., 2001, Sir Fred Hoyle. *Independent*, 23 August.
81. Demme, G., and Sarfati, J., 2001, 'Big-bang' critic dies. http://www.answersingenesis.org/news/hoyle.asp#f6, August; and *TJ*, 2001, 15(3):6–7.

82. http://www.smmirror.com/volume3/issue11/starry_skies_above.asp
83. *Yorkshire Post*, 23 August 2001.
84. http://groups.msn.com/MatterdaleMatters/cwherald2000now.msnw, August 2001.
85. Interview with Antony Hewish, Cambridge, 26 May 2004.
86. Interview with Geoffrey Burbidge, La Jolla, 7 November 2002.

Picture acknowledgements

Hoyle as a child: courtesy of St. John's College Council; *Fred Hoyle is first report*: © *The Daily Graphic*/Solo Syndication; **St. John's College**: © 2005 Top-Foto.co.uk; **first passport**: courtesy of St. John's College Council; **Hermann Bondi**: © G. Paul Bishop 1954; **Tommy Gold**: © Cornell University, all rights reserved; **George Gamow**: © Time Life Pictures/Getty Images; **Rome conference**: courtesy of St. John's College Council; *The Black Cloud*: © HarperCollins USA; **Bernard Lovell**: © Hulton-Deutsch Collection/Corbis; **Ryle's press conference report**: © *Evening Standard*/Solo Syndication; **IOTA building**: © Edward Leigh. Courtesy of the Institute of Astronomy, University of Cambridge; **IOTA staff**: AIP Emilio Segr© Visual Archives, Clayton Collection; **Richard Woolley**: © Bettmann/Corbis; **Hoyle's friends**: AIP Emilio Segr© Visual Archives, Clayton Collection; **Chandra Wickramasinghe**: courtesy of Professor Chandra Wickramasinghe; **Anglo-Australian Telescope**: © Anglo-Australian Observatory/David Malin Images; **Sir Martin Ryle**: © Corbis/Bettmann; **Jocelyn Bell Burnell and Antony Hewish**: © Hencoup Enterprises Ltd/Science Photo Library; **Hoyle writing to *The Times***: AIP Emilio Segr© Visual Archives, Clayton Collection; **60th birthday conference**: AIP Emilio Segr© Visual Archives, Clayton Collection; **Jayant Narlikar**: courtesy of Professor J. V. Narlikar; **Halton 'Chip' Arp**: courtesy of Halton Arp

The publisher and the author apologize for any errors or omissions in the above list. If contacted they will be pleased to rectify these at the earliest opportunity.

Index